Aquatic Animal Viruses and Antiviral Immunity

Aquatic Animal Viruses and Antiviral Immunity

Guest Editor

Mark Polinski

MDPI

Basel • Beijing • Wuhan • Barcelona • Belgrade • Novi Sad • Cluj • Manchester

Guest Editor
Mark Polinski
National Coldwater Marine
Aquaculture Center
U.S. Department of
Agriculture—Agricultural
Research Service
Franklin, ME
USA

Editorial Office
MDPI AG
Grosspeteranlage 5
4052 Basel, Switzerland

This is a reprint of the Special Issue, published open access by the journal *Viruses* (ISSN 1999-4915), freely accessible at: https://www.mdpi.com/journal/viruses/special_issues/8425RYXCCM.

For citation purposes, cite each article independently as indicated on the article page online and as indicated below:

Lastname, A.A.; Lastname, B.B. Article Title. *Journal Name* **Year**, *Volume Number*, Page Range.

ISBN 978-3-7258-7689-1 (Hbk)
ISBN 978-3-7258-7690-7 (PDF)
https://doi.org/10.3390/books978-3-7258-7690-7

Contents

About the Editor

Mark Polinski

Mark Polinski is a Research Immunologist at the USDA-ARS National Coldwater Marine Aquaculture Center. His work focuses on elucidating host–pathogen interactions within the context of aquatic animals for the purpose of improving aquaculture production. His laboratory focuses heavily on qPCR, RNA-seq, and protein identification methodologies aimed at understanding the complex interplay of molecular responses that underlie disease and susceptibility.

Preface

The Special Issue on Aquatic Animal Viruses and Antiviral Immunity highlights the remarkable diversity of viruses circulating in marine and freshwater systems and the equally diverse responses of their hosts. Across its contributions, the issue reveals how persistent, emerging, and previously unrecognized viruses shape the health of wild and farmed aquatic species. Several studies uncover long-term or cryptic viral infections in fish populations, showing how pathogens can endure across years and influence conservation and hatchery management. Other articles describe newly detected or genetically distinct viruses in economically important species, underscoring the continual emergence of threats in aquaculture. Advances in diagnostics and surveillance also feature prominently, including practical tools for monitoring immunity and infection at the farm level. Mechanistic work further deepens understanding of how viruses manipulate host cells, illuminating pathways that may offer future intervention potential. Collectively, the Special Issue brings together ecological, molecular, and applied perspectives to illustrate how aquatic viruses evolve, spread, and interact with their hosts. It offers a timely synthesis for researchers, managers, and industry stakeholders committed to improving fish health and sustaining resilient aquatic ecosystems.

Mark Polinski
Guest Editor

viruses

MDPI

Editorial

Aquatic Animal Viruses and Antiviral Immunity: A Closing Editorial

Mark P. Polinski

U.S. Department of Agriculture National Coldwater Marine Aquaculture Center, Orono, ME 04473, USA; mark.polinski1@usda.gov

1. Introduction

Aquatic ecosystems host the planet's greatest animal diversity, and with it, a remarkably varied virosphere. Nevertheless, across commercial and conservation aquaculture, decisionmakers are faced with three common persistent challenges: (i) distinguishing viral discovery from disease relevance, (ii) quantifying and acting on immunity at a farm/population scale, and (iii) integrating host genetics, environment, and biosecurity into coherent control strategies. This Special Issue was launched to address these challenges by advancing our knowledge of aquatic animal–virus interactions and the antiviral defenses that shape disease outcomes across fish, crustaceans, and mollusks. The assembled papers (11 in total) collectively highlight advances in virus discovery, mechanistic pathogenesis, immune monitoring, and interventions that are directly applicable to aquaculture health management, pathogen surveillance, and environmental conservation.

2. Motivations for This Special Issue

Metagenomics is redefining the diversity and evolution of fish viruses, highlighting long host-specific divergence patterns that juxtapose with cross-species transmission at the domestic–wild interfaces of aquatic systems [1]. These insights underscore how emergence is driven not only by viral genetics but also by aquaculture practices and environmental change. In parallel, molecular methods—from classical PCR to field-portable LAMP and next-gen sequencing —have matured, enabling more rapid identification and increased specificity for viral detection within aquatic hosts [2]. Progress in fish vaccinology (including recombinant and nucleic-acid platforms) is notable, although challenges remain for mucosal delivery, durability under stress, and implementation in the diverse species makeup of conservation and commercial aquaculture [3,4]. Here, comparative immunology continues to clarify innate and adaptive responses in fish and shellfish, laying foundations for targeted immunomodulation and selective breeding [5,6].

Yet, despite these advances, fundamental gaps persist. One challenge is that viral detection does not equal disease—many newly identified viruses lack clear pathogenic attribution, leaving managers unsure how to prioritize monitoring and mitigation [7,8]. A second challenge is that immunity is hard to measure at the population scale—laboratories may deliver precise assays, but farms require cost-effective, rapid immunometrics that inform real-time decision making within their budgetary capacity [9]. Third, host genetics matter. Resistance and susceptibility can vary dramatically by stock or family [10], yet validated workflows to incorporate genotype information into management—particularly in conservation aquaculture—are limited. Fourth, biosafety at conservation hatcheries is often challenged by latent infections and stress-modulated shedding, where policies and budgetary capacities are often insufficient to fully mitigate long-term reservoir risks [11].

Finally, translational therapeutics (applying laboratory discovery to practical application) beyond vaccines are under-tested in aquaculture, even when cell-level mechanisms suggest tractable antiviral targets.

3. Special Issue Synthesis

3.1. From Discovery to Surveillance and Pathogenesis Model Development

Two papers in this issue extend the U.S. marine virome and directly inform surveillance priorities. Lovy et al. [12] detected the Betanodavirus sequence and characterized a novel Nervous necrosis virus (NNV) genotype—provisionally BSBNNV—in black sea bass along the mid-Atlantic coast of North America, emphasizing the need to evaluate epidemiology and virulence in a region with expanding marine aquaculture. Raines et al. [13] identified three novel viruses from recently described nackednavirus and adomavirus lineages in a hyperpigmented largemouth bass lesion, advancing the concept that cryptogenic viruses are present "in plain sight" and should be folded into risk assessments for centrarchids. This has been further supported by independent identification of a similar nackednavirus in African cichlids this same year [14]. A third paper, by He et al. [15], moved from discovery and genomic/phenotypic characterization of multiple novel Dutch Cyprinid herpesvirus (CyHV-2) strains to establish an immersion-challenge model with consistent generation of ~40% goldfish mortality—allowing future vaccine and therapeutic evaluation tests to be conducted under conditions emulating natural transmission.

3.2. Actionable Mechanisms for Antiviral Development

Zhang et al. [16] demonstrated that an Iridovirus, ISKNV, triggers ferroptosis by suppressing GPx4 and promoting ACSL4 in CPB cells, and that pharmacological inhibition of this pathway suppresses viral yield. This finding positions ferroptosis modulation as a promising host-directed antiviral strategy in aquaculture species where vaccine solutions are incomplete or slow to deploy. Niu et al. [17] showed that the mandarin fish scTRIM44 gene positively regulated rabdoviral (SCRV) transcription despite also positively regulating RNA-virus detection pathways (RIG-I/MDA5-mediated interferon signaling). Interestingly, scTRIM44 proved inconsequential to two DNA iridoviruses (ISKNV and LMBV), revealing virus-specific host pathway effects that could be exploited for resistance or immunomodulation depending on the type of infecting virus (dsDNA vs. ssRNA).

3.3. Immunometrics and Rapid-Response Biologics for Improved Aquaculture

Yamkasem et al. [18] translated lab immunology into farm-level immune monitoring by validating a pooled-serum ELISA antibodies for the impactful Tilapia Lake Virus (TiLV). The work provides clear guidance on pooling sensitivity and cost-efficiency—an operational template that can be employed for other pathogen screening programs that could encompass individual to population-scale testing. Setthawong et al. [19] improved TiLV mitigation approaches by developing specific IgY antibodies extracted from laying hen egg-yolks targeting segment-4 of the virus. These antibodies demonstrated dose dependent neutralization and cytopathic effect inhibition in vitro for TiLV, providing a putative outbreak response tool that could serve as a bridge or adjunct to vaccination, particularly where licensing or delivery of vaccines is constrained.

3.4. Host Genetics as a Determinant of Disease

Polinski et al. [6] provided experimental evidence that Atlantic salmon stock strongly influences PRV-1/HSMI severity, independent of viral subtype (PRV-1a or PRV-1b), underscoring the value of selective breeding and stock choice for reducing disease burden. Complementing this, Toubanaki et al. [20] probed NNV re-infected transcriptomes in Euro-

https://doi.org/10.3390/v18010127

pean sea bass, identifying distinct gene networks in resistant vs. susceptible families for use in developing biomarkers and breeding targets that promote betanodavirus resistance.

3.5. Biosecurity—Recognizing Acute Reservoirs and Assessing Economic Costs

Shavalier et al. [21] documented the ~8.5-year persistence of a herpesvirus (SalHV-3) in lake trout that survived an initial disease (EED) outbreak, with stress-linked intermittent detection across fins/mucus, nervous tissues, and water. This provided evidence that survivors can act as long-term reservoirs capable of shedding under routine hatchery handling, and invites re-examination of quarantine, screening periodization, and release criteria in conservation aquaculture programs. Rahaman et al. [22] also re-examined biosecurity practices at a national scale by quantifying the economic burden of viral diseases in Australian fish and prawn aquaculture and critically assessed surveillance, diagnostics, and exclusion strategies—arguing for genetics-based resistance, modern vaccines, and integrated risk frameworks that harmonize policy with farm realities. The conclusions are transferrable to other national contexts facing similar scale-up pressures.

4. Closing

The studies of this Special Issue showcase descriptive knowledge to aid management decisions by adding realistic challenge models, identifying mechanistic antiviral levers, demonstrating actionable immunometrics, validating host genetic effects, exposing latent reservoirs, and connecting disease biology to national biosecurity planning. They show that practical impact comes from linking discovery to surveillance, aligning mechanisms with management, and embedding genetics and ecology in day-to-day decisions. These studies also highlight the continued need for forward progress to standardize immunosurveillance, develop host-directed antivirals, integrate genetic resilience, and safeguard conservation programs so that aquaculture and stewardship can better anticipate viral change rather than react to it.

Funding: M.P.P. is congressionally funded within the U.S. Department of Agriculture—Agricultural Research Service (project 8030-31000-005-00D).

Institutional Review Board Statement: Not applicable.

Informed Consent Statement: Not applicable.

Data Availability Statement: No new data were created or analyzed in this study. Data sharing is not applicable to this article.

Acknowledgments: I would like to thank all authors and reviewers and to the *Viruses* Editorial Office for their support throughout this Special Issue.

Conflicts of Interest: The author declares no conflict of interest.

References

1. Johnson, S.C.; Turcotte, L.D.M.; Siah, A.; Bradshaw, J.C.; Polinski, M.P. Analysis of Piscine orthoreovirus genotype 1 genomes collected over a 32-year period (1988–2020) suggests a fitness peak in salmon hosts with minimal evidence for temporal divergence. *Virus Evol.* **2025**, *11*, veaf033. [CrossRef] [PubMed]
2. MacAulay, S.; Ellison, A.R.; Kille, P.; Cable, J. Moving towards improved surveillance and earlier diagnosis of aquatic pathogens: From traditional methods to emerging technologies. *Rev. Aquac.* **2022**, *14*, 1813–1829. [CrossRef] [PubMed]
3. Mondal, H.; Thomas, J. A review on the recent advances and application of vaccines against fish pathogens in aquaculture. *Aquac. Int.* **2022**, *30*, 1971–2000. [CrossRef] [PubMed]
4. Ma, J.; Bruce, T.J.; Jones, E.M.; Cain, K.D. A review of fish vaccine development strategies: Conventional methods and modern biotechnological approaches. *Microorganisms* **2019**, *7*, 569. [CrossRef]
5. Hadiuzzaman, M.; Moniruzzaman, M.; Shahjahan, M.; Bai, S.C.; Min, T.; Hossain, Z. β-Glucan: Mode of action and its uses in fish immunomodulation. *Front. Mar. Sci.* **2022**, *9*, 905986. [CrossRef]

6. Polinski, M.; Gross, L.; Groman, D.; Alarcón, M.; Braceland, M.; Booman, M.; Ditlecadet, D.; May, S.; Gagné, N.; Garver, K. PRV-1 virulence in Atlantic salmon is affected by host genotype. *Viruses* **2025**, *17*, 285. [CrossRef]
7. Polinski, M.P.; Zhang, Y.; Morrison, P.R.; Marty, G.D.; Brauner, C.J.; Farrell, A.P.; Garver, K.A. Innate antiviral defense demonstrates high energetic efficiency in a bony fish. *BMC Biol.* **2021**, *19*, 138. [CrossRef]
8. Zhang, Y.; Polinski, M.P.; Morrison, P.R.; Brauner, C.J.; Farrell, A.A.P.; Garver, K.A. High-load reovirus infections do not imply physiological impairment in salmon. *Front. Physiol.* **2019**, *10*, 114. [CrossRef]
9. Manessis, G.; Gelasakis, A.I.; Bossis, I. Point-of-care diagnostics for farm animal diseases: From biosensors to integrated lab-on-chip devices. *Biosensors* **2022**, *12*, 455. [CrossRef]
10. Gervais, O.; Papadopoulou, A.; Gratacap, R.; Hillestad, B.; Tinch, A.E.; Martin, S.A.; Houston, R.D.; Robledo, D. Transcriptomic response to ISAV infection in the gills, head kidney and spleen of resistant and susceptible Atlantic salmon. *BMC Genom.* **2022**, *23*, 775. [CrossRef]
11. Breyta, R.; Samson, C.; Blair, M.; Black, A.; Kurath, G. Successful mitigation of viral disease based on a delayed exposure rearing strategy at a large-scale steelhead trout conservation hatchery. *Aquaculture* **2016**, *450*, 213–224. [CrossRef]
12. Lovy, J.; Abbadi, M.; Toffan, A.; Das, N.; Neugebauer, J.N.; Batts, W.N.; Clarke, P.J. Detection and genetic characterization of red-spotted grouper nervous necrosis virus and a novel genotype of nervous necrosis virus in black sea bass from the US Atlantic coast. *Viruses* **2025**, *17*, 1234. [CrossRef] [PubMed]
13. Raines, C.; Odenkirk, J.; Isel, M.; Mazik, P.; Biggs, M.; Iwanowicz, L. Hiding in Plain Sight: Genomic Characterization of a Novel Nackednavirus and Evidence of Diverse Adomaviruses in a Hyperpigmented Lesion of a Largemouth Bass (*Micropterus nigricans*). *Viruses* **2025**, *17*, 1173. [CrossRef] [PubMed]
14. Ben, H.; Wang, X.; Yang, P.; Li, L.; Liu, P.; Gao, Y.; Wang, Y.; Liu, Y.; Huang, C.; Chen, D. Metagenomic analysis uncovers novel hepadnaviruses and nackednaviruses. *Sci. Rep.* **2025**, *15*, 24699. [CrossRef]
15. He, B.; Sridhar, A.; Thiry, M.; Haenen, O.; Vanderplasschen, A.F.; Donohoe, O. Genomic and Phenotypic Characterization of a Novel Virulent Strain of Cyvirus cyprinidallo 2 Originating from an Outbreak in The Netherlands. *Viruses* **2025**, *17*, 658. [CrossRef]
16. Zhang, Q.; Chang, O.; Lin, Q.; Liang, H.; Niu, Y.; Luo, X.; Ma, B.; Li, N.; Fu, X. Infectious Spleen and Kidney Necrosis Virus Triggers Ferroptosis in CPB Cells to Enhance Virus Replication. *Viruses* **2025**, *17*, 713. [CrossRef]
17. Niu, Y.; Yang, X.; Liang, H.; Luo, X.; Ma, B.; Lin, Q.; Fu, X.; Li, N. scTRIM44 Positively Regulated Siniperca Chuatsi Rhabdovirus Through RIG-I-and MDA5-Mediated Interferon Signaling. *Viruses* **2024**, *16*, 1876. [CrossRef]
18. Yamkasem, J.; Tattiyapong, P.; Gardner, I.A.; Surachetpong, W. Assessment and Performance of Pooled Serum Samples for Monitoring Farm-Level Immunity in Tilapia Infected with Tilapia Lake Virus. *Viruses* **2025**, *17*, 877. [CrossRef]
19. Setthawong, P.; Yamkasem, J.; Khemthong, M.; Tattiyapong, P.; Metheenukul, P.; Prasertsincharoen, N.; Lertwanakarn, T.; Thengchaisri, N.; Surachetpong, W. Development of IgY-Based Passive Immunization Against Tilapia Lake Virus: Development and In Vitro Neutralization Assays. *Viruses* **2025**, *17*, 448. [CrossRef]
20. Toubanaki, D.K.; Tzortzatos, O.-P.; Efstathiou, A.; Bakopoulos, V.; Karagouni, E. Influence of Viral Re-Infection on Head Kidney Transcriptome of Nervous Necrosis Virus-Resistant and-Susceptible European Sea Bass (*Dicentrarchus labrax*, L.). *Viruses* **2025**, *17*, 230. [CrossRef]
21. Shavalier, M.A.; Faisal, M.; Loch, T.P. Lake Trout (*Salvelinus namaycush*) Naturally Infected with *Salmovirus salmonidallo3* (SalHV-3; Family *Alloherpesviridae*) Continue to Harbor the Virus for Nearly a Decade. *Viruses* **2025**, *17*, 1466. [CrossRef]
22. Rahaman, M.M.; Sharma, B.; Talukder, S.; Uddin, M.J.; Siddik, M.A.; Sarker, S. Viral Threats to Australian Fish and Prawns: Economic Impacts and Biosecurity Solutions—A Systematic Review. *Viruses* **2025**, *17*, 692. [CrossRef]

viruses

MDPI

Article

Lake Trout (*Salvelinus namaycush*) Naturally Infected with *Salmovirus salmonidallo3* (SalHV-3; Family *Alloherpesviridae*) Continue to Harbor the Virus for Nearly a Decade

Megan A. Shavalier [1,2], Mohamed Faisal [1,2,3] and Thomas P. Loch [1,2,3,*]

[1] Aquatic Animal Health Laboratory, Aquatic Animal Disease Ecology Program, Michigan State University, East Lansing, MI 48824, USA; shavali1@msu.edu (M.A.S.); faisal@msu.edu (M.F.)

[2] Department of Fisheries and Wildlife, College of Agriculture and Natural Resources, Michigan State University, East Lansing, MI 48824, USA

[3] Department of Pathobiology and Diagnostic Investigation, College of Veterinary Medicine, Michigan State University, East Lansing, MI 48824, USA

* Correspondence: lochthom@msu.edu; Tel.: +1-(517)-884-2019

Abstract

Salmovirus salmonidallo3 (SalHV-3) causes Epizootic Epitheliotropic Disease (EED), which has resulted in the death of millions of lake trout (*Salvelinus namaycush*) over the past 40 years. Although advancements in understanding this virus's pathogenicity and control strategies have been made, the duration and effects of chronic SalHV-3 infections remain unknown. This study focused on lake trout that survived a natural outbreak of EED in 2012 and were maintained under quarantine conditions until 2021. Following exposure to either repeated or intermittent handling stress designed to mimic typical hatchery practices, SalHV-3 was detected (via a SalHV-3-specific quantitative PCR assay) in multiple tissues from multiple fish. Non-lethally collected tissues revealed the highest prevalence and virus loads in the fin and mucus. SalHV-3 was detected in these tissues throughout the study period (49 days, 8 sampling events), with some fish having detectible virus on each study day and others only intermittently (n = 1–7 sampling days). Tissues collected lethally yielded SalHV-3 detections in multiple nervous tissues, as well as in the cornea of several fish. Experiments to evaluate virus shedding revealed that SalHV-3 was intermittently detectable in fish holding water. Collectively, results indicate that lake trout can remain SalHV-3 infected for nearly a decade and intermittently shed the virus, constituting a threat to hatchery-based lake trout conservation efforts in the Great Lakes basin.

Keywords: herpesvirus; lake trout; Great Lakes; fish disease; fishery conservation; latency; virus pathogenesis; *Alloherpesviridae*

1. Introduction

Salmovirus salmonidallo3 (SalHV-3), the causative agent of Epizootic Epitheliotropic Disease (EED), is a large, double-stranded DNA virus in the family *Alloherpesviridae* (Order *Herpesvirales*) [1] that has led to millions of mortalities in hatchery-reared lake trout (*Salvelinus namaycush*) since its discovery in the 1980s [2–4]. Disease signs associated with EED (as reported during resurgence events in 2012 and 2017) include ocular hemorrhage, corneal opacity, gill pallor, skin erythema, ulceration of the skin and fins, and hemorrhage and congestion of the visceral organs [4]. Recent studies have shed light on SalHV-3 disease course, tissue distribution, shedding, and sensitivity to a commonly used disinfectant in aquaculture [5–7]; however, the virus has yet to be cultured in vitro.

A previous study found that upon experimental bath challenge of juvenile lake trout with tissue homogenates collected and prepared from fish infected with SalHV-3, the virus could first be detected in ocular tissue as early as 9 days post exposure, then spread to multiple visceral organs by day 21 post exposure, but with a clear virus predilection of the eyes, skin, and gill tissues [5]. The median lethal dose of SalHV-3 via bath immersion was determined to be 4.7×10^4 virus copies/mL water [5]. In comparison, viral loads in dermal and ocular tissues reached titers of greater than 10^8 virus copies per milligram [5]. Gross disease signs observed in experimentally challenged fish (e.g., ocular changes, ulceration of the skin and fins) were consistent with disease signs reported during natural EED epizootics [4,5].

Another study demonstrated that SalHV-3-infected, two-year-old lake trout shed the virus for at least 9 weeks post-exposure and at relatively high loads (i.e., up to 10^9 virus copies per fish per hour), and the virus could be detected in, and shed from, surviving fish for at least a year [7]. Most remarkable was the ability of some fish to survive an infection dose that killed the majority of other fish in the same tanks. This observation raised several questions, such as: (1) can SalHV-3 persist in fish without causing overt disease, and if so, for how long?; (2) can surviving fish continue to shed the virus under culture conditions with its associated stressor episodes?; and (3) is SalHV-3 and its associated disease restricted to juvenile lake trout, and if not, what kind of tissue distribution and clinical signs can be associated with such an infection in older fish? Unfortunately, none of the previously published field or laboratory studies addressed these questions, creating a gap in knowledge that is vital to designing effective strategies for the control of EED under hatchery conditions and beyond.

In the current study, a group of adult lake trout that had survived a natural SalHV-3 outbreak associated with >25% cumulative mortality [4] were analyzed to determine if: (a) clinical EED could be induced in these fish nearly a decade following clinical recovery; (b) as in infected juvenile lake trout, the virus would be most associated with ocular and epithelial tissues; and (c) virus shedding into the surrounding water would occur at the same intensity as seen with juvenile lake trout. To answer these questions, fish were subjected to either repeated or intermittent handling stress that mimicked what is routinely experienced by captive hatchery broodstock. The ability to maintain recovered SalHV-3-infected fish for nearly a decade provided an extraordinarily rare opportunity to better understand the potential of SalHV-3 to impact lake trout populations indigenous to the Great Lakes of North America.

2. Materials and Methods

2.1. Fish and Husbandry

This study was conducted in ten-year-old lake trout (average 59.8 cm total length, 2.15 kg) that survived a naturally occurring EED outbreak in 2012 [4], were transferred from the Michigan Department of Natural Resources (MI-DNR) to the Michigan State University—Research Containment Facility (East Lansing, Michigan)—in 2013, and were subsequently reared under quarantine conditions until 2021 in accordance with the Michigan State University Institutional Animal Care and Use Committee guidelines and approval (MSU IACUC; Protocol numbers 07/12-124-99, 11/14-201-00, 11/17-197-00). During this study, fish were housed in two 680 L fiberglass aquaria receiving continuous, ultraviolet-treated, oxygenated well water (12 ± 2 °C), fed AquaMax Sport Fish 600 (Nestlé Purina, St. Louis, MO, USA) to satiation, and tanks cleaned via siphon as needed to remove detritus. The study was conducted in accordance with MSU IACUC guidelines and approval (Protocol number 202000030).

2.2. *Fish Handling and Sample Collection*

2.2.1. Water Sample Collection to Assess Virus Shedding

Prior to the start of the present study, fish were divided into two continuous flow-through 680 L tanks: nine fish were placed in Tank 1 (i.e., Group 1) and 10 fish in Tank 2 (i.e., Group 2). To determine if there was any potential virus shedding occurring, on sampling days, water flow to both tanks was turned off for one hour, while supplemental aeration was maintained and fish monitored continuously. Following fish being held for one hour in the two static holding tanks, water (~40 mL) was collected from each tank for SalHV-3 detection and quantification [7]. Additional water samples were collected each week from the holding and/or sedation tank after all fish had been sedated (for non-lethal sampling) and transferred back to their primary tanks. The collected water samples are summarized in Section 3.5.

2.2.2. Handling and PIT-Tagging

On day 0, after the one-hour holding period to assess virus shedding, and to mimic the handling stress experienced by captive broodstock in hatcheries (e.g., during gamete collection), fish were transferred via net to a 378 L holding tank containing aerated tank water, and the main tank water supply was turned back on. Each fish was sedated with 0.1 g/L tricaine methanesulfonate (MS-222; Syndel, Ferndale, WA, USA) buffered with 0.2 g/L sodium bicarbonate (Church & Dwight Co. Inc., Ewing, NJ, USA). Once sedated, the fish were injected with a 9 mm passive integrated transponder (PIT) tag (Biomark, Boise, ID, USA) in the dorsal musculature following industry standard practices for salmonid broodstocks so that each fish could be individually tracked throughout the study. Samples were then taken non-lethally as detailed below, after which the fish were returned to their flow-through tanks and monitored for recovery. Both groups of fish were likewise handled on days 0 and 49. On Days 7, 14, 21, 28, 35, and 42, Group 1 fish were handled in this manner (repeated handling stress) while Group 2 fish were not (intermittent handling stress). On Day 49, Group 2 fish were handled exactly as on day 0. In contrast, after being netted out of their holding tank and prior to tissue sample collection, Group 1 fish were euthanized with an overdose of MS-222 (0.25 g/L) buffered with sodium bicarbonate (0.5 g/L).

2.2.3. Non-Lethal Tissue Sample Collection

Non-lethal tissue samples were collected from all fish in both groups on Days 0 and 49. Additionally, non-lethal samples were also collected from Group 1 fish on Days 7, 14, 21, 28, 35, and 42. Once each fish was sedated, non-lethal samples were collected as follows:

1. Mucus—approximately 100 μL of mucus per fish was collected from along the left lateral line and around the base of the left pectoral and pelvic fins, using a 1 mL slip-tip syringe barrel (Becton, Dickinson and Company, Franklin Lakes, NJ, USA).
2. Blood—less than 2 mL of whole blood was collected from each fish (maximum of 0.3% body weight), via caudal vessel puncture using a 3 mL luer lock syringe and 20-gauge needle (Becton, Dickinson and Company, Franklin Lakes, NJ, USA). Whole blood was then centrifuged (5000 rpm, 10 min, 4 °C), after which serum was collected.
3. Fin clip—approximately 10 mg tissue was collected using sterile scissors from the left or right pectoral fin.

Following collection, all mucus, serum, and fin tissue samples were immediately frozen at −80 °C for subsequent SalHV-3 testing as detailed below.

2.2.4. Lethal Tissue Sample Collection

To assay for SalHV-3 DNA in internal organs following repeated handling stress episodes, Group 1 fish were euthanized on Day 49, and a full clinical examination and necropsy were performed on each. Likewise, tissues from the cerebellum, cerebrum, cornea, cranial nerves (Figure 1), gills, medulla oblongata, olfactory tissue, optic lobe, optic nerve, and gonads were collected from all fish, as were tissues showing additional lesions of concern, and frozen at −80 °C for further analyses.

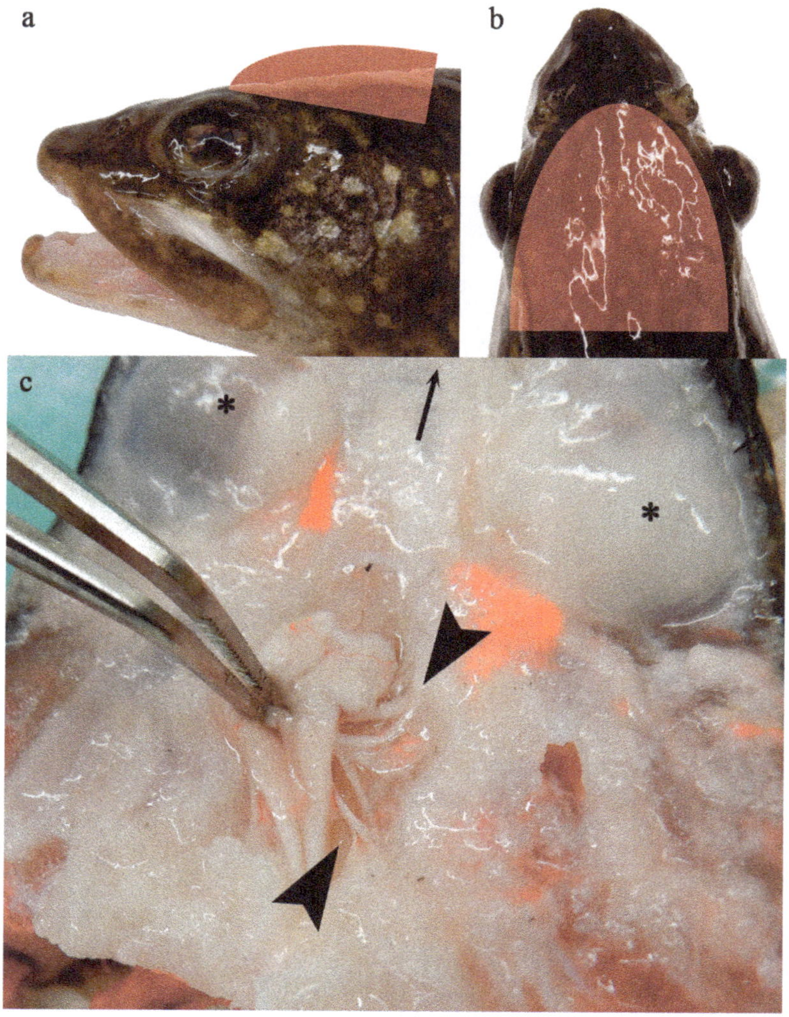

Figure 1. Group 1 fish, apparently normal nervous tissue. Red semi-ellipses indicate approximate level at which transection was made, along a dorsal plane immediately dorsal to the level of the eyes, to expose the brain and associated nervous tissues (**a**,**b**). Brain (held by forceps) and cranial nerve bundle (between arrow heads); dorsal aspect of ocular globes indicated with an *; arrow at top of image indicates rostral direction (**c**).

2.3. *Detection and Quantification of SalHV-3*

2.3.1. DNA Extraction

Salmovirus salmonidallo3 has yet to be cultured in vitro; thus, molecular methods were utilized. Nucleic acid extractions were performed on collected tissues using the Mag Bind® Blood and Tissue DNA Kit (OMEGA Bio-tek, Inc., Norcross, GA, USA), following the manufacturer's instructions (saliva protocol for mucus and serum, tissue protocol for all other samples) and with the addition of a filtering step using the E-Z 96® Lysate Clearance Plate (OMEGA Bio-tek, Inc.) [8]. Approximately 10 mg of solid tissue was used for each extraction, and a maximum of 250 μL mucus, serum or swim bladder fluid.

Nucleic acid extractions from water were performed following the Alternative PowerSoil Protocol for Low Bacterial Biomass Fluids using the Qiagen DNeasy® PowerLyzer® PowerSoil® Kit (Qiagen, Hilden, Germany) with minor modifications that also included mechanical disruption via bead-beating [7]. Frozen water samples were thawed to room temperature and vortexed briefly. The bead solution (500 μL), phenol:chloroform (200 μL; isoamyl alcohol; AMRESCO, Solon, OH, USA), and the C1 solution (60 μL) were added to the supplied bead tubes. 250 μL of the water sample was then added into this mixture, vortexed briefly, loaded into the bead beater (Mini-Beadbeater-16; Biospec Inc., Bartlesville, OK, USA) and run on high for 30 s twice with a 20 s rest period between the two bead beating cycles. The mixture was then centrifuged at 10,000× *g* for 1 min at 4 °C. The supernatant was collected and transferred into a new tube provided in the kit, and 1 μL of RNase A was added, followed by 100 μL C2 solution and 100 μL C3 solution. Tubes were then vortexed and incubated at 4 °C for 5 min. Samples were centrifuged for 1 min at 10,000× *g*, and the supernatant was transferred to a new tube. 650 μL C4 solution and 650 μL 100% ethanol were then added to each sample. The remaining steps followed the manufacturer's instructions, with the addition of the C6 solution being heated to 60 °C before being used to elute the DNA.

Extracted DNA from both protocols was quantified using a Qubit™ fluorometer (Invitrogen, Eugene, OR, USA), and samples were diluted with sterile DNase-free water to no greater than 12.5 ng/μL qPCR template DNA (50 ng per qPCR reaction).

2.3.2. Quantitative PCR Analysis

All qPCR reactions were carried out in a QuantStudio 3 real-time PCR system (ThermoFisher Scientific, Waltham, MA, USA), and were performed using the primers as described by Glenney et al. [8] and as detailed in Shavalier et al. [5]. Each 20 μL reaction contained 10 μL of SYBR® Select Master Mix, 2 μL of nuclease-free water (Promega), 1.0 μM of each primer, and template containing a maximum of 50 ng total DNA. The qPCR cycling parameters consisted of 50 °C for 2 min; 95 °C for 10 min; and 40 cycles of 95 °C for 15 s, and 60 °C for 60 s. Previously confirmed SalHV-3-positive tissue homogenate served as a positive extraction control (PEC), and sample diluent was used as a negative extraction control (NEC). SalHV-3-positive purified DNA and nuclease-free water served as the positive reaction control (PRC) and negative reaction control (NRC), respectively. Samples were considered SalHV-3-positive if amplification occurred within 35 amplification cycles as determined with the QuantStudio 3 accompanying software (ThermoFisher Connect platform, Diomni™ Design and Analysis (RUO) 3.1.0) and the manufacturer's default settings. Positive control standards for quantification were produced using known positive skin samples following the method outlined by Glenney et al. [8].

Viral loads in tissues and shedding rates (viral copies per fish per hour) were calculated using the resulting reaction copy number calculated by the QuantStudio 3 accompanying software, sample volume or weight, and sample period length.

3. Results

3.1. Clinical Observations

Over the nearly 9 years that these naturally SalHV-3-infected lake trout were held in quarantine, they experienced one additional and confirmed outbreak of EED in one of three holding tanks in use at the time. That particular tank experienced >99% mortality, and SalHV-3 was detected in 100% of fish tested from the afflicted lot (n = 20/20 gills) [4]. This event occurred one year after the initial outbreak in the hatchery, and ~six months after being moved to quarantine (i.e., in the fall of 2013). The surviving fish grew to adulthood in captivity. With the combination of this disease outbreak, occasional mortality or euthanasia of moribund fish, and periodic culling of select individuals over the years for virus screening and experimental stock production, 19 fish were available for the experiment conducted herein.

No mortality occurred throughout the present experiment in either the repeated handling group or the intermittent handling group. Likewise, neither group of fish showed observable alterations in behavior or appetite throughout the seven weeks of the study; however, some clinical signs associated with EED were observed during the postmortem clinical examination. Fish in Group 1 showed fin and skin erosion ranging from mild to severe, as well as various ocular changes, including exophthalmia (Figure 2a,b), enophthalmia (Figure 2c,d), corneal opacity (Figure 2c) and corneal scarring/thickening (Figure 2d). Although ocular changes were seen in all nine fish, chronic skin and fin changes (Figure 2e,f) appeared more pronounced in the fish with detectable levels of SalHV-3 (see below) than in those without. Internally, observed gross signs of disease included focal hemorrhage within the gonads, renal congestion, swelling of the spleen, hepatic pallor, and mild congestion of the gastric and intestinal vessels. No obvious trends were noted when comparing internal exams of fish from which SalHV-3 was detected and those that were below the detection level.

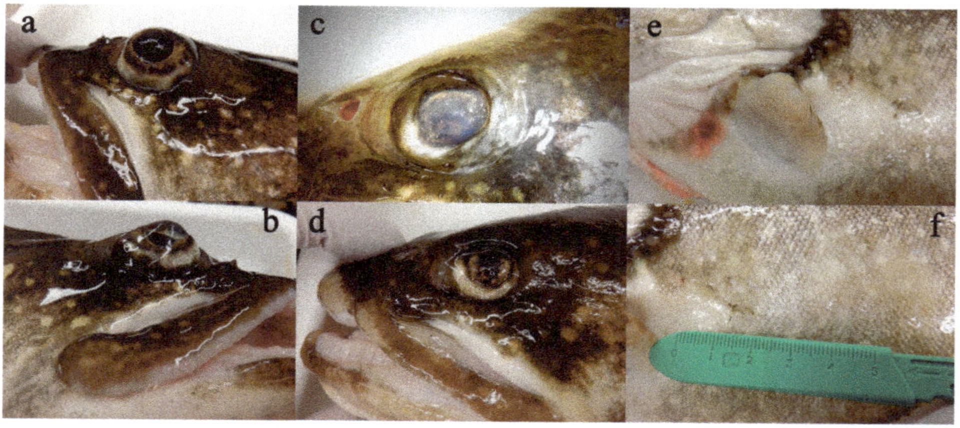

Figure 2. Representative external gross findings in Group 1 fish include exophthalmia (**a**,**b**), enophthalmia (**c**,**d**), corneal opacity (**c**), corneal scarring/thickening (**d**), and chronic fin and skin erosions (**e**,**f**).

3.2. Non-Lethal Samples—Group 1, Repeated Handling

A total of 216 tissue samples were collected from nine EED-surviving lake trout that were sampled weekly over seven weeks. Across all rounds of nucleic acid extraction and qPCR, negative controls (NEC and NRC) yielded no amplification. Positive extraction controls from previously positive skin homogenates yielded amplification within the expected range (e.g., Ct values 24–31; mean 3.2×10^4 copies per reaction), as did positive reaction control standards at 10-fold dilutions (Ct values ranging from 17–33). Five of the nine fish had at least one tissue where SalHV-3 was detected via qPCR (Tables 1 and 2), with three having detectable virus DNA at Day 0. Although three fish had virus DNA detections across multiple days, only one fish had detectable SalHV-3 on every sampling day (Tables 1 and 2). Of the 71 fin samples tested, 18 had detectable SalHV-3 (from $n = 5$ fish), with viral loads ranging from 3.88×10^2 to 8.01×10^6 virus copies per mg of tissue (mean of 1.14×10^6; median of 1.22×10^5). In comparison, SalHV-3 was detected in 13 mucus samples (from $n = 5$ fish) at loads ranging from 6.10×10^3 to 1.01×10^7 virus copies per mL tissue (mean of 2.13×10^6; median of 8.41×10^4). SalHV-3 was detected in a single serum sample throughout the study (1.78×10^4 virus copies/mL tissue; Day 14). Collectively, qPCR analyses revealed a range of patterns in individual fish, whereby some fish were consistently positive or negative throughout, and others varied over the course of repeat sampling (Tables 1 and 2).

3.3. Non-Lethal Samples—Group 2, Intermittent Handling

A total of 60 samples were collected from 10 fish in Group 2, seven weeks apart. SalHV-3 was not detected in any serum samples from intermittently sampled fish. However, there was a total of 11 SalHV-3 detections in fin or mucus samples from 5 fish across both sampling days (Tables 1 and 2). Virus loads in the fin tissues ($n = 5$) ranged from 1.95×10^2 to 5.31×10^5 viral copies per mg tissue (mean of 1.38×10^5; median of 1.08×10^5). Virus loads in the mucus samples ($n = 6$) ranged from 1.36×10^4 to 4.50×10^6 viral copies per mL tissue (mean of 9.26×10^5; median of 8.90×10^4). Four of the five fish had detectible SalHV-3 on both sampling days, although not necessarily in the same tissues (e.g., SalHV-3 was detected in two fish from fin tissues on day 0 but mucus on day 49). One fish had detectible SalHV-3 (mucus) on the second sampling day only.

Table 1. Presence (virus copies/mg tissue) or absence (-) of SalHV-3 DNA as detected by qPCR in non-lethally collected fin tissues. Blank spaces represent tissues not collected.

	Fish #	Study Day							
		0	7	14	21	28	35	42	49
Group 1—Repeated Handling	1	-	3.88×10^2	-	-	-	-	-	-
	2	1.09×10^6	1.43×10^6	3.99×10^4	2.13×10^4	1.15×10^5	2.33×10^4	6.21×10^4	3.44×10^5
	3	-	-	-	-	-	-	-	-
	4	-	-	-	5.87×10^2	-	-	-	-
	5	-	-	-	-	-	-	-	-
	6	-	-	-	-	-	-	-	-
	7	3.91×10^6	9.41×10^5	8.01×10^6	2.53×10^6	-	5.48×10^5	1.00×10^5	1.28×10^6
	8	-	-	-	-	-	-	-	-
	9	5.43×10^2	-	-	-	-	-	-	-
Group 2—Intermittent Handling	10	-							-
	11	1.95×10^2							-
	12	-							-
	13	-							-
	14	-							-
	15	5.31×10^5							-
	16	1.22×10^5							3.64×10^4
	17	-							-
	18	-							1.15×10^3
	19	-							-

Table 2. Presence (copies/mL tissue) or absence (-) of SalHV-3 DNA as detected by qPCR in non-lethally collected mucus tissues. Blank spaces represent tissues not collected.

	Fish #	Study Day							
		0	7	14	21	28	35	42	49
Group 1—Repeated Handling	1	-	-	-	-	-	-	-	-
	2	-	-	6.61×10^3	8.90×10^4	-	2.89×10^4	-	-
	3	-	-	-	-	-	-	-	-
	4	-	1.01×10^7	6.13×10^6	1.07×10^6	-	5.40×10^4	6.13×10^3	9.90×10^6
	5	-	-	-	-	-	-	-	-
	6	-	-	-	-	-	-	-	-
	7	-	-	-	8.41×10^4	8.23×10^3	-	-	-
	8	-	-	-	-	-	-	-	1.53×10^5
	9	-	-	3.69×10^4	-	-	-	-	-
Group 2—Intermittent Handling	10	-							1.05×10^5
	11	-							5.15×10^5
	12	-							-
	13	-							-
	14	-							-
	15	-							4.06×10^5
	16	-							2.04×10^4
	17	-							-
	18	1.36×10^4							4.50×10^6
	19	-							-

3.4. Lethal Samples—Group 1, Repeated Handling

Among the samples collected at the end of the study from the euthanized fish in Group 1, SalHV-3 DNA was not detected in any cerebellum, cerebrum, gill, or reproductive tissues via qPCR (Table 3). SalHV-3-positive tissues included 3/9 corneas (viral load ranging from 1.34×10^2 to 6.50×10^5 copies per mg tissue; mean of 2.37×10^5; median of 6.04×10^4), 4/9 cranial nerves (viral load ranging from 2.10×10^2 to 3.38×10^3 viral copies per mg tissue; mean of 1.01×10^3; median of 2.30×10^2), 1/9 medulla oblongata (1.14×10^2 viral copies per mg tissue), 1/8 olfactory tissues (1.49×10^2 viral copies per mg tissue), 1/9 optic lobes (2.69×10^3 viral copies per mg tissue), 1/9 optic nerves (2.94×10^1 viral copies per mg tissue), and 1/1 swim bladder fluid (9.41×10^4 viral copies per mL). Positive tissues were detected in six of the nine euthanized fish, with five of the six having at least two positive tissues. Among the six fish with detectable SalHV-3 DNA, four yielded both positive nervous/ocular (i.e., one or more of the following tissues: cranial nerve, medulla oblongata, olfactory, optic lobe, optic nerve, cornea) and external tissues (i.e., fin or mucus), with the remainder having detectable SalHV-3 DNA in only nervous (n = 1 fish) or external (n = 1 fish) tissues (Table 3).

3.5. SalHV-3 Shedding

A total of eight MS-222 water samples were collected following sedation of the Group 1 (repeated handling) fish on days 0, 7, 14, 21, 28, and 42, and the Group 2 (intermittent handling) fish on days 0 and 49. Of these, one sample had detectable SalHV-3 (Day 7, Group 1 repeated handling) at an estimated concentration of 2.36×10^3 copies/mL water. A total of 14 water samples were collected from the two fish tanks following 1 h of static holding (both Group 1 and Group 2 fish on days 0, 7, 14, 21, 28, 42, and 49). Of these, SalHV-3 was detected in one water sample (Day 7, Group 2 intermittent handling) at a concentration of 2.04×10^4 copies/mL water.

Table 3. Number of SalHV-3-positive tissue samples by fish throughout the study.

		Tissue Type												
		Non-Lethal			Lethal									
	Fish #	Fin	Mucus	Serum	Cerebellum	Cerebrum	Cranial Nerve	Medulla Oblongata	Olfactory	Optic lobe	Optic Nerve	Cornea	Gill	Gonads
Group 1—Repeated Handling	1	0/8	0/8	0/8	0/1	0/1	0/1	0/1	0/1	0/1	0/1	0/1	0/1	0/1
	2	***8/8***	***3/8***	0/8	0/1	0/1	***1/1***	0/1	0/1	***1/1***	***1/1***	***1/1***	0/1	0/1
	3	0/8	0/8	0/8	0/1	0/1	***1/1***	0/1	***1/1***	0/1	0/1	0/1	0/1	0/1
	4	***1/8***	***6/8***	***1/8***	0/1	0/1	0/1	0/1	-	0/1	0/1	0/1	0/1	0/1
	5	0/8	0/8	0/8	0/1	0/1	0/1	0/1	0/1	0/1	0/1	0/1	0/1	0/1
	6	0/8	0/8	0/8	0/1	0/1	0/1	0/1	0/1	0/1	0/1	0/1	0/1	0/1
	7	***7/8***	***2/8***	0/8	0/1	0/1	***1/1***	0/1	0/1	0/1	0/1	***1/1***	0/1	0/1
	8	0/8	***1/8***	0/8	0/1	0/1	***1/1***	***1/1***	0/1	0/1	0/1	0/1	0/1	0/1
	9	***1/8***	***1/8***	0/8	0/1	0/1	0/1	0/1	0/1	0/1	0/1	***1/1***	0/1	0/1
Group 2—Intermittent Handling	10	0/2	***1/2***	0/2	-	-	-	-	-	-	-	-	-	-
	11	***1/2***	***1/2***	0/2	-	-	-	-	-	-	-	-	-	-
	12	0/2	0/2	0/2	-	-	-	-	-	-	-	-	-	-
	13	0/2	0/2	0/2	-	-	-	-	-	-	-	-	-	-
	14	0/2	0/2	0/2	-	-	-	-	-	-	-	-	-	-
	15	***1/2***	***1/2***	0/2	-	-	-	-	-	-	-	-	-	-
	16	***2/2***	***1/2***	0/2	-	-	-	-	-	-	-	-	-	-
	17	0/2	0/2	0/2	-	-	-	-	-	-	-	-	-	-
	18	***1/2***	***2/2***	0/2	-	-	-	-	-	-	-	-	-	-
	19	0/2	0/2	0/2	-	-	-	-	-	-	-	-	-	-

4. Discussion

Although findings from the present study are based upon a relatively small number of EED-recovered fish (n = 19), their maintenance for nearly a decade allowed several heretofore unknown aspects of SalHV-3 pathogenesis to be uncovered. First, *Salmovirus salmonidallo3* DNA was detected in adult lake trout 8 years and 7 months following a natural disease outbreak. Gross signs of disease as detailed above were consistent with what is typically seen in lake trout that have previously experienced an active SalHV-3 infection, either natural or experimentally induced (chronic ocular, skin, and fin changes) [4,5]. Because this group of fish was held in quarantine for nearly a decade, the detected virus in these fish definitively originated from the natural outbreak. Whether SalHV-3 was present as a persistent infection in surviving lake trout or went through latency and became reactivated due to the handling stress is currently unknown.

Latency or viral persistence has been demonstrated in several other alloherpesviruses, including the cyprinid herpesviruses (*Cyvirus cyprinidallo1*, CyHV-1; *Cyvirus cyprinidallo2*, CyHV-2; *Cyvirus cyprinidallo3*, CyHV-3), and *Ictavirus ictalluridallo1* (IcHV-1) [9–12], and a recent review (Quijano Cardé and Soto, 2024) highlighted the current knowledge of *Alloherpesviridae* in this context [13]. Eide et al. [14] demonstrated latency and viral reactivation in koi (*Cyprinus carpio*) previously exposed to koi herpesvirus (CyHV-3), with the time interval between disease outbreak and virus reactivation reported as "several years". Baumer et al. [11] also examined latency in CyHV-3, demonstrating viral shedding in surviving fish up to 21 months post-exposure. Subclinical infection with *Cyvirus cyprinidallo2* in one-year-old gibel carp (*Carassius gibelio*) was described by Wei et al. [15], who were also able to reactivate the production of infectious virus. Latency and reactivation has been of much interest with *Ictavirus ictalluridallo1* as well; however, while virus DNA has occurred in various groups of subclinical fish, reactivation of clinical disease has proven challenging to achieve in a laboratory setting [16]. Results from the present study likewise indicate that an alloherpesvirus of lake trout (Family *Salmonidae*), SalHV-3, is capable of generating persistent and/or latent infections of a substantial duration in surviving fish, in this case at least eight and a half years after a naturally occurring EED hatchery outbreak [4].

In a similar context, SalHV-3 DNA was detected in holding tank water during this study, suggesting that eight and a half years after surviving an EED outbreak, some LAT remain capable of shedding the virus and thereby represent a potential long-term reservoir of infection. Although knowledge on shedding dynamics of other alloherpesviruses is incomplete, viral shedding and duration have indeed been investigated in several instances. For example, it has been reported (via analysis of mucosal swabs and in vivo cohabitation experiments) that CyHV-3 can be shed for up to 57 days in experimentally exposed, subclinically affected fish [17] and up to 34 days post-challenge in a separate study [18]. In both instances, longer viral shedding was associated with colder water temperatures and milder clinical disease. Kancharla and Hanson [19] showed that *Ictavirus ictaluridallo1* shedding peaks around 4 days post-infection and continues for at least 12 days. In the current study, the shedding of SalHV-3 was seemingly intermittent and, when detected, at loads estimated to be ~10^4 virus copies/mL water/hour. Although lower than the SalHV-3 shedding rates that were detected during the peak of an experimentally induced EED outbreak in juvenile lake trout [7], this virus concentration is near to the estimated median lethal dose via immersion as determined in our previous studies [5]. In the present study, SalHV-3 was detected in a subset of water samples despite a relatively short fish holding time. It is possible that increasing the fish holding duration from one hour to eight hours [7] and incorporating a filtration/ultracentrifugation step [20,21] would have revealed a higher or more consistent virus shedding rate in these EED-surviving fish. Regardless, these findings provide further evidence that EED-recovered adult lake trout constitute a potential SalHV-3 transmission risk for an extended period of time.

Virus shedding via the reproductive fluids (i.e., ovarian fluid, milt) is also of concern for salmonid herpesviruses. For example, Wolf et al. [22] identified *Salmovirus salmonidallo1* in the ovarian fluid of rainbow trout (*Oncorhynchus mykiss*); likewise, *Salmovirus salmonidallo2* has been detected in reproductive fluids, but details on the duration of shedding from infected fish via this mode are unknown [23]. SalHV-3 DNA has also been detected in the ovarian fluid of lake trout on multiple prior occasions [8,24]. Interestingly, SalHV-3 was not detected in the reproductive tissues of the fish in this current study. As these fish were not yet ripe (i.e., ready to spawn) at the time of sample collection, viral presence in gonadal tissues and fluids may relate to spawning status and warrants further study.

The present study was designed to mimic the handling and associated stress hatchery broodstock would experience during routine management procedures (i.e., transfer via netting, holding in a confined space for a period of time, anesthesia) and determine if repeated (Group 1) or intermittent (Group 2) handling affected reactivation of SalHV-3 in terms of virus loads in adult LAT that survived a previous, natural EED outbreak. In so doing, several notable observations were made. First, the mean viral load was 2.3–8.3-fold higher in the mucus and fins, respectively, of fish that were subjected to repeated handling (Group 1) compared to those that were only intermittently handled (Group 2). Second, and although only in one fish, SalHV-3 was exclusively detected in the serum of fish within the repeated handling group. The total percentage of positive fish identified by fin clip was nearly identical between the two treatment groups (24% versus 25%), whereas the percentage of SalHV-3 mucus samples was interestingly higher in the intermittently handled (Group 2) fish. Thus, it is possible that in some cases, a single handling stressor may result in just as much risk of virus reactivation as repeated handling stress. Notably, when comparing the mucus samples collected on day zero to those taken on day 49, there was an increase in the number of fish in both groups with detectable levels of SalHV-3, with the repeat sampling group going from 0/9 on day zero to 2/9 on day 49 and the intermittent group going from 1/10 on day zero to 5/10 on day 49.

Studies on other alloherpesviruses have found that stressful events favor reactivation of productive infections [25]. A study on persistently infected common carp demonstrated that netting fish was enough to reactivate CyHV-3 (measured by an increase in viral load in gill swabs) [26]. Captive broodstock can also experience unusual levels of stress during times of intense management (e.g., handling of the fish for movement, vaccination, PIT tagging, etc. [27]), as well as during spawning seasons, leading to an increased risk for herpesvirus reactivation in persistently or latently infected fish, and increased risk of virus transmission due to increased fish density and natural seasonal immune system changes [28]. Identification of infected broodstock, coupled with management planning to limit or reduce stress experienced by the fish, can help to prevent additional spread of the virus.

Results from this study also illustrate that SalHV-3 infects more tissue types than previously recognized. It is well documented that many herpesviruses target different tissues or cell types during an acute infection versus those they infect to establish latency [13,29]. Although previous studies have identified a tropism for epithelia of the skin, fin, gills, and cornea during acute SalHV-3 infections and EED events, and thus are commonly utilized for screening and diagnostic purposes [4,5], prior to this study, the potential targets during long-term SalHV-3 infections were unknown. Other herpesviruses, including many of the human herpesviruses, as well as *Cyvirus cyprinidallo1* and *Cyvirus cyprinidallo2* [25,30,31], are known to develop latency in various nervous tissues, and Tolo et al. reported that CyHV-3 DNA can commonly be detected in brain tissues despite no evidence of viral replication [32]. However, the specific neurotropism of SalHV-3 was previously unexplored. The present study expanded upon previous work that identified SalHV-3 DNA in undifferentiated "brain" tissue [5] by taking advantage of larger adult fish, which allowed for the identification, dissection, and testing of individual nervous tissues. Herein, SalHV-3 DNA was detected in the cranial nerves (including the optic nerve), the medulla oblongata, the optic lobes, and the olfactory tissues collected from inside the nares. Based on trends seen in other herpesviruses, these findings may suggest a neurotropic target for latent SalHV-3 infections pending further investigation. Interestingly, ocular tissues (i.e., cornea) have been identified as an early target during SalHV-3 infections [5], and gross ulcers around the nares are commonly reported during active disease events [4]. Thus, the presence of SalHV-3 DNA in the nervous tissues associated with these areas (olfactory tissue, optic nerve, optic lobe) as identified in this study, points toward these tissues as likely virus targets in fish during advanced stages of infection. Although additional research is needed to determine if these fish were truly latently infected or not, this study highlights the importance of targeting nervous tissues, specifically those within the head, for further investigations into SalHV-3 latency development.

In conclusion, this study demonstrates that SalHV-3 can persist in adult lake trout that survived a natural EED outbreak well beyond previously identified carrier time frames that have been determined thus far for other alloherpesviruses. As these infected fish have also been demonstrated to shed the virus, long-term survivor fish likely represent an important reservoir and virus transmission risk in hatchery environments, and appropriate control and containment measures must be put in place to prevent the spread of SalHV-3 virus to naïve fish populations.

Author Contributions: Conceptualization, M.A.S. and T.P.L.; Data curation, M.A.S.; Formal analysis, M.A.S., M.F. and T.P.L.; Funding acquisition, M.F. and T.P.L.; Investigation, M.A.S. and T.P.L.; Methodology, M.A.S. and T.P.L.; Project administration, M.F. and T.P.L.; Resources, M.F. and T.P.L.; Supervision, T.P.L.; Visualization, M.A.S.; Writing—original draft, M.A.S.; Writing–review and editing, M.A.S., M.F. and T.P.L. All authors have read and agreed to the published version of the manuscript.

Funding: Support for this project was provided by the Great Lakes Fishery Commission (Project ID —2020_LOC_440880). We also appreciate the financial support provided by Michigan State University AgBioResearch to TPL through the Stanislas F. Snieszko Endowed Scholar program.

Institutional Review Board Statement: The animal study protocol was approved by the Institutional Review Board of Michigan State University (Protocol number 202000030; 24 March 2020).

Data Availability Statement: Data is contained within the article.

Acknowledgments: The authors thank the Great Lakes Fishery Commission (Project ID—2020_LOC_440880) for providing funding support for this study. The authors thank the Michigan Department of Natural Resources, Fisheries Division, for their continued support and collaboration, as well as present and past members of the Michigan State University—Aquatic Animal Health Laboratory for their technical expertise throughout the study. The authors are especially grateful for Dana MacCallumMhor, who provided significant technical support during the sampling portion of this study.

Conflicts of Interest: The authors declare no conflicts of interest.

Abbreviations

The following abbreviations are used in this manuscript:

SalHV-3	*Salmovirus salmonidallo3*
CyHV-3	*Cyvirus cyprinidallo3*
EED	Epizootic Epitheliotropic Disease
PIT	Passive Integrated Transponder
MS-222	Tricaine Methanesulfonate
qPCR	Quantitative PCR

References

1. Walker, P.J.; Siddell, S.G.; Lefkowitz, E.J.; Mushegian, A.R.; Adriaenssens, E.M.; Alfenas-Zerbini, P.; Dempsey, D.M.; Dutilh, B.E.; García, M.L.; Hendrickson, R.C.; et al. Recent changes to virus taxonomy ratified by the International Committee on Taxonomy of Viruses (2022). *Arch. Virol.* **2022**, *167*, 2429–2440. [CrossRef] [PubMed]
2. Bradley, T.M.; Medina, D.J.; Chang, P.W.; McClain, J. Epizootic epitheliotrophic disease of lake trout (*Salvelinus namaycush*): History and viral etiology. *Dis. Aquat. Org.* **1989**, *7*, 195–201. [CrossRef]
3. Bradley, T.; Newcomer, C.; Maxwell, K. Epitheliocystis associated with massive mortalities of cultured lake trout *Saivelinus namaycush*. *Dis. Aquat. Org.* **1988**, *4*, 9–17. [CrossRef]
4. Faisal, M.; Loch, T.P.; Shavalier, M.; VanDeuren, M.G.; Standish, I.; Winters, A.; Glenney, G.; Aho, J.; Wolgamood, M.; VanAmberg, J.; et al. Resurgence of Salmonid Herpesvirus-3 Infection (Epizootic Epitheliotropic Disease) in Hatchery-Propagated Lake Trout in Michigan. *J. Aquat. Anim. Health* **2019**, *31*, 31–45. [CrossRef]
5. Shavalier, M.; Faisal, M.; Loch, T.P.; Fitzgerald, S.D.; Thaiwong, T.; Kiupel, M. Disease Progression in Lake Trout (*Salvelinus namaycush*) Experimentally Infected With Epizootic Epitheliotropic Disease Virus (Salmonid Herpesvirus-3). *Vet. Pathol.* **2020**, *57*, 687–699. [CrossRef]
6. Purbayu, M.A.; Shavalier, M.A.; Faisal, M.; Loch, T.P. Experimental Evidence of Epizootic Epitheliotropic Disease Virus (Salmoid Herpesvirus-3, *Alloherpesviridae*) Transmission via Contaminated Fomites and Subsequent Prevention Using a Disinfectant. *Pathogens* **2021**, *10*, 724. [CrossRef] [PubMed]
7. Faisal, M.; Purbayu, M.; Shavalier, M.A.; Marsh, T.L.; Loch, T.P. Shedding of the Salmonid Herpesvirus-3 by Infected Lake Trout (*Salvelinus namaycush*). *Viruses* **2019**, *11*, 580. [CrossRef]
8. Glenney, G.W.; Barbash, P.A.; Coll, J.A. A Quantitative Polymerase Chain Reaction Assay for the Detection and Quantification of Epizootic Epitheliotropic Disease Virus (EEDV; Salmonid Herpesvirus 3). *J. Aquat. Anim. Health* **2016**, *28*, 56–67. [CrossRef]
9. Wei, C.; Iida, H.; Chuah, Q.; Tanaka, M.; Kato, G.; Sano, M. Persistence of cyprinid herpesvirus 2 in asymptomatic goldfish *Carassius auratus* (L.) that survived an experimental infection. *J. Fish Dis.* **2019**, *42*, 913–921. [CrossRef] [PubMed]
10. Lepa, A.; Siwicki, A.K. Fish herpesvirus diseases: A short review of current knowledge. *Acta Vet. Brno* **2013**, *81*, 383–389. [CrossRef]
11. Baumer, A.; Fabian, M.; Wilkens, M.R.; Steinhagen, D.; Runge, M. Epidemiology of cyprinid herpesvirus-3 infection in latently infected carp from aquaculture. *Dis. Aquat. Org.* **2013**, *105*, 101–108. [CrossRef]
12. Davison, A.J.; Eberle, R.; Ehlers, B.; Hayward, G.S.; McGeoch, D.J.; Minson, A.C.; Pellett, P.E.; Roizman, B.; Studdert, M.J.; Thiry, E. The order Herpesvirales. *Arch. Virol.* **2009**, *154*, 171–177. [CrossRef]
13. Quijano Cardé, E.M.; Soto, E. A review of latency in the Alloherpesviridae family. *J. Fish Dis.* **2024**, *47*, e14016. [CrossRef]
14. Eide, K.E.; Miller-Morgan, T.; Heidel, J.R.; Kent, M.L.; Bildfell, R.J.; LaPatra, S.; Watson, G.; Jin, L. Investigation of koi herpesvirus latency in koi. *J. Virol.* **2011**, *85*, 4954–4962. [CrossRef]

15. Wei, C.; Xu, C.; Sun, Y.; Li, J.; Sano, M.; Li, Q. Investigation of the latency of *Cyprinid herpesvirus* 2 in apparently healthy farmed gibel carp, *Carassius auratus gibelio*. *Aquaculture* **2023**, *562*, 738854. [CrossRef]
16. Hanson, L.; Dishon, A.; Kotler, M. Herpesviruses that infect fish. *Viruses* **2011**, *3*, 2160–2191. [CrossRef]
17. Cano, I.; Mulhearn, B.; Akter, S.; Paley, R. Seroconversion and Skin Mucosal Parameters during Koi Herpesvirus Shedding in *Common carp, Cyprinus carpio*. *Int. J. Mol. Sci.* **2020**, *21*, 8482. [CrossRef] [PubMed]
18. Yuasa, K.; Ito, T.; Sano, M. Effect of water temperature on mortality and virus shedding in carp experimentally infected with koi herpesvirus. *Fish Pathol.* **2008**, *43*, 83–85. [CrossRef]
19. Kancharla, S.R.; Hanson, L.A. Production and shedding of channel catfish virus (CCV) and thymidine kinase negative CCV in immersion exposed channel catfish fingerlings. *Dis. Aquat. Org.* **1996**, *27*, 25–34. [CrossRef]
20. Hubbard, L.E.; Stelzer, E.A.; Poulson, R.L.; Kolpin, D.W.; Szablewski, C.M.; Givens, C.E. Development of a Large-Volume Concentration Method to Recover Infectious Avian Influenza Virus from the Aquatic Environment. *Viruses* **2024**, *16*, 1898. [CrossRef] [PubMed]
21. Ip, Y.C.A.; Chen, J.; Tan, L.Y.; Lau, C.; Chan, Y.H.; Balasubramaniam, R.S.; Wong, W.Y.J.; Ng, K.; Tan, Z.Y.B.; Fernandez, C.J.; et al. Establishing environmental DNA and RNA protocols for the simultaneous detection of fish viruses from seawater. *Environ. DNA* **2024**, *6*, e418. [CrossRef]
22. Wolf, K.; Darlington, R.; Taylor, W.; Quimby, M.; Nagabayashi, T. Herpesvirus salmonis: Characterization of a new pathogen of rainbow trout. *J. Virol.* **1978**, *27*, 659–666. [CrossRef]
23. Hanson, L.A.; Doszpoly, A.; van Beurden, S.; de Oliveira Viadanna, P.H.; Waltzek, T. Alloherpesviruses of fish. In *Aquaculture Virology*; Elsevier: Amsterdam, The Netherlands, 2024; pp. 165–189.
24. Kurobe, T.; Marcquenski, S.; Hedrick, R. PCR assay for improved diagnostics of epitheliotropic disease virus (EEDV) in lake trout *Salvelinus namaycush*. *Dis. Aquat. Org.* **2009**, *84*, 17–24. [CrossRef]
25. Grinde, B. Herpesviruses: Latency and reactivation–viral strategies and host response. *J. Oral Microbiol.* **2013**, *5*, 22766. [CrossRef]
26. Bergmann, S.; Kempter, J. Detection of koi herpesvirus (KHV) after re-activation in persistently infected common carp (*Cyprinus carpio* L.) using non-lethal sampling methods. *Bull. Eur. Assoc. Fish Pathol.* **2011**, *31*, 92–100.
27. Bordeleau, X.; Hatcher, B.G.; Denny, S.; Fast, M.D.; Whoriskey, F.G.; Patterson, D.A.; Crossin, G.T. Consequences of captive breeding: Fitness implications for wild-origin, hatchery-spawned Atlantic salmon kelts upon their return to the wild. *Biol. Conserv.* **2018**, *225*, 144–153. [CrossRef]
28. Tort, L.; Balasch, J.C.; Mackenzie, S. Fish immune system. A crossroads between innate and adaptive responses. *Inmunología* **2003**, *22*, 277–286.
29. Cohen, J.I. Herpesvirus latency. *J. Clin. Investig.* **2020**, *130*, 3361–3369. [CrossRef] [PubMed]
30. Sano, N.; Moriwake, M.; Hondo, R.; Sano, T. Herpesvirus cyprini: A search for viral genome in infected fish by infected fish by in situ hybridization. *J. Fish Dis.* **1993**, *16*, 495–499. [CrossRef]
31. Chai, W.; Qi, L.; Zhang, Y.; Hong, M.; Jin, L.; Li, L.; Yuan, J. Evaluation of Cyprinid herpesvirus 2 latency and reactivation in *Carassius gibel*. *Microorganisms* **2020**, *8*, 445. [CrossRef]
32. Tolo, I.E.; Bajer, P.G.; Mor, S.K.; Phelps, N.B. Disease ecology and host range of Cyprinid herpesvirus 3 (CyHV-3) in CyHV-3 endemic lakes of North America. *J. Fish Dis.* **2023**, *46*, 679–696. [CrossRef] [PubMed]

viruses

MDPI

Article

Detection and Genetic Characterization of Red-Spotted Grouper Nervous Necrosis Virus and a Novel Genotype of Nervous Necrosis Virus in Black Sea Bass from the U.S. Atlantic Coast

Jan Lovy [1,*], Miriam Abbadi [2], Anna Toffan [2], Nilanjana Das [3], James N. Neugebauer [4], William N. Batts [1] and Peter J. Clarke [5]

1 U.S. Geological Survey, Western Fisheries Research Center, 6505 NE 65th Street, Seattle, WA 98115, USA; bbatts@usgs.gov
2 Istituto Zooprofilattico Sperimentale delle Venezie, Viale dell'Università 10, 35020 Legnaro, PD, Italy; mabbadi@izsvenezie.it (M.A.); atoffan@izsvenezie.it (A.T.)
3 Department of Microbiology, Oregon State University, Corvallis, OR 97331, USA; dasni@oregonstate.edu
4 Office of Fish and Wildlife Health and Forensics, New Jersey Fish and Wildlife, Oxford, NJ 07863, USA
5 Bureau of Marine Fisheries, New Jersey Fish and Wildlife, Nacote Creek, NJ 08205, USA; peter.clarke@dep.nj.gov
* Correspondence: jlovy@usgs.gov; Tel.: +1-206-526-6589

Abstract

Nervous necrosis virus (NNV) causes a neurologic disease in a wide range of marine fish and poses serious disease risks to marine aquaculture worldwide. Little is known about the presence of NNV along the Atlantic coast of the United States, aside from the presence of barfin flounder nervous necrosis virus (BFNNV) in coldwater species in the northern part of this range. Herein we conducted surveillance for NNV from 2020 to 2022 in the mid-Atlantic region of the United States in black sea bass *Centropristis striata*, a serranid fish that is found throughout the eastern U.S. coast. Molecular detection methods have identified and characterized red-spotted grouper nervous necrosis virus (RGNNV) sequences at low prevalence throughout the years. Further, in 2022, a higher prevalence of a novel NNV genotype, tentatively named black sea bass nervous necrosis virus (BSBNNV), was characterized for the first time. Though virus isolation was unsuccessful, this study was the first to genetically identify NNV in this region and in this species. These findings highlight the need for further research on NNV to understand epidemiology and virulence in the context of marine fisheries and an emerging marine aquaculture industry in the United States.

Keywords: nervous necrosis virus; viral encephalopathy and retinopathy; viral nervous necrosis; genetics; black sea bass; United States; Atlantic

1. Introduction

Viral encephalopathy and retinopathy, also known as viral nervous necrosis (VNN), is caused by nervous necrosis virus (NNV), a neurotropic RNA virus belonging to the genus *Betanodavirus* within the family Nodaviridae [1,2]. The disease VNN causes clinical signs in fish, including whirling, hyperactivity, and/or lethargy, which is associated with vacuolating lesions within the brain and retina [3]. NNV has a relatively small genome (total of 4.5 kb) comprising two single-stranded, positive-sense molecules, including the RNA1 (3.1 kb) encoding the RNA-dependent RNA polymerase and the RNA2 (1.4 kb) encoding the viral capsid [4–6]. NNV has a wide host range in marine fish species with four unique genotypes recognized, based on genetic sequences within a highly variable region (T4) of the RNA2 gene [7]. These include red-spotted grouper nervous necrosis virus (RGNNV), striped jack nervous necrosis virus (SJNNV), tiger puffer nervous necrosis virus (TPNNV), and barfin flounder nervous necrosis virus (BFNNV) [7]. These genotypes are associated with certain host species [8] and optimal replication temperatures [9–11]; as such, they may be confined to certain geographic ranges. For example, BFNNV occurs in cooler regions including Japan, northern Europe, and northern North America [12–15], whereas RGNNV occurs in tropical and temperate fishes across a wider range, as reviewed by [16]. Though these four genotypes have been well described and accepted as viral species by the ICTV [1], a high level of genetic diversity and additional genotypes are still being recognized. This is evidenced by other recently described genotypes that are awaiting formal recognition as viral species [17,18].

Viral nervous necrosis is considered one of the most serious disease threats to global marine aquaculture. For example, in all regions of the Mediterranean Sea, VNN is viewed as the most important contributor to disease in the farming of European sea bass *Dicentrarchus labrax* and gilthead sea bream *Sparus aurata* [19–21]. Serious disease in marine aquaculture has also been reported in Asia, including Japan [15,22,23] and Taiwan [24], and in Norwegian Atlantic halibut *Hippoglossus hippoglossus* juvenile rearing facilities [25]. In North America, BFNNV caused disease issues in farmed Atlantic cod *Gadus morhua* and haddock *Melanogrammus aeglefinus* from Atlantic Canada and the northeastern United States [14,26], while in the western United States, RGNNV was associated with an outbreak in farmed white sea bass *Atractoscion nobilis* from the Pacific coast in California [27] (Figure 1). Little is known of NNV prevalence outside these specific regions in the U.S. Considering the high priority to expand marine offshore aquaculture in the U.S. to offset the capture of wild fisheries [28], understanding risks from NNV in the region may help develop fish health planning as the industry is established. Because of limited biosecurity control in the offshore environment, viruses and other pathogens may be introduced to marine aquaculture from wild fish reservoirs [29], emphasizing the need to understand pathogen distribution and prevalence in wild fish.

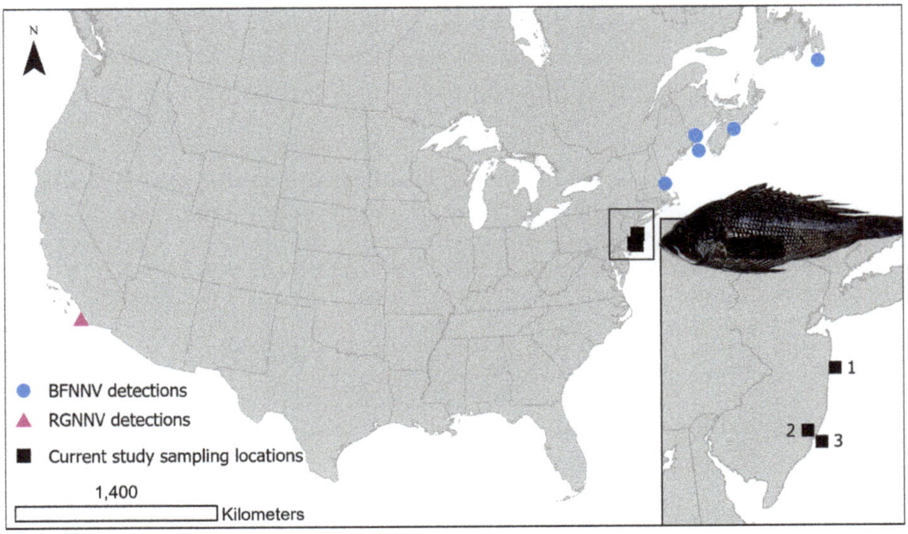

Figure 1. Sites of black sea bass *Centropristis striata* collection from coastal New Jersey for this study, including 1. Sea Girt Reef; 2. Rutgers Marine Field Station; 3. Little Egg Reef. Previous reports of nervous necrosis virus detected in North America are indicated; barfin flounder nervous necrosis virus (BFNNV) [13,14,26]; red-spotted grouper nervous necrosis virus (RGNNV) [27].

Much of the attention to NNV has been within aquaculture settings, while the impacts of this virus on wild fish are less known. Wild fish are considered natural reservoirs for NNV, with BFNNV detected in wild winter flounder *Pleuronectes americanus*, in Atlantic Canada [13] and RGNNV in wild groupers *Epinephelus* spp. and European sea bass in regions of marine aquaculture in the Mediterranean Sea [30,31]. Virulence of the virus is known to be highest in larval and juvenile stages [32,33], though mortality in these early stages of fish may be easily missed in the wild. Disease in wild adult fish has been associated with erratic swimming, lethargy, skin and fin erosions, eye lesions, and hyperinflated swim bladders in adult groupers *Epinephelus* spp. and European sea bass [30,31]. In the present study, we investigated NNV in wild black sea bass *Centropristis striata*, an important species that is fished commercially and recreationally in the mid-Atlantic U.S. In addition to supporting fisheries, black sea bass are considered an ideal commercially ready species for U.S. marine aquaculture [34].

Black sea bass, within the family Serranidae (sea basses and groupers), are demersal fishes with a preference for warm temperate waters with a range from Maine to Florida in the northwestern Atlantic Ocean [35]. The species is managed in two distinct populations, including a mid-Atlantic population that spans from the Gulf of Maine to Cape Hatteras, North Carolina, and a southern population that spans from Cape Hatteras to Florida [36,37]. The mid-Atlantic population feed and spawn in coastal areas between May and October and in the fall migrate offshore toward the continental shelf and south to the Chesapeake Bight to overwinter in warmer waters [38,39]. With a preference for structured habitat, and the middle Atlantic Bight comprising a relatively flat topology, black sea bass have benefited from man-made reefs in this region, such as shipwrecks and artificial reefs [40]. The success in artificial reefs supporting higher fish abundance and species diversity may also influence transmission factors for certain infectious agents. For example, *Lernaeenicus radiatus*, a pennellid copepod with a complex life cycle, was found at a higher prevalence and infection intensity in black sea bass from artificial reef sites compared to non-structured habitats. This is likely a result of increased transmission factors related to increased fish density and biodiversity favored in reef environments [41].

In the present investigation, surveillance for NNV has been conducted alongside an artificial reef black sea bass monitoring survey led by the Bureau of Marine Fisheries, New Jersey Fish and Wildlife. Between 2020 and 2022, 865 adult black sea bass were screened with molecular methods for NNV, documenting the virus in this species for the first time. Genetic sequence analysis conducted herein documents the presence of RGNNV, not previously detected on the Atlantic coast of North America, and a previously undescribed NNV genotype, indicating the role of NNV as a potential pathogen within the region that is highly relevant to fisheries and emerging aquaculture.

2. Materials and Methods

2.1. Fish Collection and Environmental Data

Across 2020–2022, adult black sea bass were collected in collaboration with the artificial reef monitoring trap survey conducted by the Bureau of Marine Fisheries, New Jersey Fish and Wildlife. Artificial reefs are made of concrete, dredge rock, decommissioned barges, old ships, tanks, subway cars, and steel demolition debris to provide structured habitat for demersal fishes. Fish were captured from two artificial reefs, Sea Girt Reef and Little Egg Reef, both located off the shore of New Jersey (Figure 1). Little Egg Reef is 3.2 km^2 in size and located 6.1 nautical km offshore with a depth of 15–18 m. Sea Girt Reef is 2.25 km^2 in size, located 5.6 nautical km offshore, and has a depth range of 17–22 m.

For fish collection, 22 un-baited ventless traps measuring 112 L × 58 W × 38 H (cm) were randomly placed throughout the reef sites between April and November. The traps were checked within 7 d of deployment. After capture, fish were placed on ice and transferred to the Pequest Aquatic Animal Health Laboratory, Oxford, New Jersey for tissue sampling within 24 h of capture. Acceptable methods for fish handling/sampling were under the direction of the New Jersey Fish and Wildlife and the Federal Sportfish and Restoration Project (FW-69-R21). Total length, weight, and sex was recorded for each fish. In 2020, in addition to NNV screening, fish were additionally screened for other general systemic viruses. For this general screening, spleen and kidney were aseptically collected from 146 fish with up to five fish pooled per sample to screen for viruses by isolation using viral cell culture assays, as further described below. Between 2020 and 2022, 865 adult black sea bass were screened for NNV using a two-step reverse transcription real-time PCR (rRT-PCR), further described below. Brain was aseptically collected into a 2 mL microcentrifuge tube. Instruments were disinfected with bleach, water, ethanol, and flame between fish samples. Tissue samples were immediately frozen at −80 °C. The remaining brain and eye tissue was collected into a sterile sample bag and kept frozen at −80 °C.

In 2020 only, young-of-the-year (YOY) black sea bass were also collected between March 24 and August 4 in collaboration with the Rutgers University Marine Field Station. Un-baited Gee wire mesh traps were deployed at the marine field station boat basin and sampled twice weekly, as previously described [42]. A total of 116 YOY black sea bass were collected, euthanized with an overdose (200 mg/L) of buffered MS-222 (Syndel, Ferndale, WA, USA), and frozen at −20 °C until further processing. Fish were sampled while frozen by excising and pooling brain tissue from 5 fish into 2 mL microcentrifuge tubes. The samples were immediately returned to the freezer, maintaining the tissues frozen to avoid an additional freeze–thaw. Brain samples were processed as described below for homogenization and genetic screening for NNV.

2.2. Viral Cell Culture Assays

In 2020, for general virus screening not related to NNV, pooled spleen and kidney were processed for virus isolation at the Animal Health Diagnostic Laboratory, Department of Agriculture, Ewing, New Jersey. Tissues were homogenized in Hanks' balanced salt solution (HBSS) at a volume of 9:1 (HBSS:tissue), followed by centrifugation. The supernatant was diluted 1:1 with antibiotic media and incubated for at least 2 h at 4 °C. Three cell lines were used for general virus screening, including Bluegill fry-2 (BF-2), Epithelioma papulosum cyprini (EPC), and Chinook salmon embryo-214 (CHSE-214). The cells were grown in 24-well plates and within 24 h of confluence were seeded in duplicate with 100 µL of the centrifuged supernatant and incubated at 20 °C. Cells were monitored every 3 days for 2 weeks, followed by blind passaging of the samples onto freshly grown cells for a further 2-week monitoring period. If no cytopathic effects (CPE) were noted in the cells within the 4-week period, then they were considered negative.

For samples that were positive for NNV by molecular methods (described below), the original tissue homogenate or parallel eye/brain tissue stored at −80 °C was used to attempt virus isolation specific to NNV. In 2020 and 2021, the frozen tissue samples were submitted to the National Veterinary Services Laboratory (NVSL), United States Department of Agriculture, Animal and Plant Health Inspection Service (Ames, IA, USA). Tissue homogenates from each fish were individually inoculated onto striped snakehead-1 cell line (SSN-1) and a clone of the striped snakehead cell line (E-11), cell lines known to be permissible to NNV isolation [43], and incubated at 20 °C. A minimum of two blind passages were made and cells were monitored for CPE for a 28-day incubation period. In 2022, samples were processed at the Western Fisheries Research Center (Seattle, WA, USA). Previously frozen supernatant from processed tissue homogenates from 15 samples that showed the lowest cycle threshold (CT) values in rRT-PCR, ranging from 17.5 to 24.8, were selected for virus isolation by cell culture assays. Each sample was tested individually, without pooling. These homogenates were centrifuged at 12,000× g for 5 min and held on ice. The supernatant was collected and diluted 1:10 in L-15 Leibovitz media (Cytiva, Logan, UT, USA), followed by two additional serial dilutions (1:100 and 1:1000) in L-15. A volume of 200 µL of the three dilutions of each sample were inoculated onto freshly seeded SSN-1 cells pretreated with 7% Polyethylene glycol (PEG; Sigma-Aldrich, St. Louis, MO, USA) in L-15. Following inoculation, plates were subject to an adsorption step comprising centrifugation of cell culture plates (600 g) at 20 °C for 45 min. Following the adsorption spin, 1 mL of L-15 was added to each well. This was duplicated, such that one set of sample plates was incubated at 20 °C and a second set of plates incubated at 25 °C. Cells were monitored regularly for CPE for 14 days, followed by a blind passage onto fresh SSN-1 cells and monitoring for an additional 14 days. If no CPE was detected within the 28-day incubation period, then the samples were considered negative in cell culture.

2.3. Molecular Screening and Confirmation of Nervous Necrosis Virus

Upon thawing brain tissue samples, 900 µL of Gibco© minimum essential medium (Gibco, Grand Island, NY, USA) and a 5 mm bead were added to each sample tube before homogenizing using a TissueLyser (Qiagen, Germantown, MD, USA) for 2 min at 20 hz. The homogenate was left on ice for 5 min and centrifuged at 3000× g for 5 min at 6 °C. The supernatant was removed to a clean 2 mL microcentrifuge tube. For RNA extraction, 50 µL of each sample was transferred to a 96-well plate and extracted using the KingFisher MagMax-96 Viral RNA Isolation Kit (Thermo Fisher Scientific, AM1836) run on the KingFisher Flex RNA extraction robot (Thermo Fisher Scientific), according to manufacturer's directions. A positive and negative extraction control was run along with each plate. Positive control material included BFNNV amplified in the SSN-1 cell line, provided from an outside laboratory (Kennebec River Biosciences, Richmond, ME, USA). For NNV screening, a rRT-PCR assay [44] was used with slight modifications. Specifically, cDNA synthesis was conducted using a high-capacity cDNA reverse transcription kit (Thermo Fisher Scientific, Carlsbad, CA, USA, #4368814) according to manufacturer's instructions and maintained at −20 °C until further use. For rRT-PCR, the QuantiNova Probe PCR Kit (Qiagen #208252) was used with primers known to amplify the four established NNV species [44], RNA2 FOR: 5′-CAA CTG ACA RCG AHC ACA C-3′ and RNA2 REV: 5′-CCC ACC AYT TGG CVA C-3′ at a concentration of 0.4 µM and the RNA2 PROBE: 5′ 6-FAM TY CAR GCR ACT CGT GGT GCV G-BHQ1-3′ at a concentration of 0.2 µM. Plates were run on an Applied Biosystems 7500 or an ABI 7500 Fast real-time PCR system (Thermo Fisher Scientific) with a positive amplification control and negative template control. Cycling conditions were either 50 °C for 30 min, 95 °C for 15 min, followed by 40 cycles of 94 °C for 15 s and 60 °C for 60 s, or a fast mode that included 95 °C for 2 min, followed by 40 cycles of 95 °C for 5 s and 60 °C for 5 s. Samples crossing the automatically designated threshold were identified as presumptively positive.

An independent molecular assay, an endpoint reverse transcription PCR (RT-PCR) assay targeting a conserved region of the RNA2 gene using VNNV1 and VNNV2 primers [45] was used with slight modifications (detailed below) to confirm presumptive positive findings. Samples were run on a Veriti thermocycler (Applied Biosystems). A Platinum Taq DNA polymerase master mix kit (Thermo Fisher Scientific, Carlsbad, CA, USA) was used where each reaction contained 1× PCR buffer, 0.2 mM dNTP mixture, 1.5 mM $MgCl_2$, 3 µL of template cDNA, and 0.4 µM of each primer, with molecular grade water added up to a 50 µL reaction volume. Cycle conditions were 94 °C for 2 min, followed by 40 cycles of 94 °C for 15 s, 50 °C for 30 s, and 68 °C for 1 min, followed by final extension at 68 °C for 5 min and holding at 4 °C. PCR products were visualized via gel electrophoresis on 1.2% agarose E-gels (Thermo Fisher Scientific, Carlsbad, CA, USA) containing SYBR Safe DNA gel stain alongside an E-gel 1KB Plus Ladder to estimate amplicon size and a GeneRuler Low Range Ladder DNA Ladder (Thermo Fisher Scientific) to roughly estimate DNA quantity. PCR products that contained amplicons around 605 bp in size were selected for Sanger sequencing. In situations when no amplicons were detected, then the RT-PCR product was subject to nested RT-PCR using the primers VNNV3 and VNNV4 to amplify a 255 bp product according to a previously published protocol [45]. When the expected amplicon size occurred in either the first or nested round of RT-PCR, then the PCR product was processed for genetic sequencing, as detailed below. Genetic sequences were confirmed as NNV by comparing the sequences to those deposited in the National Library of Medicine, National Center for Biotechnology Information (NCBI), using the nucleotide basic local alignment search tool (BLASTn, https://blast.ncbi.nlm.nih.gov). Samples that were positive by rRT-PCR and confirmed with the independent endpoint RT-PCR assay combined with genetic sequencing were considered positive for NNV.

2.4. Genetic Sequencing

Various primer combinations (Table 1) were used to PCR amplify and sequence the RNA1 and RNA2 genomes of select samples. When necessary, new primers were designed using the Primer3 web-based software (Primer3web, version 4.1.0) [46], based on available genetic sequences from our study. For RT-PCR amplification, the master mix and primer concentrations described above were used. Cycling conditions described above or those matching the respective reference for the primers listed in Table 1 were used. When appropriately sized amplicons were detected by gel electrophoresis, the RT-PCR product was enzymatically purified using ExoSAP-IT (Thermo Fisher Scientific) according to the manufacturer's directions. Sequencing reactions were prepared by diluting the purified DNA to approximately 2 ng/µL with molecular grade water, followed by the addition of 5 µL of the sequencing primer at a 5 µM concentration to make a total volume of 15 µL. The primers used for RT-PCR amplification were also used for sequencing. Sequencing was performed in both directions by Azenta (South Plainfield, NJ, USA) using ABI BigDye version 3.1 (Applied Biosystems) and run on an ABI 3730xl DNA analyzer (Applied Biosystems). Sequence quality was examined using Chromas Lite (Technelysium, Version 2.6.6) and sequence alignment was conducted using BioEdit (Version 7.2.5) to assemble sequences for the RNA1 and RNA2 segments.

To amplify and sequence the 5′ segment of the RNA2 in the new genotype from 2022, a representative sample (Sample ID: 22-137) was used to perform 5′RACE (Thermo Fisher Scientific) to obtain the genomic terminal sequences, as previously described [47]. The 5′RACE Kit protocol was followed for synthesis, purification, and for TdT tailing of the cDNA according to manufacturer's directions and as previously reported [47].

Table 1. Primers used for the amplification and sequencing of the RNA1 and RNA2 segments of nervous necrosis virus. Unshaded were used for red-spotted grouper nervous necrosis virus, black shading was used only for the unique genotype in 2022, and gray shading was used for both genotypes. Nested PCR is shown by a first round PCR * followed by a nested PCR **. Intended amplicon size (Amp size) in base pairs.

Primer Name	Sequence	Target and Amp Size	Reference
RNA1_283-5′F RNA1_856-874R	TAA CAT CAC CTT CTT GCT GGT GCT CAC CCA TCT TGA	RNA1 874	[48]
RNA1_684-705F RNA1_1346-1367R	GAA CTA CAA CCA GGA TAC CAT G GAC TCA CTT GGA AAT ACA	RNA1 684	[48]

Table 1. *Cont.*

Primer Name	Sequence	Target and Amp Size	Reference
JRNV1_F1 JRNV1_R1	TCA CTT ACG CAA GGT TAC CG GAC CGG CGA ACA GTA TCT GAC	RNA1 1122	[49]
JRNV1_F1	TCA CTT ACG CAA GGT TAC CG	RNA1 1973	[49]
RNA1_1955-1973R	TGA CAG CAG GTG CTT GG		[48]
JRNV1_F2 JRNV1_R2	AGT CTG GGY YTG GAR GGC GAC GAA AGC RTT DGC AAT GC	RNA1 1032	[49]
VNNV5 VNNV6	GTT GAG GAT TAT CGC CAA CG ACC GGC GAA CAG TAT CTG AC	RNA1 953	[50]
FOR521	ACG TGG ACA TGC ATG AGT TG	RNA1 630	[51]
VNNV6	ACC GGC GAA CAG TAT CTG AC		[50,51]
RNA1_1248-1267F RNA1_2068-2087R	CTT GCK CGT CAT TAC CAA GC GCG ACC AGC AAG GTA TGA GA	RNA1 839	This study
JRNV1_F3 JRNV1_R3	TCC AAG CAC CWG CTG T GGG GTG GGA GCR GGC A	RNA1 1099	[49]
BF Pol 1698-1715F BF Pol 2465-2482R	GTC CAG CTA CAC CTA CGC AGT CTG CGT ATT GGA CCA	RNA1 785	[48]
RNA2_283-5′F RNA2_578-597R	TAA TCC ATC ACC GCT TTG GCT GCC AAC ACA CAG GA	RNA2 593	[48]
RNA2_8-21F RNA2_410-429R	TCA YCG CTT TGC MAT CAC AA CGT TGT CAG TTG GAT CAG GC	RNA2 421	This study
VNNV1 * VNNV2 *	ACA CTG GAG TTT GAA ATT CA GTC TTG TTG AAG TTG TCC CA	RNA2 605	[45]
VNNV3 ** VNNV4 **	ATT GTG CCC CGC AAA CAC GAC ACG TTG ACC ACA TCA GT	RNA2 255	[45]
RNA2_818-837F RNA2_1206-1227R	CAT TGA CTA CAA CCT TGG AG CAA TGG TAC CAA CAA TAG	RNA2 409	[48]

2.5. Phylogenetic Analysis of Nervous Necrosis Virus Sequences

Partial RNA1 and RNA2 consensus sequences were aligned and compared to reference nucleotide sequences retrieved from GenBank using the MEGA 7.0 package [52]. Reference sequences were selected according to the BLAST results obtained for each gene of each sample. Representative strains of each genotype were also included. For both RNA1 and RNA2 nucleic and amino acid alignments, phylogenetic relationships among strains were inferred using the maximum likelihood (ML) method available in the IQ-Tree software v1.6.9 [53]. The best fitting model of nucleotide (nt) and amino acid (AA) substitution was determined with ModelFinder [54]. One thousand bootstrap replicates were performed to assess the robustness of individual nodes of the phylogeny, and only values ≥70% were considered significant. Phylogenetic trees were visualized with the FigTree v1.4 software (http://tree.bio.ed.ac.uk/software/figtree/, accessed on 31 May 2025).

Phylogenetic trees based on partial nucleic sequence alignments were also developed for both genetic segments using the neighbor-joining (NJ) method with 1000 bootstrap re-samplings using the MEGA 7.0 package [52]. Pairwise similarities were also determined.

3. Results

3.1. Virology and Nervous Necrosis Virus Findings in 2020

Viral cell culture assays from spleen and kidney pools for general virus screening were negative for virus isolation in all three cell lines and incubation temperatures (Table 2). For NNV screening by rRT-PCR, all YOY black sea bass were negative. From adult fish, NNV was detected from one fish (Sample ID: 20-045) out of 258 fish screened (Table 2). This fish was collected from the Sea Girt Reef and had a total length of 220 mm and weight of 148.6 g. The detection was very low, having a CT of 35. Confirmation and genetic sequencing of this sample yielded a 1338 bp sequence of the RNA1 segment and 1226 bp sequence of the RNA2 segment, both deposited in GenBank (Table 3). A BLASTn search indicated that both segments have the closest identities (approximately 95–98% similarity; searched in July 2025) with various RGNNV sequences. An attempt to isolate this virus from previously frozen homogenates in the E-11 cell line failed to induce CPE.

Table 2. Black sea bass *Centropristis striata* collected from the coast of New Jersey for general viral screening by virus isolation in cell culture (VI) or two-step reverse transcription real-time PCR (rRT-PCR) for nervous necrosis virus. Total length (TL) in mm and fish weight in g ± standard deviation; young-of-the-year (YOY). The number of confirmed detections/total fish are indicated.

Sampling Dates	Life Stage	Fish TL	Fish Weight	Method	Results
2020—August–October	Adult	246 ± 46	204 ± 113	VI cell culture	0/146
2020—March–August	YOY	62 ± 11	NA	rRT-PCR	0/116
2020—August	Adult	246 ± 46	203 ± 112	rRT-PCR	1/132
2020—October–November	Adult	239 ± 50	203 ± 120	rRT-PCR	0/126
2021—April	Adult	259 ± 25	220 ± 52	rRT-PCR	0/32
2021—July	Adult	244 ± 38	193 ± 96	rRT-PCR	2/180
2021—October	Adult	253 ± 38	225 ± 87	rRT-PCR	3/91
2022—July	Adult	262 ± 44	258 ± 150	rRT-PCR	23/304

Table 3. Samples confirmed positive for nervous necrosis virus by sequencing of the RNA1 and/or RNA2 genes. Associated GenBank accession numbers are provided. Short* are short or discontinuous sequence lengths not submitted to GenBank, but data are available in a U.S. Geological Survey data release [55]; (-) indicates no sequence was generated.

Year	Strain ID	RNA1	RNA2	Year	Strain ID	RNA1	RNA2
2020	20-045	PV877385	PV877386	2022	22-167	PV993951	PV993960
2021	21-058	PV877388	-		22-170	PV993952	PV993961
	21-182	PV877387	PV877389		22-175	PV993953	PV993962
	21-252	Short*	Short*		22-178	Short*	-
	21-261	-	Short*		22-191	PV877392	PV993963
	21-266	-	Short*		22-203	Short*	-
2022	22-021	PV993948	PV993956		22-213	Short*	-
	22-028	PV877390	PV877394		22-228	-	PV877397
	22-045	PV877391	PV877395		22-236	PV993954	PV993964
	22-053	PV993949	PV993957		22-243	Short*	PV993965
	22-062	PV993950	Short*		22-259	-	PV993966
	22-082	-	PV993958		22-273	PV993955	PV993967
	22-086	Short*	-		22-279	-	PV993968
	22-137	Short*	PV877396		22-298	PV877393	-
	22-163	-	Short*				

3.2. Nervous Necrosis Virus Findings in 2021

A total of six presumptive NNV detections resulted from a total of 303 fish screened using rRT-PCR. Five fish, three male, and two females were confirmed positive with end-point RT-PCR and sequencing (Table 2). Samples 21-252, 21-261, and 21-266 were confirmed positive with nested PCR [45] and partial sequences of the RNA2 gene (200–240 bp) confirmed their identity within RGNNV (Table 3). Fish 21-223 with a CT value of 38.9 failed to produce a reliable sequence, thus was not confirmed positive. Three of the detections came from the Sea Girt Reef and the other two from Little Egg Reef. The average total length and weight of fish testing positive for the virus was 263 ± 39 mm and 225 ± 77 g, respectively.

Partial RNA1 and RNA2 sequences were generated from these samples and deposited to GenBank (Table 3). Sequences spanning a large portion (93%) of the RNA1 sequence (2896 bp) and a majority (87%) of the RNA2 gene (1243 bp) were obtained for sample 21-182, while a large portion of only the RNA1 gene (95%) was obtained for sample 21-058 (2930 bp). BLASTn analysis of the RNA1 sequence for sample 21-058 showed the closest identity to RGNNV from red grouper *Epinephelus morio* in Taiwan with 97.47% identity (accession# FJ809938; BLAST search in July 2025) and around 96–97% identity with other RGNNV sequences in NCBI. Comparison of the RNA1 segments between samples 21-058 and 21-182 showed 97.76% similarity, with a total of 65 nucleotide substitutions between them, indicating considerable sequence variation in samples from the same fish population at the same point in time. BLASTn analysis of the RNA2 segment of 21-182 demonstrated a 97.1% match to an RGNNV isolate from China (accession# EF558369; BLAST search in July 2025). Comparison of the RNA2 sequences between samples 20-045 and 21-182 showed a 98.61% identity with 17 nucleotide differences.

Previously frozen tissue homogenates from all five positive samples processed for virus isolation failed to induce CPE in SSN-1 and E-ll cell lines. Molecular detection of the virus from the first passage in the cell line was confirmed in sample 21-58 using endpoint RT-PCR [45], though this was likely not from virus amplification in cell culture, but rather detection of the virus carried over in the cell culture inoculum. The genetic sequence matched that of the previous sequence generated for that sample.

3.3. *Nervous Necrosis Virus Findings in 2022*

A total of 82 presumptive positive detections of NNV occurred using rRT-PCR from a total of 304 fish collected in July. Of the 82 detections, 28 samples had a CT value under 30, 18 samples were between 30 and 35, and 36 samples had a CT value over 35 (Figure 2). The two lowest CT values were 17.5 and 18.4. A total of 11 out of those 28 samples with CT under 30 originated from the Sea Girt Reef, while 17 were from Little Egg Reef. This included 16 female and 12 male fish with a mean total length (TL) of 240 ± 38 mm and weight of 204 ± 96 g. Only a subset of samples was selected to confirm with endpoint RT-PCR and sequencing, including all 28 samples that had a CT under 30 and 2 additional samples. Of the 30 samples, a total of 23 were confirmed positive with endpoint RT-PCR and genetic sequencing. Representative sequences were deposited in GenBank (Table 3).

Genetic sequencing of the RNA1 yielded five samples (22-028, 22-045, 22-053, 22-191, and 22-273) with continuous sequences covering much of the RNA1 (about 2390 bp: about 77% complete). Sequences for 22-045, 22-053, and 22-273 were identical, whereas two nucleotide substitutions occurred in the other sequences. BLASTn comparison in NCBI indicates the closest identity of this RNA1 segment is with an isolate from Atlantic cod from Norway (Accession# EF577395; BLAST search performed in July 2025) with 91.24% identity and BFNNV from barfin flounder *Verasper moseri* from Japan (Accession# NC013458; BLAST search performed in July 2025) with 90.9% identity. The remaining 13 RNA1 sequences (Table 3) were partial or discontinuous sequences, ranging between 757 and 1950 bp in length. Alignment and comparison of the sequences indicated that all the partial RNA1 sequences had high identity to each other with rare single nucleotide substitutions or single nucleotide polymorphisms. Alignment of 12 partial sequences spanning 1022 nucleotides of the RNA1 correspond to nucleotide positions 80–1102 when aligned with the full RNA1 reference BFNNV sequence (accession# EU236146). Comparison of the sequences showed two single nucleotide polymorphisms at position 798 in three sequences and position 1005 in two sequences. Both nucleotide polymorphisms were a C-T substitution. Additionally, single nucleotide substitutions occurred in three sequences at positions 531, 723, and 912, which comprised a C-T, T-C, and A-G substitution, respectively.

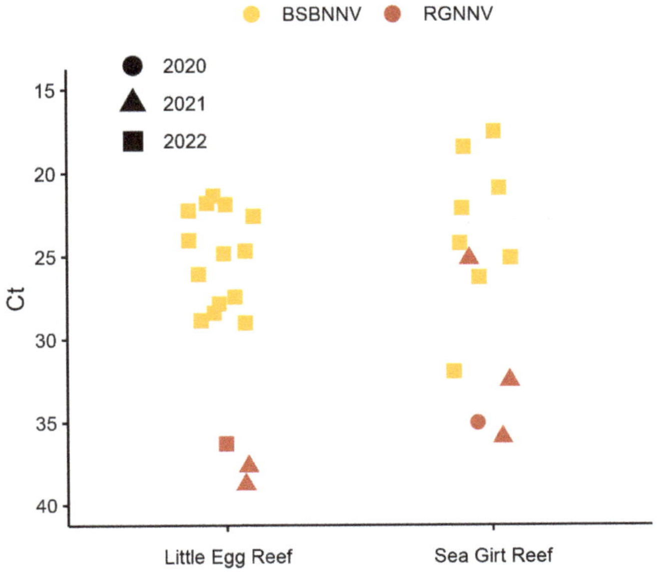

Figure 2. 2020–2022 confirmed nervous necrosis virus reverse transcription real-time PCR (rRT-PCR) detections with cycle threshold (CT) values from black sea bass *Centropristis striata* collected from two reef sites. Note the reverse Y-axis scale showing cycle threshold values in rRT-PCR. Red-spotted grouper nervous necrosis virus (RGNNV); black sea bass nervous necrosis virus (BSBNNV).

Genetic sequencing of the RNA2 gene yielded partial sequences around 577 bp in length from 18 fish samples with the exception of 2 samples that only yielded sequences 250 bp in length. Two distinct sequence types were detected: a predominant sequence type (type I) that occurred in 17 samples and an infrequent sequence type II that occurred in 1 sample (22-228). The type I sequences were nearly identical to each other with rare single nucleotide substitutions detected. Fifteen of the partial RNA2 type I sequences (575 bp) were aligned along with the complete RNA2 gene from BFNNV (accession # EU236147). The partial sequences aligned with nucleotide positions 365 to 940 in this reference RNA2 sequence. This comparison showed five sequences that had a single nucleotide substitution, which included either an A-G, G-T, C-T, G-A, or T-C substitution. The other 10 sequences were identical. GenBank accession numbers are summarized in Table 3. 5′RACE techniques provided the sequence on the 5′ end of a single representative sample of the type I RNA2 (sample 22-137, accession # PV877396), which provided the most complete sequence of the RNA2 from this set of sequences (955 bp). BLASTn comparison of this sequence to other sequences published in NCBI showed closest identity (88.6%) to two submissions of TPNNV (accession # NC013461 and D38637; search performed in July 2025).

The type II RNA2 sequence from sample 22-228 (Table 3) was 546 bp in length. BLASTn analysis in NCBI indicates that 22-228 shares closest identity with various sequences including a viral isolate from Japanese flounder *Paralichthys olivaceus* (D38527), SJNNV (AY600956), viral isolate from European sea bass *Dicentrarchus labrax* (AF175510), and RGNNV (accession NC008041), all with around 95–96% identity. When this sequence was compared with the two RGNNV sequences from 2020 and 2021, sequence divergence was about 2.6%.

All viral cell culture assays on previously frozen tissue homogenates failed to induce CPE in the SSN-1 cell lines incubated at either 20 or 25 °C.

3.4. Phylogenetic Analysis of Nervous Necrosis Virus Sequences

Maximum likelihood phylogenetic analysis based on a selection of the longest nucleotide sequences for both RNA1 and RNA2 clearly grouped black sea bass viruses in two distinct viral clusters (Figure 3). The first cluster fell within the RGNNV genotype and included viral sequences obtained in 2020 (20-045—RNA1 and RNA2) in 2021 (21-058 and 21-182—RNA1 only) and 2022 (22-228—RNA1 and RNA2). The second cluster included only samples detected in 2022, which formed a new, homogeneous, and strongly supported group of sequences clustering separately from all previously described genotypes. This group is most closely related to BFNNV and TPNNV according to RNA1 and TPNNV and SJNNV according to RNA2 (Figure 3).

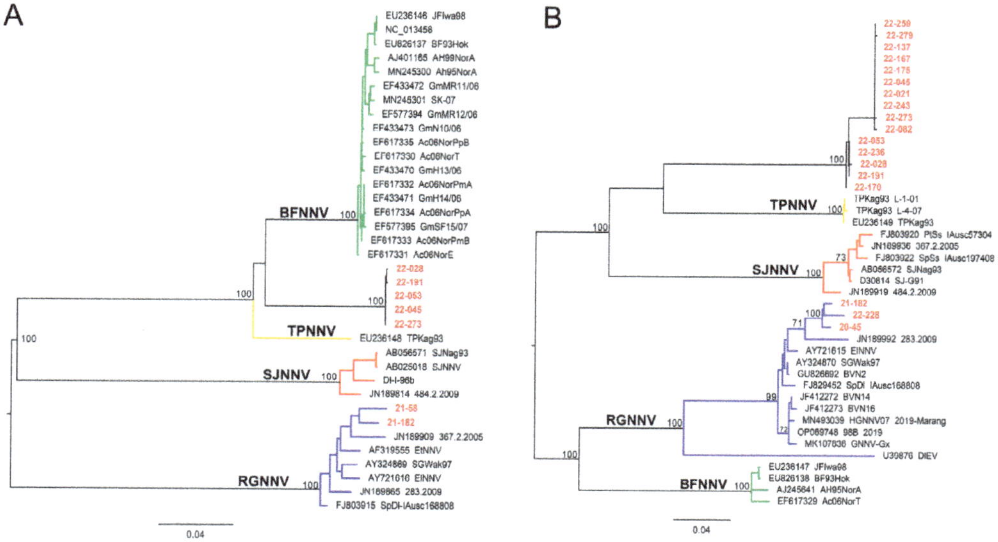

Figure 3. Maximum likelihood (ML) phylogenetic tree based on nucleotide sequences of the RNA1 (**A**) and RNA2 (**B**). Betanodaviruses collected from black sea bass *Centropristis striata* reported in this work are in red. The numbers at branch points represent bootstrap values expressed as percentages (only values ≥ 70 are reported). The genotype subdivision according to [7] is shown at the main branches. Scale bar represents the number of substitutions per site. Barfin flounder nervous necrosis virus (BFNNV) in green; tiger puffer nervous necrosis virus (TPNNV) in yellow; striped jack nervous necrosis virus (SJNNV) in orange; red-spotted grouper nervous necrosis virus (RGNNV) in blue.

Indeed, when compared to TPNNV, RNA1 nucleotide identity ranged from 90.96% to 91.29% and amino acid (AA) identity ranged between 95.50 and 95.96%, while nucleotide identity to BFNNV ranged from 90.81% to 91.40% and AA identity ranged between 95.56 and 95.95%.

RNA2 nucleotide identity to TPNNV ranged from 82.74% to 83.43% and AA identity ranged between 83.33 and 83.51%. Nucleotide identity to SJNNV ranged from 81.03% to 81.54% and AA identity was 81.93–82.44%. Nucleotide identity to BFNNV was 77.55–78.07% and AA identity was 78.12–79.12%. The obtained findings were corroborated by the ML amino acid phylogenetic analysis on the same samples (Figure 4) and by the NJ phylogeny performed on a larger dataset (Supplementary Figure S1).

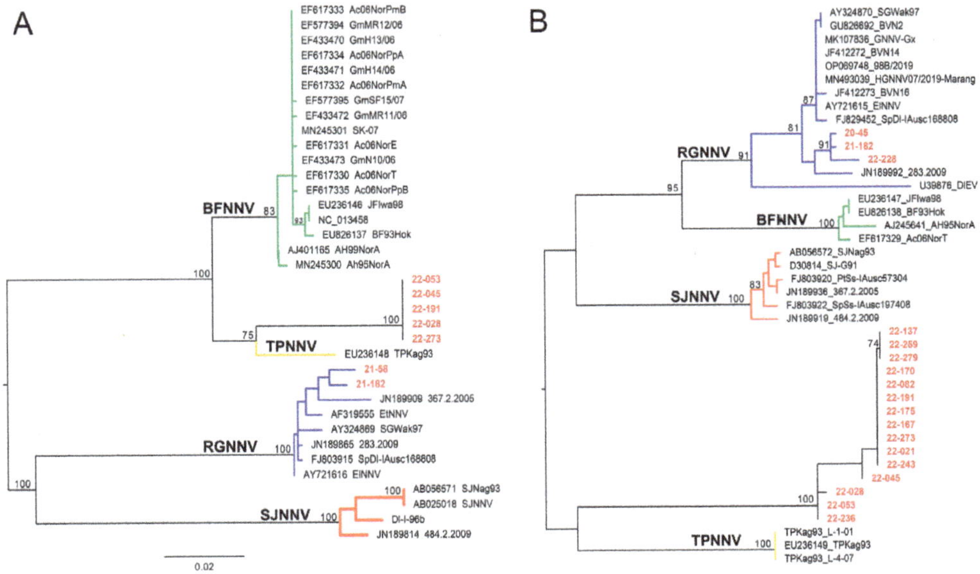

Figure 4. Maximum likelihood (ML) phylogenetic tree based on amino acid sequences of the RNA1 (**A**) and RNA2 (**B**). Betanodaviruses collected from black sea bass *Centropristis striata* reported in this work are in red. The numbers at branch points represent bootstrap values expressed as percentages (only values $\geq$ 70 are reported). The genotype subdivision according to [7] is shown at the main branches. Scale bar represents the number of substitutions per site. Barfin flounder nervous necrosis virus (BFNNV) in green; tiger puffer nervous necrosis virus (TPNNV) in yellow; striped jack nervous necrosis virus (SJNNV) in orange; red-spotted grouper nervous necrosis virus (RGNNV) in blue.

4. Discussion

These findings mark the first molecular detection of RGNNV within the North American Atlantic coast. This finding in black sea bass is consistent with previous detections of RGNNV, which are often found in groupers within the family Serranidae [22,24,31,49]. To our knowledge, the only other previous report of RGNNV in the United States was off the California coast associated with mortality in farmed white sea bass [27]. A notable finding herein was that the RGNNV sequences were unique from each other, having diverged between 1.4% and 2.6% within partial RNA2 sequences. This pattern probably does not indicate a single virus strain that has recently been transmitted or one that is circulating within the population, but rather could suggest individual fish carrying unique virus strains that likely originated from separate transmission events. The nucleotide substitution rate between samples is similar to that described for RGNNV from Europe [56] and Asia [57]. Though this is the first detection of RGNNV in the U.S. mid-Atlantic, the sequence diversity within these individual detections does suggest that it has likely been present long before this detection. Although no histology was available in this study to look for disease signs, the low prevalence of detection in the population, ranging from 0.66 to 3.3%, combined with the high CT values in rRT-PCR suggest very low virus levels in fish, as expected in "carrier" host stages that are not associated with clinical signs. The low prevalence and lack of sequence similarity also indicates that this virus was not transmitted actively among the population of black sea bass during the time of sampling. Further work to collect additional RGNNV sequence types in this region would be needed to learn more about RGNNV diversity to better confirm nucleotide substitution rates and to understand if there is regional clustering of sequence types as shown in other regions [56,57].

An unexpected finding in 2022 was a unique genotype that was not detected in the preceding two years. Indeed, phylogenetic analysis for both genetic segments, based on nucleotide and amino acid alignments, highlighted the presence of a new and strongly supported cluster, tentatively called black sea bass nervous necrosis virus (BSBNNV). This genetic uniqueness was also confirmed from the estimated pairwise nucleotide and amino acid distances showing clear divergence of this cluster from all other genotypes, while showing a relative closeness to the BFNNV and TPNNV genotypes. According to ICTV, in order to claim a new genotype of NNV, the nucleotide sequence of the newly detected genomic RNAs should be compared with those of other nodaviruses and show encoding capsid proteins that differ at >15% of nucleotides and >12% of amino acid positions [1]. In our case, the RNA2 of BSBNNV showed identities ranging from 77 to 83% for the nt sequence and from 78 to 83% for the AA sequence with the other known betanodaviruses—strongly supporting the discovery of a new viral genotype. Therefore, in addition to the turbot nodavirus (TNV) [17] and the SKNNV in shellfish [58], the herein described BSBNNV genotype from black sea bass could be considered in the next revision of the genus *Betanodavirus* taxonomy.

The unique BSBNNV showed a distinct detection pattern suggesting active infection and transmission of the virus in the black sea bass population. This is supported by the high prevalence in the population and detections by rRT-PCR with low CT values that were more indicative of heavy viral infections (two lowest CT values were 17.5 and 18.4). Although the rRT-PCR assay used was not a quantitative method, the relative differences in the CT values observed in 2022 compared to those found in RGNNV infected fish during the earlier two years is notable and likely related to virus concentration in the tissue samples. Although standard curves associating virus concentration and CT value were not established herein, this same rRT-PCR assay was previously validated and a standard curve showed a linear relationship of CT and serial dilutions of in vitro transcribed RNA [44]. In this study, when brain tissue from fish showing clinical signs of VNN were tested using this assay then CT values ranged from 9.9 to 18.6 [44]. In addition to the relatively low CT values observed in 2022, when viral sequences from individual fish were compared, they were highly uniform indicating a single virus strain being transmitted within the population. This sequence uniformity is consistent with other reported NNV outbreaks within hatcheries when sequences are identical in samples collected over a similar point in time [57]. This indicates that the 2022 sampling had overlapped with a natural epizootic with this novel genotype. An important question that remains unanswered is related to the virulence of BSBNNV in wild black sea bass. Viral nervous necrosis has been reported to cause disease in adult groupers in the wild, as seen by loss of equilibrium and lethargy, fin erosion, skin ulcerations around the head, swim bladder hyperinflation, and corneal opacity [30,31]. Samples for histology were unavailable to evaluate for neurologic lesions related to VNN and behavioral clinical signs would have easily been missed during sampling with the trapping method used. Further, it is not uncommon to observe some ulcerations and fin erosions from the traps, thus it may have been difficult to discern sampling injuries from viral specific lesions. The origin of this virus genotype is unknown, though it is notable to document its high prevalence and relatively high levels in the adult black sea bass population. Further, whereas RGNNV was detected at low levels in all three years, this novel genotype was only detected in 2022, which poses the question if this was a more recently introduced virus into this population. These questions may be further addressed with continued monitoring to learn more about the prevalence and sequence diversity of this genotype, along with the collection of parallel samples for histology and fresh tissues to further attempt virus isolation and evaluate the virulence of this genotype.

The presumptive and confirmatory detection methods for NNV in this study were molecular based, using independent rRT-PCR, endpoint RT-PCR, and genetic sequencing to confirm the presence of viral genetic sequences directly from fish brain tissue. These methods do not confirm the presence of replicating intact virus, which requires isolation of the virus in cell culture. Herein, previously frozen tissue homogenates from rRT-PCR positive fish and confirmed positive samples for RGNNV failed to isolate the virus in the SSN-1 and E-11 cell lines. This may be explained by multiple factors, including the expected low levels of virus based on rRT-PCR CT values (25.1–38.7) combined with the use of frozen tissue homogenates that were exposed to at least two freeze–thaw cycles. The novel genotype detected in 2022 also failed to be isolated in SSN-1 cells. Considering that the CT values in rRT-PCR suggested heavy virus infection, other factors need to be considered for isolating this virus. The novel genotype here is closely related to BFNNV based on sequence identity and phylogenetic analysis. It has previously been reported that BFNNV from Atlantic halibut *Hippoglossus hippoglossus* failed to replicate in the SSN-1 cell line even when fresh tissues from clinically affected fish were used. This was evidenced by no CPE and a decrease in virus quantity as estimated by rRT-PCR CT values during cell culture incubation, which increased from 20.9 at inoculation to 34.8 by the end of the incubation period [59]. Further, another study on NNV from turbot *Scophthalmus maximus* (TNV) failed to isolate the virus from clinically affected fish within the SSN-1 cell line [17]. Though the SSN-1 cell line and associated clones (E-11) are considered permissive to the various genotypes of NNV [43], it is clear that some virus strains are not effectively isolated in this cell line. Considering that these are the first detections of NNV in this species and in this geographic range, isolating a replicating virus can help further confirm risks associated with this virus. Based on the results herein, further work to identify a suitable cell line to isolate these NNV genotypes may aid in performing additional work to understand characteristics and virulence of this virus. Incubation temperature is an important consideration for in vitro isolation and cultivation of *Betanodavirus* spp. [9–11]. Considering that BSBNNV is closely related to BFNNV and TPNNV, which both replicate at relatively lower temperatures [43], 15–20 °C may be a good starting point for cultivating this virus.

As thermal constraints limit fish species and population ranges, these similar constraints influence NNV replication and prevalence, as seen by temperature preferences for the different genotypes [43]. Temperature influence has been shown to be regulated by the RNA1 segment encoding an RNA-dependent RNA polymerase [10], with optimal replication temperatures for BFNNV being 15–20 °C, RGNNV at 25–30 °C, and SJNNV and TPNNV in the middle of this range [43]. The black sea bass collected from this study are part of the mid-Atlantic population, which range from Cape Hatteras, North Carolina, north to the Gulf of Maine. The mid-Atlantic population historically reached only as far north as Massachusetts, though because of warming sea temperatures their range has recently expanded to the Gulf of Maine [60,61]. This recent range expansion for black sea bass into the Gulf of Maine is significant, as the Gulf of Maine is where BFNNV has been reported in coldwater species, including winter flounder, haddock, and Atlantic cod [13,14,26]. It is possible that fish range expansions could lead to new fish species interactions, which may influence virus transmission and evolution. These fish species interactions and life history are important to consider, particularly for NNV, as this is an RNA virus with a wide host range that may adapt through evolution and reassortment of RNA1 and RNA2 gene segments, as has been shown between RGNNV and SJNNV genotypes [8,62]. Considering that the novel BSBNNV described herein is closely related to BFNNV, it would be curious to see if the ranges of these genotypes overlap in the Gulf of Maine and if that could influence the interaction or evolution of these genotypes. Additionally on the southern end of the range, Cape Hatteras, North Carolina, is considered a major biogeographical boundary due to a sharp thermal change in this region [37]. This boundary divides the mid-Atlantic black sea bass populations from the south-Atlantic population, which are genetically distinct and range from Cape Hatteras, North Carolina, south to Florida, though some limited mixing between the populations may occur in North Carolina [37,63]. It would be interesting to know if similar NNV genotypes exist in the south-Atlantic population, or if the thermal barriers and limited interactions prevent exchange of viruses between these populations. Considering that RGNNV replicates well in warm conditions (25–30 °C), it would not be surprising if that genotype occurs in the south-Atlantic population. Monitoring that population and other fish species from these regions will shed light on the molecular epidemiology of NNV within the U.S. Atlantic coast.

As black sea bass are a prime candidate species for marine aquaculture [34,64], the potential of NNV infection in this species suggests that biosecurity may be important to avoid future disease issues. As this is a species relatively new to aquaculture, practices involve the use of wild-captured broodstock held in recirculation systems [34,65,66]. Vertical transmission has been suggested for NNV and this can lead to heavy mortality of larval progeny [67,68], suggesting that these disease risks may exist when utilizing wild collected black sea bass broodstock. As with other marine fish larvae, black sea bass larvae are sensitive to environmental factors in early life stages and mortality is not uncommon [69]. Poor early survival believed to be related to environmental factors may obscure mortality related to NNV if not properly investigated. With the findings herein, NNV could be a consideration in the list of rule-outs related to larval or juvenile mortality of black sea bass. As NNV has been shown to impact marine aquaculture in diverse fish species in many regions, the findings here suggest that implementing biosecurity plans to prevent NNV in marine aquaculture practices in this region could help prevent disease issues. Further, continued monitoring of wild fish and developing methods to isolate these viruses in cell culture could help to better understand NNV virulence in commercially important fish species.

5. Conclusions

Two genotypes of NNV have been detected using molecular methods for the first time in black sea bass collected off the coast of New Jersey, USA. These findings suggest an expanded range for red-spotted grouper nervous necrosis virus (RGNNV) to include the U.S. Atlantic coast. Further, a novel genotype tentatively named black sea bass nervous necrosis virus (BSBNNV) has been genetically characterized from this region. As NNV is known to cause major disease issues in marine aquaculture worldwide, these findings highlight the need for additional research to better understand virulence and epidemiology of NNV in this region.

Supplementary Materials: The following supporting information can be downloaded at: https://www.mdpi.com/article/10.3390/v17091234/s1, Figure S1: Neighbour-joining (NJ) phylogenetic trees.

Author Contributions: Conceptualization, J.L.; methodology, J.L., N.D., J.N.N., P.J.C. and W.N.B.; formal analysis, J.L., A.T. and M.A.; investigation, N.D., J.N.N., J.L., W.N.B. and P.J.C.; resources, P.J.C. and J.L.; data curation, J.L.; writing—original draft preparation, J.L.; writing—review and editing, A.T., M.A., N.D., J.N.N., W.N.B. and P.J.C.; visualization, J.L., A.T., M.A. and N.D.; supervision, J.L.; project administration, J.L.; funding acquisition, J.L. All authors have read and agreed to the published version of the manuscript.

Funding: This research was funded by the Federal Sportfish and Restoration Grant Project, grant numbers FW-69-R20 and FW-69-R21 and the U.S. Geological Survey Ecosystems Mission Area.

Institutional Review Board Statement: No experimental research on animals occurred in this study. Euthanized fish were provided to the laboratory. Field sampling protocols were in accordance with the state New Jersey Department of Environmental Protection Agency, Fish, and Wildlife Section, as detailed in the Federal Sportfish and Restoration Grant Project.

Data Availability Statement: Genetic sequences have been deposited to the National Center for Biotechnology Information (NCBI) under accession numbers PV877385–PV877398 and PV993948–PV993968. Metadata and raw data from this study are available through the U.S. Geological Survey ScienceBase at the following: Lovy, J. and Batts, W.N., 2025, Surveillance for nervous necrosis virus in black sea bass from the U.S. Atlantic coast: U.S. Geological Survey data release [55], https://doi.org/10.5066/P17MAMLV.

Acknowledgments: The authors are grateful to Kaitlyn Barone and Sarah Friend (New Jersey Fish and Wildlife) for their assistance in processing tissue samples for molecular biology, as well as Jeremy Snyder and Lana Castellano (New Jersey Department of Agriculture) for their support and assistance with PCR processing of samples and viral cell culture assays. We would like to thank the U.S. Department of Agriculture's National Veterinary Services Laboratory, specifically Janet Warg and Mia Kim Torchetti for their technical support, and Justin Greer (U.S. Geological Survey, Western Fisheries Research Center) for his technical assistance. Further we would like to thank the staff from the Bureau of Marine Fisheries, N.J. Fish and Wildlife for their assistance in collecting fish for this study.

Conflicts of Interest: The authors declare no conflicts of interest. Any use of trade, firm, or product names is for descriptive purposes only and does not imply endorsement by the U.S. Government. The funders had no role in the design of the study; in the collection, analyses, or interpretation of data; in the writing of the manuscript; or in the decision to publish the results.

Abbreviations

The following abbreviations are used in this manuscript:

NNV	Nervous necrosis virus
VNN	Viral nervous necrosis
BFNNV	Barfin flounder nervous necrosis virus
RGNNV	Red-spotted grouper nervous necrosis virus
SJNNV	Striped jack nervous necrosis virus
TPNNV	Tiger puffer nervous necrosis virus
BSBNNV	Black sea bass nervous necrosis virus
rRT-PCR	Real-time reverse transcription polymerase chain reaction
RT-PCR	Endpoint/conventional reverse transcription polymerase chain reaction
ICTV	International Committee on Taxonomy of Viruses
CPE	Cytopathic effects
BF-2	Bluegill fry-2 cells
EPC	Epithelioma papulosum cyprini cells
CHSE-214	Chinook salmon embryo-214 cells
SSN-1	Striped snakehead-1 cells
E-11	Clone of striped snakehead cells
CT	Cycle threshold
PEG	Polyethylene glycol
HBSS	Hanks' balanced salt solution
MEM	Minimum essential medium
nt	Nucleotide
AA	Amino acid
NCBI	National Center for Biotechnology Information
TL	Total length
YOY	Young of the year
BLAS	Nucleotide basic local alignment search tool

References

1. Sahul Hameed, A.; Ninawe, A.; Nakai, T.; Chi, S.; Johnson, K.; Consortium, I.R. ICTV virus taxonomy profile: Nodaviridae. *J. Gen. Virol.* **2019**, *100*, 3–4. [CrossRef] [PubMed]
2. Mori, K.-I.; Nakai, T.; Muroga, K.; Arimoto, M.; Mushiake, K.; Furusawa, I. Properties of a new virus belonging to nodaviridae found in larval striped jack (*Pseudocaranx dentex*) with nervous necrosis. *Virology* **1992**, *187*, 368–371. [CrossRef]
3. Peducasse, S.; Castric, J.; Thiery, R.; Jeffroy, J.; Le Ven, A.; Laurencin, F.B. Comparative study of viral encephalopathy and retinopathy in juvenile sea bass Dicentrarchus labrax infected in different ways. *Dis. Aquat. Org.* **1999**, *36*, 11–20. [CrossRef]
4. Nagai, T.; Nishizawa, T. Sequence of the non-structural protein gene encoded by RNA1 of striped jack nervous necrosis virus. *J. Gen. Virol.* **1999**, *80*, 3019–3022. [CrossRef]
5. Nishizawa, T.; Mori, K.-i.; Furuhashi, M.; Nakai, T.; Furusawa, I.; Muroga, K. Comparison of the coat protein genes of five fish nodaviruses, the causative agents of viral nervous necrosis in marine fish. *J. Gen. Virol.* **1995**, *76*, 1563–1569. [CrossRef] [PubMed]
6. Johnson, K.N.; Johnson, K.L.; Dasgupta, R.; Gratsch, T.; Ball, L.A. Comparisons among the larger genome segments of six nodaviruses and their encoded RNA replicases. *J. Gen. Virol.* **2001**, *82*, 1855–1866. [CrossRef]
7. Nishizawa, T.; Furuhashi, M.; Nagai, T.; Nakai, T.; Muroga, K. Genomic classification of fish nodaviruses by molecular phylogenetic analysis of the coat protein gene. *Appl. Environ. Microbiol.* **1997**, *63*, 1633–1636. [CrossRef]
8. Iwamoto, T.; Okinaka, Y.; Mise, K.; Mori, K.-I.; Arimoto, M.; Okuno, T.; Nakai, T. Identification of host-specificity determinants in betanodaviruses by using reassortants between striped jack nervous necrosis virus and sevenband grouper nervous necrosis virus. *J. Virol.* **2004**, *78*, 1256–1262. [CrossRef]
9. Chi, S.-C.; Lin, S.-C.; Su, H.-M.; Hu, W.-W. Temperature effect on nervous necrosis virus infection in grouper cell line and in grouper larvae. *Virus Res.* **1999**, *63*, 107–114. [CrossRef]
10. Hata, N.; Okinaka, Y.; Iwamoto, T.; Kawato, Y.; Mori, K.-I.; Nakai, T. Identification of RNA regions that determine temperature sensitivities in betanodaviruses. *Arch. Virol.* **2010**, *155*, 1597–1606. [CrossRef] [PubMed]
11. Souto, S.; Olveira, J.G.; Bandín, I. Influence of temperature on Betanodavirus infection in Senegalese sole (*Solea senegalensis*). *Vet. Microbiol.* **2015**, *179*, 162–167. [CrossRef]
12. Aspehavg, V. The phylogenetic relationship of nervous necrosis virus from Halibut. *Bull. Eur. Assoc. Fish Pathol.* **1999**, *19*, 196.
13. Barker, D.E.; MacKinnon, A.-M.; Boston, L.; Burt, M.D.; Cone, D.K.; Speare, D.J.; Griffiths, S.; Cook, M.; Ritchie, R.; Olivier, G. First report of piscine nodavirus infecting wild winter flounder Pleuronectes americanus in Passamaquoddy Bay, New Brunswick, Canada. *Dis. Aquat. Org.* **2002**, *49*, 99–105. [CrossRef]
14. Johnson, S.C.; Sperker, S.A.; Leggiadro, C.T.; Groman, D.B.; Griffiths, S.G.; Ritchie, R.J.; Cook, M.D.; Cusack, R.R. Identification and characterization of a piscine neuropathy and nodavirus from juvenile Atlantic cod from the Atlantic coast of North America. *J. Aquat. Anim. Health* **2002**, *14*, 124–133. [CrossRef]
15. Nguyen, H.D.; Mekuchi, T.; Imura, K.; Nakai, T.; Nishizawa, T.; Muroga, K. Occurrence of viral nervous necrosis (VNN) in hatchery-reared juvenile Japanese flounder *Paralichthys olivaceus*. *Fish. Sci.* **1994**, *60*, 551–554. [CrossRef]
16. Bandín, I.; Souto, S. Betanodavirus and VER disease: A 30-year research review. *Pathogens* **2020**, *9*, 106. [CrossRef]
17. Johansen, R.; Sommerset, I.; Tørud, B.; Korsnes, K.; Hjortaas, M.; Nilsen, F.; Nerland, A.; Dannevig, B. Characterization of nodavirus and viral encephalopathy and retinopathy in farmed turbot, *Scophthalmus maximus* (L.). *J. Fish Dis.* **2004**, *27*, 591–601. [CrossRef]
18. Kim, Y.C.; Kwon, W.J.; Min, J.G.; Kim, K.I.; Jeong, H.D. Complete genome sequence and pathogenic analysis of a new betanodavirus isolated from shellfish. *J. Fish Dis.* **2019**, *42*, 519–531. [CrossRef] [PubMed]
19. Le Breton, A.; Grisez, L.; Sweetman, J.; Ollevier, F. Viral nervous necrosis (VNN) associated with mass mortalities in cage-reared sea bass, *Dicentrarchus labrax* (L.). *J. Fish Dis.* **1997**, *20*, 145–151. [CrossRef]
20. Toffan, A.; Pascoli, F.; Pretto, T.; Panzarin, V.; Abbadi, M.; Buratin, A.; Quartesan, R.; Gijón, D.; Padrós, F. Viral nervous necrosis in gilthead sea bream (*Sparus aurata*) caused by reassortant betanodavirus RGNNV/SJNNV: An emerging threat for Mediterranean aquaculture. *Sci. Rep.* **2017**, *7*, 46755. [CrossRef] [PubMed]
21. Vendramin, N.; Zrncic, S.; Padrós, F.; Oraic, D.; Le Breton, A.; Zarza, C.; Olesen, N.J. Fish health in Mediterranean Aquaculture, past mistakes and future challenges. *Bull. Eur. Assoc. Fish Pathol.* **2016**, *36*, 38–45.
22. Fukuda, Y.; HD, N.; Furuhashi, M.; Nakai, T. Mass mortality of cultured sevenband grouper, Epinephelus septemfasciatus, associated with viral nervous necrosis. *Fish Pathol.* **1996**, *31*, 165–170. [CrossRef]
23. Yoshikoshi, K.; Inoue, K. Viral nervous necrosis in hatchery-reared larvae and juveniles of Japanese parrotfish, *Oplegnathus fasciatus* (Temminck & Schlegel). *J. Fish Dis.* **1990**, *13*, 69–77.
24. Chi, S.; Lo, C.; Kou, G.; Chang, P.; Peng, S.; Chen, S. Mass mortalities associated with viral nervous necrosis (VNN) disease in two species of hatchery-reared grouper, *Epinephelus fuscogutatus* and *Epinephelus akaara* (Temminck & Schlegel). *J. Fish Dis.* **1997**, *20*, 185–193.

25. Grotmol, S.; Nerland, A.H.; Biering, E.; Totland, G.K.; Nishizawa, T. Characterisation of the capsid protein gene from a nodavirus strain affecting the Atlantic halibut Hippoglossus hippoglossus and design of an optimal reverse-transcriptase polymerase chain reaction (RT-PCR) detection assay. *Dis. Aquat. Org.* **2000**, *39*, 79–88. [CrossRef] [PubMed]
26. Gagné, N.; Johnson, S.; Cook-Versloot, M.; MacKinnon, A.; Olivier, G. Molecular detection and characterization of nodavirus in several marine fish species from the northeastern Atlantic. *Dis. Aquat. Org.* **2004**, *62*, 181–189. [CrossRef]
27. Curtis, P.; Drawbridge, M.; Iwamoto, T.; Nakai, T.; Hedrick, R.; Gendron, A. Nodavirus infection of juvenile white seabass, Atractoscion nobilis, cultured in southern California: First record of viral nervous necrosis (VNN) in North America. *J. Fish Dis.* **2001**, *24*, 263–271. [CrossRef]
28. Lester, S.E.; Gentry, R.R.; Kappel, C.V.; White, C.; Gaines, S.D. Offshore aquaculture in the United States: Untapped potential in need of smart policy. *Proc. Natl. Acad. Sci. USA* **2018**, *115*, 7162–7165. [CrossRef]
29. Kurath, G.; Winton, J. Complex dynamics at the interface between wild and domestic viruses of finfish. *Curr. Opin. Virol.* **2011**, *1*, 73–80. [CrossRef]
30. Vendramin, N.; Patarnello, P.; Toffan, A.; Panzarin, V.; Cappellozza, E.; Tedesco, P.; Terlizzi, A.; Terregino, C.; Cattoli, G. Viral Encephalopathy and Retinopathy in groupers (*Epinephelus* spp.) in southern Italy: A threat for wild endangered species? *BMC Vet. Res.* **2013**, *9*, 20. [CrossRef]
31. Kara, H.; Chaoui, L.; Derbal, F.; Zaidi, R.; Boisséson, C.d.; Baud, M.; Bigarré, L. Betanodavirus-associated mortalities of adult wild groupers *Epinephelus marginatus* (Lowe) and *Epinephelus costae* (Steindachner) in Algeria. *J. Fish Dis.* **2014**, *37*, 273–278. [CrossRef] [PubMed]
32. Jaramillo, D.; Hick, P.; Whittington, R. Age dependency of nervous necrosis virus infection in barramundi *Lates calcarifer* (Bloch). *J. Fish Dis.* **2017**, *40*, 1089–1101. [CrossRef] [PubMed]
33. Toffan, A.; Biasini, L.; Pretto, T.; Abbadi, M.; Buratin, A.; Franch, R.; Dalla Rovere, G.; Panzarin, V.; Marsella, A.; Bargelloni, L. Age dependency of RGNNV/SJNNV viral encephalo-retinopathy in Gilthead Sea Bream (*Sparus aurata*). *Aquaculture* **2021**, *539*, 736605. [CrossRef]
34. Watanabe, W.O.; Carroll, P.M.; Alam, M.S.; Dumas, C.F.; Gabel, J.E.; Davis, T.M.; Bentley, C.D. The status of black sea bass, Centropristis striata, as a commercially ready species for US marine aquaculture. *J. World Aquac. Soc.* **2021**, *52*, 541–565. [CrossRef]
35. Steimle, F.W.; Zetlin, C.A.; Berrien, P.L.; Chang, S. *Essential Fish Habitat Source Document. Black Sea Bass, Centropristis Striata, Life History and Habitat Characteristics*; Northeast Fisheries Science Center (U.S.); NOAA Technical Memorandum NMFS-NE, 1999; Volume 143.
36. Roy, E.; Quattro, J.; Greig, T. Genetic management of Black Sea Bass: Influence of biogeographic barriers on population structure. *Mar. Coast. Fish.* **2012**, *4*, 391–402. [CrossRef]
37. McCartney, M.A.; Burton, M.L.; Lima, T.G. Mitochondrial DNA differentiation between populations of black sea bass (*Centropristis striata*) across Cape Hatteras, North Carolina (USA). *J. Biogeogr.* **2013**, *40*, 1386–1398. [CrossRef]
38. Musick, J.A.; Mercer, L.P. Seasonal distribution of black sea bass, Centropristis striata, in the Mid-Atlantic Bight with comments on the ecology and fisheries of the species. *Trans. Am. Fish. Soc.* **1977**, *106*, 12–25. [CrossRef]
39. Moser, J.; Shepherd, G.R. Seasonal distribution and movement of black sea bass (*Centropristis striata*) in the Northwest Atlantic as determined from a mark-recapture experiment. *J. Northwest Atl. Fish. Sci.* **2009**, *40*, 17–28. [CrossRef]
40. Steimle, F.W.; Zetlin, C. Reef habitats in the middle Atlantic bight: Abundance, distribution, associated biological communities, and fishery resource use. *Mar. Fish. Rev.* **2000**, *62*, 24–42.
41. Lovy, J.; Lewis, N.L.; Friend, S.E.; Able, K.W.; Shaw, M.J.; Hinks, G.S.; Clarke, P.J. Host, seasonal and habitat influences on incidence of *Lernaeenicus radiatus* (Copepoda: Pennellidae) in the mid-Atlantic Bight. *Mar. Ecol. Prog. Ser.* **2020**, *642*, 83–101. [CrossRef]
42. Able, K.W.; Valenti, J.L.; Grothues, T.M. Fish larval supply to and within a lagoonal estuary: Multiple sources for Barnegat Bay, New Jersey. *Environ. Biol. Fishes* **2017**, *100*, 663–683. [CrossRef]
43. Iwamoto, T.; Nakai, T.; Mori, K.-i.; Arimoto, M.; Furusawa, I. Cloning of the fish cell line SSN-1 for piscine nodaviruses. *Dis. Aquat. Org.* **2000**, *43*, 81–89. [CrossRef]
44. Panzarin, V.; Patarnello, P.; Mori, A.; Rampazzo, E.; Cappellozza, E.; Bovo, G.; Cattoli, G. Development and validation of a real-time TaqMan PCR assay for the detection of betanodavirus in clinical specimens. *Arch. Virol.* **2010**, *155*, 1193–1203. [CrossRef] [PubMed]
45. Dalla Valle, L.; Zanella, L.; Patarnello, P.; Paolucci, L.; Belvedere, P.; Colombo, L. Development of a sensitive diagnostic assay for fish nervous necrosis virus based on RT-PCR plus nested PCR. *J. Fish Dis.* **2000**, *23*, 321–327. [CrossRef]
46. Untergasser, A.; Cutcutache, I.; Koressaar, T.; Ye, J.; Faircloth, B.C.; Remm, M.; Rozen, S.G. Primer3—New capabilities and interfaces. *Nucleic Acids Res.* **2012**, *40*, e115. [CrossRef] [PubMed]
47. Batts, W.N.; LaPatra, S.E.; Katona, R.; Leis, E.; Ng, T.F.F.; Brieuc, M.S.; Breyta, R.B.; Purcell, M.K.; Conway, C.M.; Waltzek, T.B. Molecular characterization of a novel orthomyxovirus from rainbow and steelhead trout (*Oncorhynchus mykiss*). *Virus Res.* **2017**, *230*, 38–49. [CrossRef]

48. Panzarin, V.; Cappellozza, E.; Mancin, M.; Milani, A.; Toffan, A.; Terregino, C.; Cattoli, G. In vitro study of the replication capacity of the RGNNV and the SJNNV betanodavirus genotypes and their natural reassortants in response to temperature. *Vet. Res.* **2014**, *45*, 56. [CrossRef]
49. Jia, P.; Chen, X.; Fu, J.; Yi, M.; Chen, W.; Jia, K. Near-complete genome sequence of a fish nervous necrosis virus isolated from hybrid grouper in China. *Microbiol. Resour. Announc.* **2020**, *9*, e01453-19. [CrossRef]
50. Toffolo, V.; Negrisolo, E.; Maltese, C.; Bovo, G.; Belvedere, P.; Colombo, L.; Dalla Valle, L. Phylogeny of betanodaviruses and molecular evolution of their RNA polymerase and coat proteins. *Mol. Phylogenetics Evol.* **2007**, *43*, 298–308. [CrossRef]
51. Bovo, G.; Gustinelli, A.; Quaglio, F.; Gobbo, F.; Panzarin, V.; Fusaro, A.; Mutinelli, F.; Caffara, M.; Fioravanti, M. Viral encephalopathy and retinopathy outbreak in freshwater fish farmed in Italy. *Dis. Aquat. Org.* **2011**, *96*, 45–54. [CrossRef]
52. Kumar, S.; Stecher, G.; Tamura, K. MEGA7: Molecular evolutionary genetics analysis version 7.0 for bigger datasets. *Mol. Biol. Evol.* **2016**, *33*, 1870–1874. [CrossRef]
53. Nguyen, L.-T.; Schmidt, H.A.; Von Haeseler, A.; Minh, B.Q. IQ-TREE: A fast and effective stochastic algorithm for estimating maximum-likelihood phylogenies. *Mol. Biol. Evol.* **2015**, *32*, 268–274. [CrossRef]
54. Kalyaanamoorthy, S.; Minh, B.Q.; Wong, T.K.; Von Haeseler, A.; Jermiin, L.S. ModelFinder: Fast model selection for accurate phylogenetic estimates. *Nat. Methods* **2017**, *14*, 587–589. [CrossRef]
55. Lovy, J.; Batts, W.N. *Surveillance for Nervous Necrosis Virus in Black Sea Bass from the U.S. Atlantic Coast*; U.S. Geological Survey Data Release: Reston, VA, USA, 2025. [CrossRef]
56. Panzarin, V.; Fusaro, A.; Monne, I.; Cappellozza, E.; Patarnello, P.; Bovo, G.; Capua, I.; Holmes, E.C.; Cattoli, G. Molecular epidemiology and evolutionary dynamics of betanodavirus in southern Europe. *Infect. Genet. Evol.* **2012**, *12*, 63–70. [CrossRef]
57. Knibb, W.; Luu, G.; Premachandra, H.; Lu, M.-W.; Nguyen, N.H. Regional genetic diversity for NNV grouper viruses across the Indo-Asian region–implications for selecting virus resistance in farmed groupers. *Sci. Rep.* **2017**, *7*, 10658. [CrossRef]
58. Kim, Y.; Kwon, W.; Min, J.; Jeong, H. Isolation and initial characterization of new betanodaviruses in shellfish. *Transbound. Emerg. Dis.* **2018**, *65*, 1557–1567. [CrossRef]
59. Korsnes, K.; Devold, M.; Nerland, A.H.; Nylund, A. Viral encephalopathy and retinopathy (VER) in Atlantic salmon Salmo salar after intraperitoneal challenge with a nodavirus from Atlantic halibut *Hippoglossus* hippoglossus. *Dis. Aquat. Org.* **2005**, *68*, 7–16. [CrossRef] [PubMed]
60. Bell, R.J.; Richardson, D.E.; Hare, J.A.; Lynch, P.D.; Fratantoni, P.S. Disentangling the effects of climate, abundance, and size on the distribution of marine fish: An example based on four stocks from the Northeast US shelf. *ICES J. Mar. Sci.* **2015**, *72*, 1311–1322. [CrossRef]
61. McMahan, M.D.; Sherwood, G.D.; Grabowski, J.H. Geographic variation in life-history traits of black sea bass (*Centropristis striata*) during a rapid range expansion. *Front. Mar. Sci.* **2020**, *7*, 567758. [CrossRef]
62. Olveira, J.G.; Souto, S.; Dopazo, C.P.; Thiéry, R.; Barja, J.L.; Bandin, I. Comparative analysis of both genomic segments of betanodaviruses isolated from epizootic outbreaks in farmed fish species provides evidence for genetic reassortment. *J. Gen. Virol.* **2009**, *90*, 2940–2951. [CrossRef] [PubMed]
63. Gray, I.E.; Cerame-Vivas, M.J. The Circulation of Surface Waters in Raleigh Bay, North Carolina 1. *Limnol. Oceanogr.* **1963**, *8*, 330–337. [CrossRef]
64. Watanabe, W.O.; Dumas, C.F.; Carroll, P.M.; Resimius, C.M. Production economic analysis of black sea bass juveniles to support finfish mariculture growout industry development in the southeastern United States. *Aquac. Econ. Manag.* **2015**, *19*, 226–250. [CrossRef]
65. Watanabe, W.O.; Smith, T.I.; Berlinsky, D.L.; Woolridge, C.A.; Stuart, K.R.; Copeland, K.A.; Denson, M.R. Volitional spawning of Black Sea Bass Centropristis striata induced with pelleted luteinizing hormone releasing hormone-analogue. *J. World Aquac. Soc.* **2003**, *34*, 319–331. [CrossRef]
66. Cotton, C.F.; Walker, R.L.; Recicar, T.C. Effects of temperature and salinity on growth of juvenile black sea bass, with implications for aquaculture. *N. Am. J. Aquac.* **2003**, *65*, 330–338. [CrossRef]
67. Azad, I.; Jithendran, K.; Shekhar, M.; Thirunavukkarasu, A.; De la Pena, L. Immunolocalisation of nervous necrosis virus indicates vertical transmission in hatchery produced Asian sea bass (*Lates calcarifer* Bloch)—A case study. *Aquaculture* **2006**, *255*, 39–47. [CrossRef]
68. Ransangan, J.; Manin, B.O. Mass mortality of hatchery-produced larvae of Asian seabass, *Lates calcarifer* (Bloch), associated with viral nervous necrosis in Sabah, Malaysia. *Vet. Microbiol.* **2010**, *145*, 153–157. [CrossRef] [PubMed]
69. Berlinsky, D.L.; Taylor, J.C.; Howell, R.A.; Bradley, T.M.; Smith, T.I. The effects of temperature and salinity on early life stages of black sea bass *Centropristis striata*. *J. World Aquac. Soc.* **2004**, *35*, 335–344. [CrossRef]

viruses

MDPI

Article

Hiding in Plain Sight: Genomic Characterization of a Novel Nackednavirus and Evidence of Diverse Adomaviruses in a Hyperpigmented Lesion of a Largemouth Bass (*Micropterus nigricans*)

Clayton Raines [1,2], John Odenkirk [3], Michael Isel [3], Patricia Mazik [2], Morgan Biggs [1] and Luke Iwanowicz [1,4,*]

1 Leetown Research Laboratory, Eastern Ecological Science Center, U.S. Geological Survey, 11649 Leetown Road, Kearneysville, WV 25430, USA; craines@usgs.gov (C.R.)

2 West Virginia Cooperative Fish and Wildlife Research Unit, West Virginia University, U.S. Geological Survey, 322 Percival Hall, Morgantown, WV 26506, USA; pmazik@wvu.edu

3 Virginia Department Wildlife Resources, 1320 Belman Road, Fredericksburg, VA 22401, USA; john.odenkirk@dwr.virginia.gov (J.O.); mike.isel@dwr.virginia.gov (M.I.)

4 National Center for Cool and Coldwater Aquaculture, Agriculture Research Service, United States Department of Agriculture, 11861 Leetown Road, Kearneysville, WV 25430, USA

* Correspondence: luke.iwanowicz@usda.gov

Abstract

Largemouth bass (LMB; *Micropterus nigricans*) are popular both as a sportfish and an aquaculture species. At present, six described viruses are associated with LMB, of which two are typically considered in cases of LMB mortality events. Advances in discovery and diagnostic capabilities using next-generation sequencing have augmented surveillance efforts and subsequently led to the discovery of novel cryptogenic viruses. Here, we present evidence of three novel viruses from a single skin sample collected from a hyperpigmented melanistic lesion of an LMB with blotchy bass syndrome associated with MnA-1 co-infection. These viruses represent recently described groups of viruses (adomaviruses and nackednaviruses) that infect fish. Both are markedly understudied and of unknown significance to fish health. This work highlights the diversity of viruses associated with LMB and further advances our understanding of the LMB virome. Application of de novo sequencing approaches presents an opportunity to explore a new frontier of host–pathogen relationships and microbes associated with changing environments.

Keywords: fish virus; nackednavirus; adomavirus; largemouth bass

1. Introduction

Largemouth bass (*Micropterus nigricans*; LMB), so named for its oversized jaws, is a member of the sunfish family Centrarchidae which are endemic to North America. Centrarchidae consists of seven genera, including the genus Micropterus, commonly referred to as black basses. Initially, the genus contained only two black bass species: largemouth bass (historically identified as *M. salmoides*) and smallmouth bass (*M. dolomieu*), though the taxonomy is constantly evolving [1,2]. The closely related Florida bass (recognized as *Micropterus salmoides*; FLB) was long considered a subspecies of LMB, though it is now recognized as a standalone species [3,4]. FLB and LMB are frequently treated as a single management unit, as their ranges now substantially overlap and are frequently found to be syntopic [5]. Despite decades old calls for cessation of FLB stocking outside the state of Florida [3], present day management frequently includes overstocking of FLB into the historic range of LMB to produce hybrids with increased trophy potential [6,7]. In doing so, native basses are then subject to the negative effects of population mixing, including loss of local adaptations or co-adapted gene complexes [5,8]. Failure to differentiate overlapping species management units, or creation of new overlaps, can result in overexploitation and a reduction in underrepresented genetic traits [9] or even the extinction of rare, undiscovered, or overlooked species [10]. Even so, LMB stocking and management exists as one of the primary cornerstones of freshwater fisheries in North America. Consequently, due to the ongoing management and taxonomy overlap within the study area, FLB and LMB are considered as a single unit in this manuscript, referred to hereafter as LMB.

Of the black basses, LMB have the widest native distribution, ranging as far north as the James River in Virginia (Figure 1), east to the Atlantic seaboard, and south into northern Mexico [4,11–13]. With managed fisheries in both warm and cool water, black bass inhabit all fifty states [1,14]. In warmwater fisheries, LMB are the most sought-after species of the black basses and are more available to U.S. anglers than any other species of fish [15,16]. While largemouth bass are native to North America, translocation of fish either by individuals or government entities enabled spread far outside of their endemic range. As a result, black bass are now among the most widely distributed aquatic organisms in the world and constitute established non-indigenous populations in Africa, Asia, South America, and Europe [17]. As a result of increased demand for sportfish stocking, intensive aquaculture of LMB soon became a necessity. LMB-specific aquaculture in the United States began in the mid-1800s [18,19] and quickly expanded to 2,000,000 ponds and 500 agency-managed hatcheries by the 1960s [19,20]. Domestic reports from 2013 reported sales of LMB approaching 1,000,000 kg/year in a single state [19], with the worldwide market LMB production rates at 152,200 metric tons in the same year [21,22].

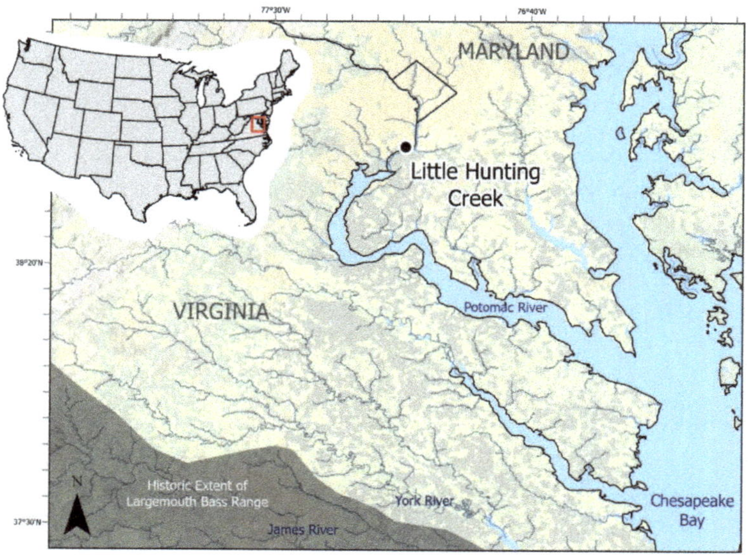

Figure 1. Sampling location of wild-caught largemouth bass (*Micropterus nigricans*; LMB) sampled in 2021 within Little Hunting Creek, Virginia, USA. The black dot denotes GPS coordinates of the centroid of the electrofishing transects. Historic range of LMB is modified from Miranda et al. [4].

Despite their popularity as a sportfish and aquaculture species, and that the obligate and opportunistic pathogens of LMB are better studied than most non-intensively cultured species, a substantial knowledge gap remains. While a number of bacterial pathogens of bass have been described, there is little experimental viral research in this species [23–25]. At present, there are six viruses described that infect largemouth bass. These include largemouth bass reovirus (LMBRV), *Micropterus salmoides* reovirus (MeReV), the iridoviruses largemouth bass virus (LMBV) and ISKNV-like virus (ISKNV-ZY), *Micropterus salmoides* rhabdovirus (MSRV), and the recently described *Micropterus nigricans* adomavirus 1 (MnA-1) [26–31]. The causal relationship between the reoviruses and clinical disease is ambiguous, which is typical of this family of orphan viruses [32]. LMBV was first discovered in 1991 and is routinely ascribed as the causative pathogen of LMB mortality events in the United States (US) [26,33,34]. The acutely lethal megalocytivirus, ISKNV-ZY, was documented in LMB aquaculture in China in 2017 [31]. Later, in 2018 MSRV was identified as a novel primary pathogen of LMB, having caused substantial economic losses in Chinese aquaculture since 2011 [29,35,36]. Most recently, MnA-1 has been identified as the causative agent of blotchy bass syndrome (BBS) [30].

The first documented report of BBS in LMB was in the late 1980s [30,37]. The association of a virus, MnA-1, with this condition was only recently confirmed [30]. This disease is characterized by the manifestation of pathognomonic hyperpigmented melanistic lesions (HMPLs) on the skin of LMB and is typically observed between the fall and spring [30]. The significance of the HPMLs is unknown and they purportedly resolve during the summer months [30]. Fish affected by this seasonal disease have been reported in natural waters from Texas to the East Coast. Here, we screened largemouth bass for MnA-1 in Little Hunting Creek, VA to further establish the relationship between the virus and the clinical manifestation of BBS and better understand the diversity of MnA-1 in LMB in this region. Application of next-generation sequencing led to the discovery of a novel nackednavirus and two previously undescribed adomaviruses (ADmVs). These findings highlight the black box containing the universe of uncharacterized viruses of this host.

2. Materials and Methods

Live LMB were collected from Little Hunting Creek (Figure 1), in Fairfax County, VA (USA). A tidal tributary of the Potomac River, Little Hunting Creek is a fifth-order stream located in the Chesapeake Bay watershed, the largest estuary in the United States [38]. It lies within the Atlantic Coastal Plain terrane [39,40] and encompasses 7067 acres, 82% of which is currently developed and 1762 acres of that developed landscape being totally impervious [41]. Little Hunting Creek is also of historical and cultural importance. Much of the land located within the catchment was once owned by George Washington, and the historic plantation now known as Mount Vernon was initially known as Little Hunting Creek Plantation [42].

LMB were collected during routine daytime DC electrofishing surveys performed by the Virginia Department of Wildlife Resources. Boat electrofishing was conducted from a vessel equipped with twin anodes, each affixed with six droppers and powered by a 5000 W generator. The control box was a Smith-Root (Vancouver, WA, USA) Type VI-A set to 884 V and run at 7 A. Shoreline electrofishing efforts were concentrated in shallow water (<2 m) at the channel margins and around structures or floating vegetation. Fish visually identified as having HPMLs (n = 20) were subject to scale or tissue collection for downstream DNA extraction prior to release. If HPMLs were present on the body surface, 1–2 scales were removed via sterilized forceps. When possible, we collected HPMLs from tissue that could be less-invasively sampled (e.g., fin margins) instead of scales. Clinically normal tissues were also collected from bass with BBS for molecular screening. All collected samples were immediately stored in RNAlater™ (ThermoFisher, Waltham, MA, USA). Samples were stored at room temperature for 24 h prior to being stored at −20 °C, where they remained until extraction.

DNA from preserved scale tissue was then extracted using a DNeasy Blood and Tissue Kit (Qiagen, Valencia, CA, USA), following manufacturer instructions for Purification of Total DNA from Animal Tissues (Spin-Column Protocol; DNeasy Blood & Tissue Handbook version 07/2020). As scales were dentinous, collected tissues were consumptively digested overnight (~10 h) in 180 μL of lysis buffer and 20 μL proteinase K (600 mAU/mL) prior to proceeding with the specified protocol. Extracted DNA was quantified using a Qubit dsDNA HS Assay Kit and a Qubit 4.0 Fluorometer (Invitrogen, Carlsbad, CA, USA).

Given that the intention here was to screen for ADmVs, we used rolling circle amplification (RCA) to enrich a single sample for circular dsDNA viral genomes. RCA was conducted using an Illustra TempliPhi 100 Amplification Kit (Cytiva, Wilmington, DE, USA) in accordance with manufacturer instructions specified for M13 Phage DNA. The process was initiated by combining 0.5 μL template DNA (10 ng/μL) with the included sample buffer and using included random hexamer primers. Using a thermocycler, the mix was denatured for 3 min at 95 °C and immediately cooled to 4 °C. A TempliPhi™ reaction master mix was made by combining 5 μL of reaction buffer and 0.2 μL enzyme mix, with a total of 5 μL of the cocktail added to the denatured samples upon cooling. The combined mixture was then incubated on a thermocycler at 30 °C for 18 h, followed by heat-inactivating the enzyme at 65 °C for 10 min. The resultant RCA product was diluted with 40 μL of nuclease free water before preparation for next-generation sequencing.

The diluted amplification product from the RCA was quantified using a Qubit dsDNA HS Assay Kit and a Qubit 4.0 Fluorometer. The product was normalized to 0.2 ng/μL using nuclease free water. A total of 1 ng (5 μL) of normalized product was then used as a template for next-generation library preparation. An Illumina Nextera XT Library Preparation Kit (Illumina, San Diego, CA, USA) was used in accordance with Nextera XT Library Preparation Reference Guide (Doc # 15031942 ver. 5) for MiSeq preparation. The final library was normalized with Illumina's Bead-Based Normalization (BBN) method and pooled as described in the BBN Loading Concentrations Exceptions Table 2 (MiSeq System Denature and Dilute Libraries Guide (Doc # 15039740 ver. 10). Pooled libraries were then sequenced for 2 × 301 cycles and loaded with a 15% PhiX (12.5 pM) spike.

Paired reads were quality trimmed and assembled using Megahit (v1.2.9) [43] and screened for putative viral sequences using Cenote Taker 2 [44]. Genome coverage and mapping metrics were determined using CLC Genomics Workbench v.25.0.1. This led to the identification of three novel viruses including a nackednavirus and two ADmVs. No additional DNA viruses were identified in these efforts. We predicted open reading frames (ORFs) using Geneious Prime (v.2022.2.2). Predicted proteins were queried against the NCBI Conserved Domain, PDB, Pfam-A, UniProt-SwissProt-viral70 databases via HH-pred [45]. Genomic sequences were queried using blastx against the NCBI non-redundant protein and transcriptome assembly proteins databases to identify similar uncharacterized sequences. MnA-1 contigs were aligned to the reference MnA-1 genome (PV430023) [30]. We analyzed for variations and single nucleotide polymorphisms (SNPs) using Geneious Prime and the Find Variations/SNPs tool. We compared phylogenetic relationships of nackednaviruses (NDVs) to other known fish NDVs using the polymerase (P) protein. Sequences of the P protein were obtained using NCBI and GenBase [46], as well as supplemental data from nackednavirus publications [47–49]. Novel ADmVs were compared to NCBI reference ADmVs using RepE1 protein (Superfamily 3 helicase; S3H) [50,51]. Sequences were aligned using MUSCLE and default settings [52]. We used IQ-TREE to determine model selection and ultrafast bootstrapping to identify tree topology [53] for both NDVs and ADmVs. We used iTOL v.6 to visualize tree topology and phylogenetic relationships [54]. Pairwise relationships of NDV P proteins and ADmV RepE1 proteins across fish hosts were determined and visualized using the Sequence Demarcation Tool 1.3 [55].

We designed endpoint PCR assays to facilitate screening of the novel viruses obtained during the 2021 sampling effort (Table 1). Primers were designed using Primer3 v2.3.7 bundled in Geneious Prime. We targeted the RepE1 locus for the ADmVs and the polymerase ORF for MnNDV-1. PCR was run on extracted DNA from both HPMLs and paired clinically normal tissues from the same individuals. All PCR reactions were conducted on a Bio-Rad T100 thermal cycler as 25 μL reactions consisting of 1 μL of template, 1 μL of 10 μM forward primer, 1 μL of 10 μM reverse primer, 10 μL of nuclease free water, and 12 μL of 2× GoTaq® Green Master Mix (Promega Corporation, Madison, WI, USA). Cycling conditions for each reaction can be found in Table 1. Obtained PCR products were visualized on a 2% agarose gel and using a Bio-Rad ChemiDoc Imaging System to confirm amplification.

Table 1. PCR Conditions and Primer Sequences. All PCR reactions were conducted as 25 μL reactions consisting of 1 μL of template, 1 μL of 10 uM forward primer, 1 μL of 10 uM reverse primer, 10 μL of nuclease free water, and 12 μL of 2× GoTaq® Green Master Mix. Cycling profiles were as follows: Denature at 95 °C for 3 min and then 30 cycles of 30 s at 95 °C, 30 s at 60 °C, and 60 s at 72 °C. Final extension at 72 °C for 5 min, final hold at 4 °C for ADmVs. The cycling profile was identical for the NDV except for a 5 min denaturation step and 40 s extension time. Product size is indicated in the primer name.

Primer Bind	Primer Name	Sequence (5′ → 3′)	Target
		MnA-1	
F	Mna1-E01_851F	TGCTCGTGCCCTTAACAGAG	RepE1 (NCBI accession: PV469406)
R	Mna1-E01_851R	TCTCTCAGACGGTTCGT	
		MnA-2	
F	Mna2_E01_802F	GCAGCTAAATCGCAGACAGC	RepE1 (NCBI accession: PV469407)
R	Mna2_E01_802R	TGTTCACTGGCACCTCATCC	
		MnA-3	
F	Mna3_E01_699F	ATCTGAAACCCGGAACCGTC	RepE1 (NBCI accession: PV469408)
R	Mna3_E01_699R	GATGGGATGCACCAGTGACA	
		MnNDV-1	
F	MnNV462F	CCCGAGATTCAACGATGGGT	Polymerase (NCBI Accession: PV448632)
R	MnNV462R	GCATGCAAAAAGGGGAGCTC	

3. Results

3.1. Genetic Data Availability

Short reads used for viral assemblies obtained from the HPML sample VA12B were deposited in the Sequence Read Archive (SRX13896994) associated with BioProject (PRJNA785556), BioSample (SAMN23566442), and accessions (PV469405, PV469406, PV469407, PV469408, PV430023, and PV448632).

3.2. LMB Nackednavirus Genome Organization

The complete genome of a novel dsDNA nackednavirus MnNDV-1 (Figure 2) of 3206 bp in length was obtained (NCBI accession: PV448632). The average coverage of this genome was 164,861×, consisting of 3,212,600 reads. The GC content of the genome was 46.5%. Organizationally, the genome structure of MnNDV-1 (Table 2) was consistent with that of other nackednaviruses: <4 kb in length, possessing 14 bp direct repeats, two predicted partially overlapping ORFs >500 bp, and three non-overlapping ORFs <500 bp (Figure 3) [56].

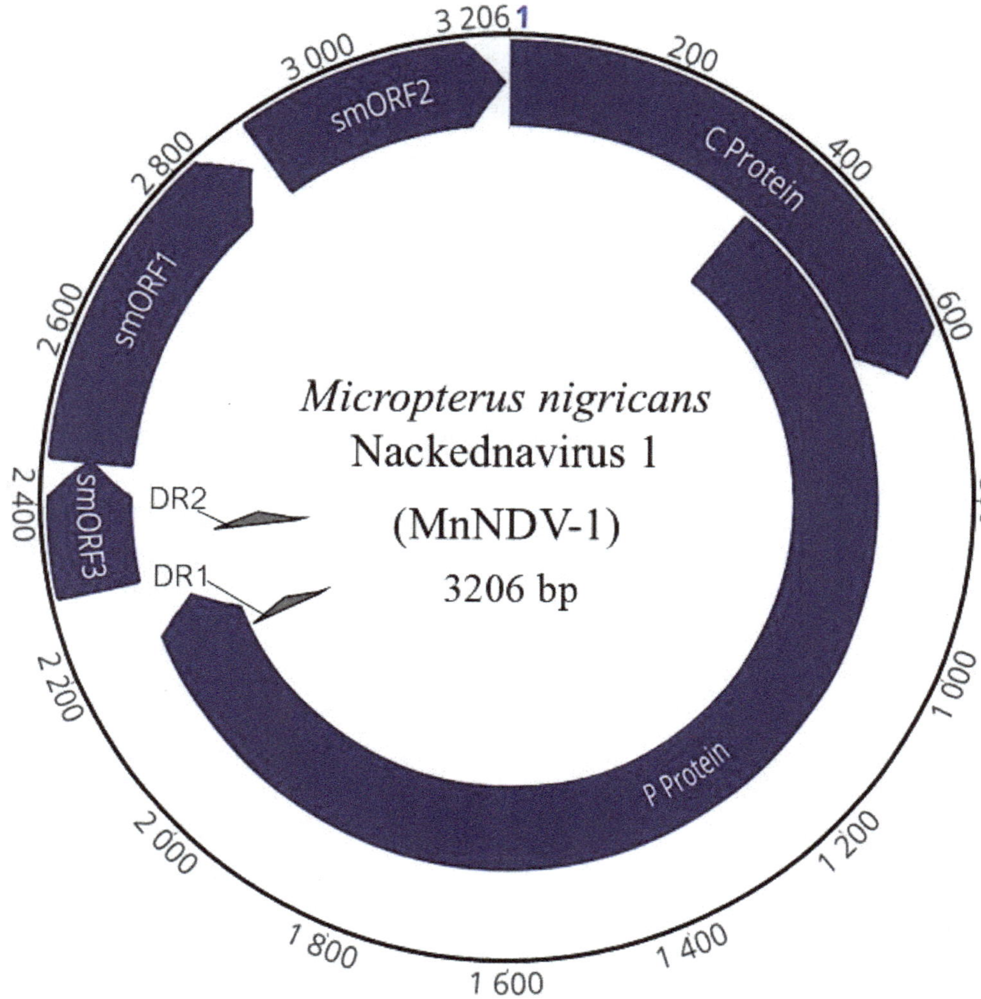

Figure 2. Genome organization of the novel *Micropterus nigricans* nackednavirus (MnNDV-1). The complete genome is 3206 bp and includes partially or completely overlapping ORFs encoding for the core (C) and polymerase (P) proteins, as well as three small open reading frames smORF1, smORF2, and smORF3 (indigo). The blue numeral denotes the origin of nucleotide numbering for the viral genome, corresponding to position 1 in the reference sequence.

Table 2. Genome organization of novel ADmVs and NDVs obtained from largemouth bass sample VA12B. Coding sequences (CDSs) denoted in bold and italics were identified in this sample.

CDS	Highest Structural Similarity (HHpred)	GC%	Length (bp)	Interval (nt)	pI	Amino Acids	Molecular Weight (Da)
			MnNDV-1 (3206 bp, 46.5% GC)				
C Protein	Core protein HBV	53.5	648	1–648	10.24	215	24,343
P Protein	Polymerase protein HBV	49.5	2298	353–2257	10.17	765	72,880
smORF 3	Phage minor head protein GP7	45.1	162	2293–2454	9.13	53	5886
smORF 1	Uncharacterized protein	46.7	420	2451–2870	9.78	139	15,990
smORF 2	Uncharacterized protein	41.7	312	2892–3203	5.26	103	11,958

Table 2. *Cont.*

CDS	Highest Structural Similarity (HHpred)	GC%	Length (bp)	Interval (nt)	pI	Amino Acids	Molecular Weight (Da)
			MnA-2 (17,869 bp, 46.1% GC)				
1940	Y1940 ATV Protein Acidianus two-tailed virus	49.6	1071	1464–2534	11.38	357	39,252
Wasp	Neural Wiskott–Aldrich syndrome protein	44.7	237	2476–2712	10.55	79	9051
Cahalt	Dynein intermediate chain, motor protein	52	327	2739–3065	4.15	109	12,153
Penton	Signal transduction histidine kinase, adenovirus penton	46	783	3079–3861	4.73	261	28,150
Macc	Late L2 mu core protein	47.5	888	3851–4738	10.31	296	31,895
Hexon	ATP synthase subunit b (MdA-2 LO6 analog)	48.2	1494	4750–6243	5.35	498	55,108
Adenain	Adenain; alpha and beta protein	46	633	6353–6985	8.37	211	24,253
Bcoroid	BCOR, BCL-6 co-repressor	50.3	1968	7274–9241	5.02	656	71,264
Prim	DNA primase small subunit	49.9	2256	9244–11,499	5.98	751	82,566
Otomem	Proton channel, otopetrin, membrane protein	44.2	591	11,496–12,086	4.67	197	22,283
Spamem	Serine palmitoyltransferase small subunit A membrane	36.1	291	12,114–12,404	5.91	97	11,238
Broz	3-ketodihydrosphingosine reductase, oxidoreductase	37.6	303	13,478–13,780	7.52	101	11,645
RepE1	Replication protein E1	45.8	2787	14,451–17,237	6.05	928	104,573
Alt	Movement protein TGBp3	47.4	228	16,397–16,624	7.9	76	8592
SET	SET domain Methyltransferase	45.2	897	16,973–17,869	9.89	298	34,089
Sealt	Tospovirus NSs protein	44.4	207	17,620–17,826	8.5	69	8469
			MnA-3 (partial sequence)				
Prim	DNA primase small subunit	51.2	2250	1578–3827	7.22	749	83,240
Otomem	Structural protein VP2 Pyrobaculum filamentous virus 1	48.0	483	3704–4186	5.79	161	17,277
RepE1	Replication protein E1	47.2	2214	5685–7808	10.79	708	74,945

The first of the major ORFs in reading frame 1 (RF +1) encodes the core (C) protein (nt 847—1494, 648 bp) of 215 amino acids (aa) with a theoretical average molecular weight of 24,343 daltons. The size of the C protein was comparable to that of other nackednaviruses and identical (in size) to that of the Rainwater killifish Nackednavirus Lp-2 (KNDV-Lp-2). Similarly, the predicted isoelectric point (pI) of 10.24 is similar to other NDV C proteins. The C ORF partially overlaps open reading frame 2 (RF +2), which encodes the viral polymerase (P) protein (nt 1119–3103, 2298 bp). This ORF contains conserved domains associated with NDVs and hepadnaviruses including viral DNA polymerases (DNA polymerase C and N terminal domains) and reverse transcriptase (DIRS1 group of retrotransposons and RVT_1). This 765 aa protein has a pI of 10.17. Also present within the genome were three predicted small ORFs (smORFs), two of which were partially overlapping. The smORF1 (nt 91–510, 420 bp) encodes a small 139 aa hypothetical protein. While data are unavailable to confirm the expression of this protein, an 80 aa locus similar to that of an uncharacterized protein from Acanthamoeba polyphaga mimivirus (BAV62857.1) was identified in this ORF. Partially overlapping smORF1 is the 53 aa smORF3 (nt 3139–94, 162 bp). Like the hypothetical protein from smORF1, the nearest predicted protein comparison for smORF3 was a 38 aa locus of a minor head protein from a Bacillus subtilis bacteriophage (Q38442.2). Lastly, smORF2 (nt 532–843, 312 bp) consisting of 103 aa shared weak but greatest identity (35–46.5%) with syntenic proteins from other NDVs. Comparisons of MnNDV-1 using blastx identified the greatest identity (36.28%) to an uncharacterized *Lepomis macrochirus* virus (WXG22992.1) [56].

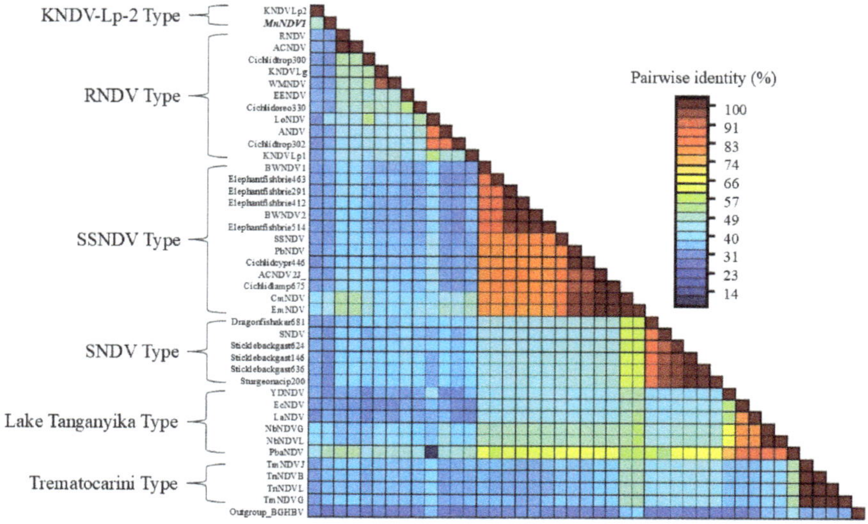

Figure 3. Pairwise identity (%) of nackednavirus (NDVs) polymerase proteins. *Micropterus nigricans* nackednavirus (MnNDV-1) is denoted in bold and italics. Brackets denote associations based on similarities to "Types" identified previously. NCBI and GenBase accession numbers are available in Table 3.

Table 3. Pairwise identity (%) of the *Micropterus nigricans* nackednavirus 1 (MnNDV-1). Accession number indicates NCBI or Genbase accession numbers, and asterisks denote sequences obtained from NCBI SRA experiments in lieu of accession numbers. Samples with the prefix "C_" were obtained from GenBase. MnNDV-1 accession numbers identified in bold and italics. Host and tissue of isolation are indicated.

Accession Number	Host	Tissue Source	Figure Label	% Identity
PV448632	*Micropterus nigricans*	Fin tissue	***MnNDV-1***	-
WLN26308	*Pseudosimochromis babaulti*	Gill	PbaNDV	49.6
SRX340853 *	*Lucania parva*	Pooled organs	KNDVLp2	49.0
WLN26317	*Enantiopus melanogenys*	Brain	EmNDV	45.3
WLN26318	*Callochromis macrops*	Gill	CmNDV	43.8
WLN26322	*Lamprologus ocellatus*	Gonads	LoNDV	42.3
OR350361	*Neolamprologus buescheri*	Gill	NbNDVG	39.0
OR350369	*Trematocara nigrifrons*	Brain	TnNDVB	36.7
WLN26321	*Trematocara nigrifrons*	Liver	TnNDV-L	36.7
OR350368	*Trematocara marginatum*	Lower pharyngeal jaw	TmNDV-J	36.0
WLN26320	*Trematocara marginatum*	Gill	TmNDV-G	36.0
SRX340836	*Lucania parva*	Pooled organs	KNDV-Lp-1	35.6
OR350333	*Lepidiolamprologus attenuatus*	Gill	LaNDV	35.3
ERX240954	*Astatotilapia sp.*	Unknown	ANDV	35.2
C_AA050294	Brienomyrus brachyistius	Unknown	Elephantfishbrie463	35.1
C_AA050263	Tropheus sp. "Ikola"	Unknown	Cichlidtrop302	35.0
C_AA050253	Paracyprichromis brieni	Unknown	Cichlidcypr446	34.9
C_AA050256	Lamprologus ocellatus	Unknown	Cichlidlamp675	34.5
C_AA050291	Brienomyrus brachyistius	Unknown	Elephantfishbrie291	34.4
C_AA050292	*Brienomyrus brachyistius*	Unknown	Elephantfishbrie412	34.2
SRX553136	*Brienomyrus brachyistius*	Muscle	BWNDV1	34.2
OR350367	*Paracyprichromis brieni*	Brain	PbNDV	34.1
SRX700630	*Anguilla anguilla*	Olfactory epithelium	EENDV	34.1

Table 3. *Cont.*

Accession Number	Host	Tissue Source	Figure Label	% Identity
C_AA050295	*Brienomyrus brachyistius*	Unknown	Elephantfishbrie514	33.9
C_AA050270	*Akarotaxis nudiceps*	Unknown	Dragonfishakar681	33.9
OR350336	*Eretmodus cyanostictus*	Lower pharyngeal jaw	EcNDV	33.8
SRX265393	*Oncorynchus nerka*	Pooled organs	SSNDV	33.7
SRX340220	*Lucania goodei*	Pooled organs	KNDV-Lg	33.6
SRX376926	*Gambusia affinis*	Ovary	WMNDV	33.4
SRX573075	*Brienomyrus brachyistius*	Electric organ	BWNDV-2	33.4
WLN26316	*Lamprologus ocellatus*	Lower pharyngeal jaw	ACNDV-2J	33.4
AZP02119	*Sebastes nigrocinctus*	Brain	RNDV	33.1
MH158727	*Ophthalmotilapia ventralis*	Pooled organs	ACNDV	33.1
C_AA050351	*Gasterosteus aculeatus*	Unknown	Sticklebackgast624	32.8
C_AA050259	*Oreochromis sp. (strain BJ-2021)*	Unknown	Cichlidoreo330	32.7
C_AA050262	*Tropheus sp. 'Ikola'*	Unknown	Cichlidtrop300	32.5
SRX367575	*Nibea albiflora*	Unknown	YDNDV	32.4
C_AA050350	*Gasterosteus aculeatus*	Unknown	Sticklebackgast146	32.3
C_AA050352	*Gasterosteus aculeatus*	Unknown	Sticklebackgast636	32.3
C_AA050353	*Acipenser ruthenus x Huso huso*	Unknown	Sturgeonacip200	32.3
SRX1037831 *	*Gasterosteus aculeatus*	Unknown	SNDV	31.3
YP_009259541	*Lepomis macrochirus*	Lip	BGHBV	28.3

Pairwise Comparisons and Phylogenetic Analysis of Nackednavirus P Proteins

The amino acid identity of the MnNDV P protein with that of other nackednaviruses ranged from 31.3 to 49.6% by pairwise alignment (Figure 3) and shared the highest identity with Pseudosimochromis babaulti nackednavirus (PbaNDV) and Rainwater killifish Nackednavirus Lp-2 (KNDV-Lp-2).

Phylogenetic analyses of NDV P proteins (Figure 4) supported the assignment of the MnNDV-1 to the nackednaviruses at the family level (Nudnaviridae). MnNDV-1 was most similar to KNDV-Lp-2 (Figure 4). Clades of NDVs grouped in accordance with either previously established "Types" (Supplemental Table S1) or by criteria including geographic region (Lake Tanganyika) or a singular genus (Trematocarini) [47,48].

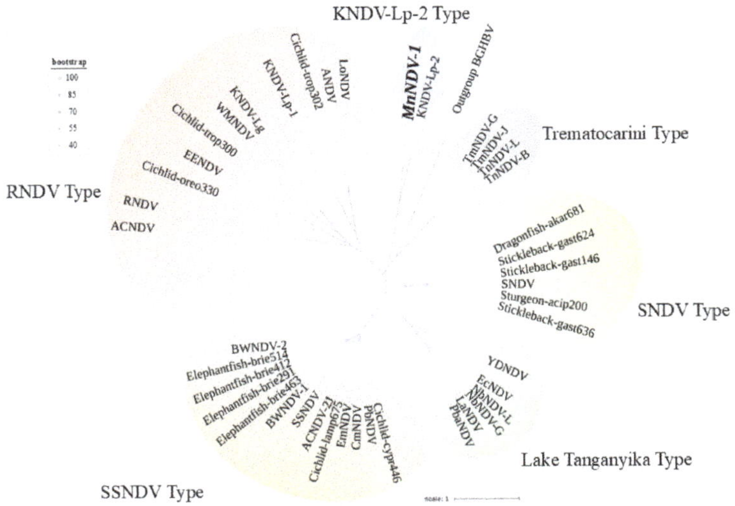

Figure 4. Radial phylogram depicting the relationships of the polymerase protein from 30 nackednaviruses. The *Micropterus nigricans* nackednavirus (MnNDV-1) is depicted in bold and italics and was most similar to KNDV-Lp-2 (Rainwater killifish Nackednavirus Lp-2). Resolved clades are idicated by color and label according to their corresponding NDV lineages.

3.3. Adomaviruses

We identified three ADmVs in the HPML sample from which the nackednavirus was recovered (Table 2). These ADvMs included MnA-1 and two novel ADmVs. MnA-1 was represented by two contigs that contained nine ORFs and 90.6% of the total genome. Contig 1 (PV469405) was 8527 bp corresponding to nucleotides 3795–12,321 of the reference genome and included Wasp, Cah, Penton, Macc, Hexon, Adenain, and Prim ORFs (Table 2). This contig was represented by 4614 reads (162× coverage). The pairwise identity of this contig to the reference genome was 99.965%. Three SNPs were identified within the Wasp (two non-synonymous) and Prim (one non-synonymous) ORFs. Contig 2 (PV469406) was 5965 bp corresponding to nucleotides 12,439–15,989 and contained the viral helicase RepE1 and the SET ORFs. This contig was more highly represented in the sample than contig 1 (7977 reads; 401×coverage). Contig 2 differed from the reference genome by a single synonymous nucleotide difference in RepE1.

We recovered the complete genome of a novel ADmV (PV469407) that we designate as MnA-2 (Figure 5). Sequencing led to 741× coverage of this genome consisting of 98,846 reads. The GC content was 46%, and it was organizationally similar to MnA-1 but with significant differences in the number of coding domains. This genome contained sixteen discrete ORFs (Table 2). MnA-2 was notably larger than MnA-1 (17,869 bp) and more similar in length to MdA-2 that infects smallmouth bass [57]. Also present within MnA-2 are five previously unobserved ORFs we identified and designated as 1940, Otomem, Spamem, Broz, and Sealt. Additional homologous ADmV ORFs [30,57], Wasp, Cahalt, Penton, Macc, Hexon, Adenain, Bcoroid, Prim, RepE1, and Alt, are described in Table 2.

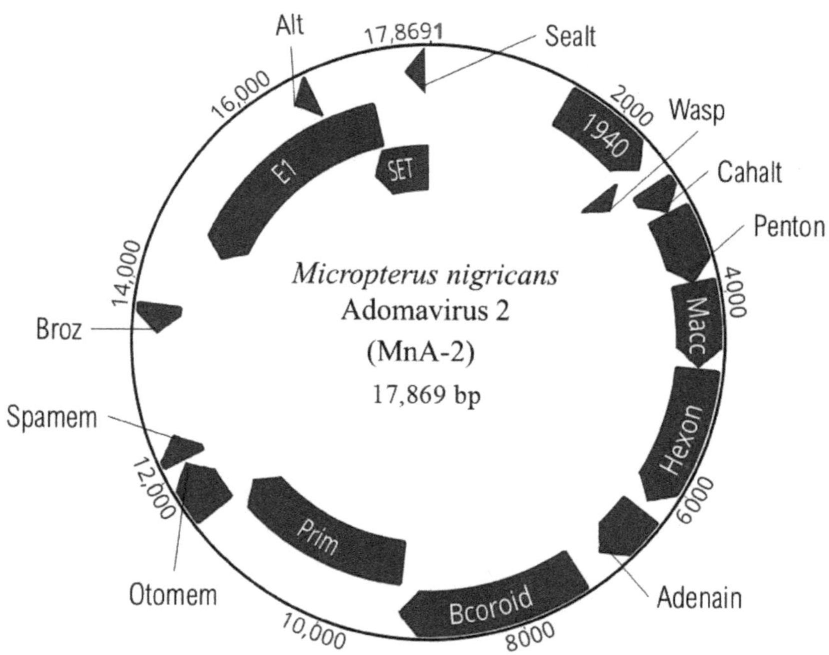

Figure 5. Genome organization of *Micropterus nigricans* adomavirus 2 (MnA-2). The complete genome is 17,869 bp of dsDNA and includes 14 non-overlapping and 2 overlapping ORFs (indigo). Detailed genome organization can be found in Table 2. The blue numeral denotes the origin of nucleotide numbering for the viral genome, corresponding to position 1 in the reference sequence.

A single 7966 contig corresponding to a third novel ADmV we identified as MnA-3 (NBCI accession: PV469408) was also recovered. A total of 496,318 reads were mapped to the contig, resulting in 13,355× mean coverage. The contig included three ORFs, two of which were homologous to other ADmVs (Table 2), Prim and RepE1. MnA-3 Prim (2250 bp) was comparable in size (749aa) to its MnA-2 homolog (751aa). However, the predicted isoelectric point of MnA-3 Prim was circumneutral (7.22), compared to the 5.98 pI observed in MnA-2. We also noted an additional ORF we designated by synteny as Otomem (483 bp). Unlike the other LMB ADmVs, the nearest predicted protein for MnA-3 Otomem was structural protein VP2 from Pyrobaculum filamentous virus 1 (AML61167.1). While we did not recover complete genomes, E1 ORFs for all three ADmVs (Figure 6) were recovered, allowing for multiple alignment for sequence comparison.

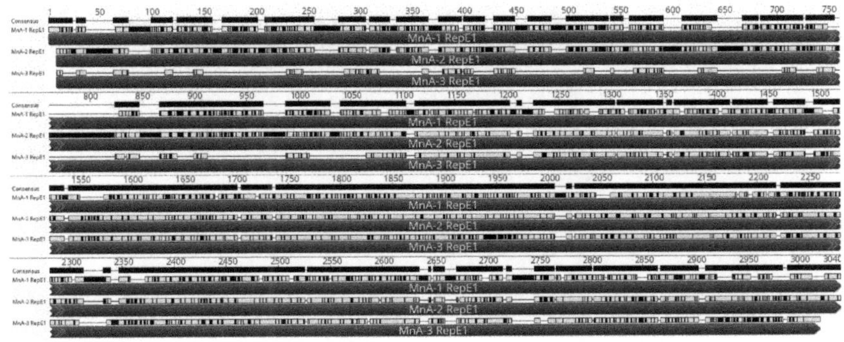

Figure 6. Multiple alignment of MnADmV E1 (Superfamily 3 helicase) ORFs. Black bars visible on the top line denote variations, and E1 CDSs are represented by indigo bars and labeled with specific ADmV. Gray rectangles above indigo bars denote identity to consensus alignment, and black rectangles above indigo bars identify dissimilarity to consensus sequence.

Pairwise Comparisons and Phylogenetic Analysis of ADmV Replicase Proteins

Comparative pairwise alignment of ADmV replicase proteins identified 15.07–99.10% amino acid identity between evaluated viruses (Figure 7). The greatest level of protein identity was observed in ADmVs from hosts of the same genus or family. This included ADmVs from two cichlid species from Lake Tanganyika (*Perissodus microlepis* and *Lamprologus lemairii* (99.09%)), MnA-1 and Micropterus dolomieu adomavirus 1 (85.5%) from micropterans, and ADmVs from *Betta splendens* and *B. smaragdina* (81.7%). MnA-2 and MnA-3 shared weak but the greatest identity with one another (51.9%). Comparison of both MnA-2 and MnA-3 revealed the greatest similarities with Blueface angelfish adomavirus (AdomaBK011013_RepE1), Tilapia adomavirus 2 (AdomaBK010892_RepE1), and Symphysodon discus adomavirus 1 (AdomaMF946549_RepE1), respectively.

Phylogenetic analyses (Figure 8) of replicase proteins recapitulated phylogenetic tree topology based on replicase type [58]. All micropteran ADmVs grouped most closely with ADmVs with RepE1 replicases. MnA-2 and MnA-3 resolved to a clade separate from that on MnA-1 that included the Blueface angelfish adomavirus, Tilapia adomavirus 2, and Symphysodon discus adomavirus 1 (Figure 8). Co-infection of RepE1 and RepLT ADmVs has been previously reported; however, the LMB ADmVs are exclusively represented by RepE1 replicases.

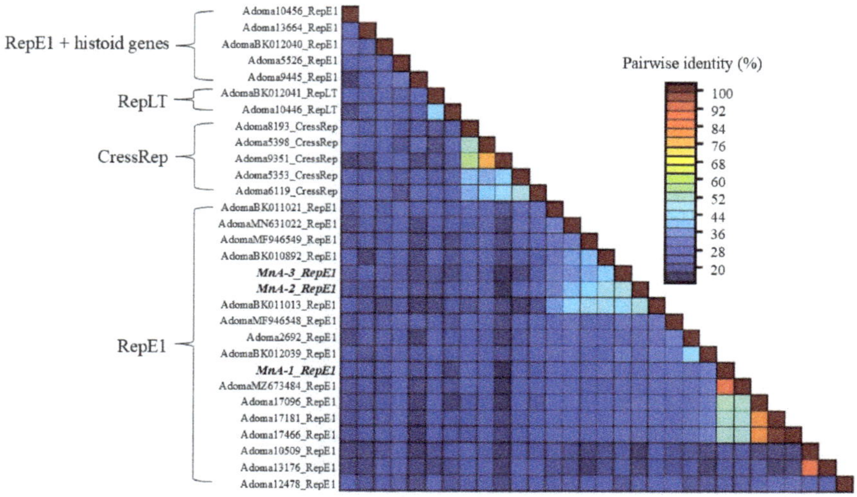

Figure 7. Pairwise sequence alignment identity (percentage) of 30 adomavirus (ADmVs) replicase (Rep) proteins based on pairwise alignment of ADmVs. The novel ADmVs obtained from *Micropterus nigricans* in this study are labeled in bold and italics. Brackets denote de facto associations based on assignment to adomavirus lineages based on replicase structures previously identified. NCBI accession numbers or NCBI SRA experiments (prefix SRX/ERX) are detailed in Supplemental Table S2.

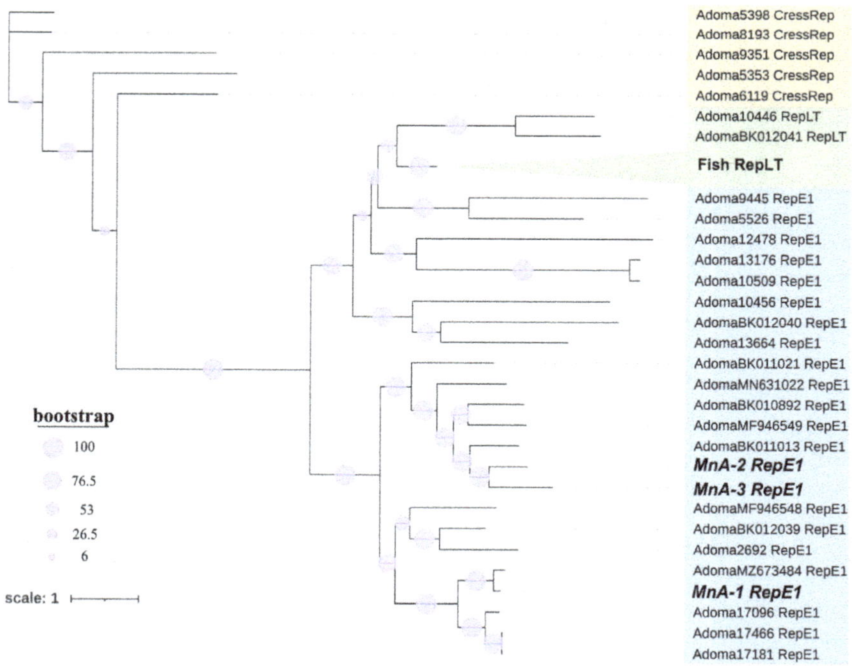

Figure 8. Rectangular phylogram depicting the relationships of replicase (Rep) proteins from 47 adomaviruses (ADmVs). Replicases were classified as CressRep (yellow), RepLT (green), or RepE1 (blue). RepLT tip labels from multiple fish hosts (n = 17) were collapsed for visual clarity and are collectively labeled in bold. The ADmVs obtained from *Micropterus nigricans* in this study are labeled in bold and italics. Detailed information and metadata can be found in Supplementary Table S2.

We detected MnA-1 in 85% of the HPMLs. It was never detected in clinically normal skin from fish with BBS. MnNDV-1, MnA-2, and MnA-3 were only detected in a single sample, and that sample was also positive for MnA-1 (Figure 9). HPML samples in which MnA-1 was not detected had a low starting template (<1 ng/μL) compared to the MnA-1 positive HPML samples (>10 ng/μL), which may explain the lack of detection [30].

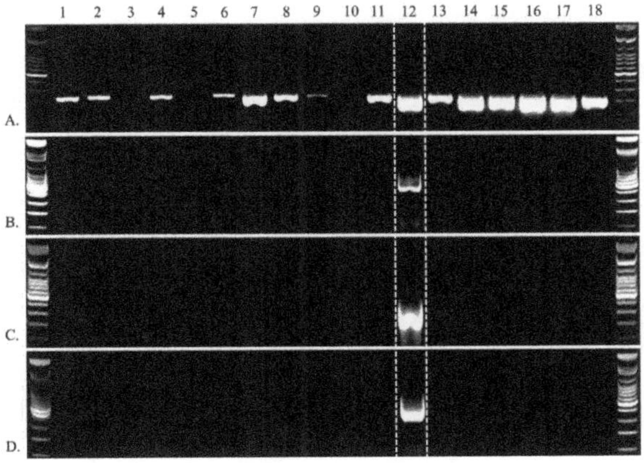

Figure 9. Endpoint PCR results documenting the presence of multiple viruses within a single sample. Enumeration at top identifies samples 1 through 18 collected from hyperpigmented lesions of discrete largemouth bass (*Micropterus nigricans*) during a singular sampling effort. Virus identification is denoted by the capital letters on the left column: (**A**) (MnA-1), (**B**) (MnA-2), (**C**) (MnA-3), and (**D**) (MnNDV-1). Additional positive detections for MnA-1 (n = 2) and negative non-template controls (n = 4) were included in the assay but are not pictured.

4. Discussion

Here, we identified a novel nackednavirus and evidence of viral co-infection with three adomaviruses, of which two are undescribed, all from a single hyperpigmented skin lesion on a largemouth bass. This study contributes to the list of LMB viruses and informs future disease research. At present, only three primary viral pathogens have been described that impact LMB. This includes LMBV which is routinely ascribed as the causative pathogen in most LMB mortality events within the United States [26,34]. In China, ISKNV-ZY and MSRV have also been identified as viral pathogens that impact LMB aquaculture [29,31,35]. Most recently, blotchy bass syndrome in LMB has been ascribed to MnA-1. In the current viral surveillance effort, MnA-1 was detected in 85% of HPMLs and of the viruses described here was the only virus identified in multiple individuals. Horizontal gene transfer (HGT) has been posited as a mechanism of ADmV evolution, diversity, and adaptation [58]. The co-infection of three ADmVs within a single skin lesion observed here documents an infection opportunity where HGT could occur.

This is the first report of MnA-2, MnA-3, and MnNDV-1. While they were associated with an HPML, their association with this lesion type is unclear. Moreover, they were not detected in other HPMLs. The identification of MnA-1 and its relationship with the HPMLs that characterize blotchy bass syndrome in LMB has only recently been documented despite the observation of this condition in bass in the Chesapeake Bay watershed for decades [30]. Investigations into SMB skin lesions in this watershed have also identified divergent RepE1 ADmVs. The MdA-1 is associated with blotchy bass syndrome in SMB, while MdA-2 is associated with raised mucoid lesions [30,57]. SMB co-infection of MdA-1 and MdA-2 has been observed [37,57]. Other examples of diverse ADmVs associated with a single host include those that infect *Xiphophorus birchmanni* and *Oreochromis niloticus*. The presence of previously undescribed adomaviruses, explicitly associated with melanistic lesions in other black basses as well as the documentation of a novel nackednavirus, highlights a need to better understand the significance of viral co-infection.

With so much still unknown, more cryptogenic diseases and viruses of unknown consequence are only now being addressed in earnest. Those associated with grossly observable anomalies like skin lesions are more likely to be investigated. Adomaviruses, a family of circular double-stranded DNA (dsDNA) viruses which are known to cause pathogenesis in vertebrates including skin lesions, fall into this category [59–61]. Many viruses including adomaviruses are now being discovered via metagenomic methods. These viruses seem to be present in clinically normal individuals. NDVs represent another understudied novel group of dsDNA fish viruses discovered in clinically normal hosts. They are a sister family of hepadnaviruses (HBVs) [47,49,62–65]. Unlike HBVs that infect birds and mammals, NDVs have no specific liver tropism and lack a surface protein, and diseases of NDVs have yet to be described [66]. Thus far, no specific NDV tissue tropism has been observed, which may hinder targeted detection in host organisms. Without a preference for particular cells or tissues, NDVs may circulate more broadly within the host or result in variable viral loads across different physiological states or environmental conditions [47–49].

The detection of multiple viruses in a single sample here suggests viral co-infection or a similarly complex microbial disease ecology [67–69], which may indicate "pre-fragilization" of the host [70] that would make them more susceptible to infection. Mortality events and lowered recruitment of SMB within the Chesapeake Bay watershed since the early 2000s has led to intensive research in the context of fish health, given the evidence of immunomodulation and ecosystem impairment [71–79]. Such pre-fragilization is common in cases of ecosystem impairment [80–82] and chronic environmental stress [80,83–85]. Disease susceptibility may be incited by disrupting beneficial microbiota on external mucosa [86–88], causing "dysbiosis" [86,89]. Deleterious effects of dysbiosis may be further catalyzed by increased environmental temperatures [90,91] which can act synergistically to accelerate the intensity of co-infections and fish immune disruption [92]. Similarly, virus abundance and virulence in fish hosts is influence by thermal stress [93,94].

Viral pathogens beyond LMBV should be considered in the context of LMB management. Sportfishing and associated handling and stress serve as the primary focus of human and LMB interactions. Virus transmission associated with the handling of fish has not been adequately studied, but damage to the external glycocalyx slime coat may provide entry for pathogens [95,96]. High-intensity fishing tournaments directly contribute to the mortality of angled fish [97,98] and may serve as a vector for disease transmission [99–101]. Likewise, the movement of recreational boats from separate water bodies has also been documented as posing transmission threats [102,103]. The growing popularity of sportfishing and interstate boat travel will likely continue to increase the risk of disease spread into the future [104]. LMB aquaculture for both stocking programs and as food fish serve as a remaining interface. Intensive aquaculture often leads to disease by pathogens that are not notably problematic under natural environmental settings given the associated crowding and other stressors [105–107]. However, pathogens can often be eradicated within aquaculture environments, whereas control of an established pathogen in a naturalized system is not always possible [108].

The results of this study emphasize the utility of massively parallel sequencing (MPS) technologies as a meaningful and perhaps necessary approach for biosurveillance and the identification of new or multiple pathogens that may be present in fish hosts without clinical signs of disease. Substantial advances in virus discovery have been made in both fish [48,56,93,109] and non-fish [110–115] via MPS approaches. Despite the transition into an age of virus discovery and molecular biological inventorying, viral co-infections are still comparatively understudied. Co-infections including ADmVs are currently limited to this study, the isolation of an ADmV and an aquareovirus from marbled eel (*Anguilla marmorata*) [116], and co-infection of giant guitarfish polyomavirus 1 (GfPyV1) and guitarfish adomavirus (GAdoV) [60]. Reoviruses are not routinely associated with clinical disease [27,28], although co-infections which incorporate reoviruses have been documented intermittently [116–118]. Viral co-infections including polyomaviruses are also repeatedly noted, including numerous associations with skin lesions [60,119–121]. Perhaps more importantly, polyomaviruses may provide another avenue for HGT, particularly within co-infected hosts [58,122]. Likewise, the proposed evolutionary origin of ADmV proteins Cah and Adenain is hypothesized to be rooted in HGT from reoviruses [58], further complicating these specific associations. Additional investigation targeting viral co-infections is crucial, as they may offer multiple insights simultaneously, though they are underrepresented in the literature.

The general paucity of the mixed viral infections described is likely due to the historic use of outcome-based sampling approaches, where a single specific viral pathogen is usually the target of interest. PCR methods or conventional culture methods are effective for the detection of known pathogens, but no consistent methodology has been ascribed for the detection of novel or mixed infections. Traditionally, virus discovery and characterization in fishes have relied on inoculating cell lines with a presumptive pathogen. However, some well-studied viral diseases, including Viral Erythrocytic Necrosis (VEN) and Piscine Orthoreovirus-1 (PRV-1), have thus far have not been cultivatable on fish cell lines [123–125]. While hundreds of immortal fish cell lines have been developed and characterized during the last half century, the number of commercially available cell lines is substantially lower [126], and those permissive to viruses of non-model organisms are lower still. MPS methods that have now become standard technology in most molecular diagnostic laboratory workflows can provide a contemporary approach to discovering the diverse and otherwise unculturable viruses that affect fish species of management interest [126–130].

Massively parallel sequencing approaches have been widely adopted as standard workflows which can provide a contemporary approach to discovering otherwise unculturable viruses that affect fish species of management interest [126–128,131]. While characterization of unknown viruses and identification of notable pathogens are often separate processes, it is still useful to expand our inventory of the known virome. Specific considerations should be expanded to include the potential undescribed viruses when considering stocking and translocation of LMB. The potential interplay between environmental stressors and the LMB virome and how these factors may disrupt microbial homeostasis require further investigation. Looking forward, the application of MPS technologies will be critical for elucidating these complex interactions and enhancing our understanding of viral diversity within aquatic ecosystems. Additional viroinformatics advancements can be anticipated in the coming years and will likely reduce current inadequacies [132,133]. For example, many viral sequences obtained via MPS methods are uncharacterized simply due to the lack of reference sequences in common databanks [128]. This undefined viral "dark matter" can comprise as much 90% of obtained viral sequences and serves as one of the greatest challenges in the emerging field of viroinformatics [134]. As viroinformatics advances, additional reference genomes will be available for comparison, facilitating the identification of viruses associated with fishes. Nevertheless, detection of novel viruses by means of MPS substantially enhances our capacity for biosurveillance of unknown viruses and should remain a crucial component of any strategic risk assessments.

Supplementary Materials: The following supporting information can be downloaded at: https://www.mdpi.com/article/10.3390/v17091173/s1, Table S1: Supplemental Table S1 (sequence metadata for NDVs); Table S2: Supplemental Table S2 (sequence metadata for ADmVs).

Author Contributions: Conceptualization, C.R., L.I. and P.M.; formal analysis, C.R., M.B. and L.I.; investigation, J.O., M.I., M.B., L.I. and C.R.; resources, J.O., M.I. and P.M.; data curation, M.B. and C.R.; writing—original draft preparation, C.R. and L.I.; writing—review and editing, C.R., J.O., M.I., P.M., M.B. and L.I. All authors have read and agreed to the published version of the manuscript.

Funding: This work was supported with funding from the U.S. Geological Survey, Environmental Health (Contaminant Biology Program GX25LB00UU11000) and Ecosystems (Environments Chesapeake Bay and Biological Threats and Invasive Species Research Program- GX23LB00U7KC100) Mission Areas.

Institutional Review Board Statement: This study was performed in compliance with the USGS Eastern Ecological Science Center Institutional Animal Care and Use Committee (IACUC) (ID: 2021-17L, 04 Mar 2024).

Informed Consent Statement: Not applicable.

Data Availability Statement: The original data presented in the study are openly available in the sequence read archive (SRA) associated with Bioproject (PRJNA785556), Biosample (SAMN23566442), and National Center for Biotechnology Information accessions (PV469405, PV469406, PV469407, PV469408, PV430023, and PV448632). At the time of publication, fish annual survey data were not publicly available from the Virginia Division of Wildlife Resources. Figure 1 Map Metadata: State layers downloaded from: the United States Census Bureau 2018 TIGER/Line Shapefiles (machine readable data files); U.S. Department of Commerce https://www.census.gov/cgi-bin/geo/shapefiles/index.php. Streamlines available from the U.S. Environmental Protection Agency and the U.S. Geological Survey National Hydrography Dataset Plus—NHDPlus Version 2.1. http://www.horizon-systems.com/nhdplus/nhdplusv2_home.php. Range estimates identified by proximity to the existing literature [4].

Acknowledgments: We thank Stephanie Gordon for map creation and Cynthia Holt and Anna Sjodin for their constructive comments that improved the manuscript. Any use of trade, firm, or product names is for descriptive purposes only and does not imply endorsement by the U.S. Government. The USDA is an equal opportunity employer and provider.

Conflicts of Interest: The authors declare no conflicts of interest.

References

1. Long, J.M.; Allen, M.S.; Porak, W.F.; Suski, C.D. A Historical Perspective of Black Bass Management in the United States. In Proceedings of the American Fisheries Society Symposium, Portland, OR, USA, 19 November 2015; Volume 82, pp. 99–122.
2. Taylor, A.T.; Long, J.M.; Tringali, M.D.; Barthel, B.L. Conservation of Black Bass Diversity: An Emerging Management Paradigm. *Fisheries* **2019**, *44*, 20–36. [CrossRef]
3. Kassler, T.W.; Koppelman, J.B.; Near, T.J.; Dillman, C.B.; Levengood, J.; Swofford, D.L.; VanOrman, J.L.; Claussen, J.E.; Philipp, D.P. Molecular and Morphological Analyses of the Black Basses: Implications for Taxonomy and Conservation. In Proceedings of the American Fisheries Society Symposium, Bethesda, MD, USA, 19 August 2002; Volume 2002, pp. 291–322.
4. Miranda, L.E.; Bettoli, P.W. Largemouth Bass Natural History. In *Largemouth Bass Aquaculture*; CABI: Oxford, UK, 2019.
5. Kim, D.; Taylor, A.T.; Near, T.J. Phylogenomics and Species Delimitation of the Economically Important Black Basses (Micropterus). *Sci. Rep.* **2022**, *12*, 9113. [CrossRef]
6. Philipp, D.P. Genetic Implications of Introducing Florida Largemouth Bass, Micropterus Salmoides Floridanus. *Can. J. Fish. Aquat. Sci.* **1991**, *48*, 58–65. [CrossRef]
7. Barthel, B.; Allen, M.; Porak, W.; Kerns, J. Florida Bass Micropterus Floridanus (LeSueur, 1822). In *Black Bass Diversity: Multidisciplinary Science for Conservation*; American Fisheries Society, Symposium: Bethesda, MD, USA, 2015; Volume 82, pp. 43–53.
8. Sammons, S.M.; Dorsey, L.G.; Loftis, C.S.; Chrisman, P.; Scott, M.; Hammonds, J.; Jolley, M.; Hatcher, H.; Odenkirk, J.; Damer, J.; et al. Alabama Bass Alter Reservoir Black Bass Species Assemblages When Introduced Outside Their Native Range. *N. Am. J. Fish. Manag.* **2023**, *43*, 384–399. [CrossRef]

9. Johansen, T.; Besnier, F.; Quintela, M.; Jorde, P.E.; Glover, K.A.; Westgaard, J.-I.; Dahle, G.; Lien, S.; Kent, M.P. Genomic Analysis Reveals Neutral and Adaptive Patterns That Challenge the Current Management Regime for East Atlantic Cod *Gadus morhua* L. *Evol. Appl.* **2020**, *13*, 2673–2688. [CrossRef] [PubMed]
10. Morrison, W., III; Lohr, J.; Duchen, P.; Wilches, R.; Trujillo, D.; Mair, M.; Renner, S. The Impact of Taxonomic Change on Conservation: Does It Kill, Can It Save, or Is It Just Irrelevant? *Biol. Conserv.* **2009**, *142*, 3201–3206. [CrossRef]
11. MacCrimmon, H.R.; Robbins, W.H. Distribution of the Black Basses in North America. In *Black Bass Biology and Management*; Sport Fishing Institute: Washington, DC, USA, 1975; pp. 56–66.
12. Lee, D.S.; Gilbert, C.R.; Hocutt, C.H.; Jenkins, R.E.; McAllister, D.E.; Stauffer, J.R., Jr. *Atlas of North American Freshwater Fishes*; North Carolina State Museum of natural history: Raleigh, NC, USA, 1980.
13. Shaw, S. Black Bass Diversity and Conservation: An Overview. In *Black Bass Diversity: Multidisciplinary Science for Conservation*; American Fisheries Society, Symposium: Bethesda, MD, USA, 2015; Volume 82, pp. 3–8.
14. Robbins, W.H.; MacCrimmon, H.R. *The Blackbass in America and Overseas*; Publications Division, Biomanagement and Research Enterprises: Sault Ste. Marie, ON, Canada, 1974.
15. Brown, T.; Runciman, B.; Pollard, S.; Grant, A. Biological Synopsis of Largemouth Bass (*Micropterus salmoides*). *Can. Manuscr. Rep. Fish. Aquat. Sci.* **2009**, *2884*, 1–26.
16. USFWS. *2016 National Survey of Fishing, Hunting and Wildlife-Associated Recreation*; U.S. Fish & Wildlife Service: Washington, DC, USA, 2018.
17. Seguy, L.; Long, J.M. Perceived Ecological Threats and Economic Benefits of Non-Native Black Bass in the United States. *Fisheries* **2020**, *46*, 56–65. [CrossRef]
18. Moffitt, C.M. *Reflections: A Photographic History of Fisheries and the American Fisheries Society in North America*; American Fisheries Society: Bethesda, MD, USA, 2001.
19. Tidwell, J.H.; Nickum, J.G. History of Largemouth Bass Production. In *Largemouth Bass Aquaculture*; CABI: Oxford, UK, 2019.
20. Swingle, H. History of Warmwater Pond Culture in the United States. In *A Century of Fisheries in North America*; American Fisheries Society, Special Publication: Bethesda, MD, USA, 1970; Volume 7, pp. 95–105.
21. Zhou, Y.; Liu, C. Largemouth Bass Production in China. In *Largemouth Bass Aquaculture*; 5M Publishing Ltd Benchmark House: Sheffield, UK, 2019; pp. 37–47.
22. Bai, J.; Li, S. Current Status and Development Trend on China Largemouth Bass Industry. *Chin. Fish. Econ.* **2013**, *31*, 104–108.
23. Steeger, T.M.; Grizzle, J.M.; Weathers, K.; Newman, M. Bacterial Diseases and Mortality of Angler-Caught Largemouth Bass Released after Tournaments on Walter F. George Reservoir, Alabama–Georgia. *N. Am. J. Fish. Manag.* **1994**, *14*, 435–441. [CrossRef]
24. Francis-Floyd, R.; Reed, P.; Bolon, B.; Estes, J.; McKinney, S. An Epizootic of *Edwardsiella tarda* in Largemouth Bass (*Micropterus salmoides*). *J. Wildl. Dis.* **1993**, *29*, 334–336. [CrossRef]
25. Camus, A.; Griffin, M.; Armwood, A.; Soto, E. A Spontaneous Outbreak of Systemic *Edwardsiella piscicida* Infection in Largemouth Bass *Micropterus salmoides* (Lacépède, 1802) in California, USA. *J. Fish Dis.* **2019**, *42*, 759–763. [CrossRef]
26. Grizzle, J.M.; Altinok, I.; Fraser, W.A.; Francis-Floyd, R. First Isolation of Largemouth Bass Virus. *Dis. Aquat. Org.* **2002**, *50*, 233–235. [CrossRef]
27. Chen, Z.-Y.; Gao, X.-C.; Zhang, Q.-Y. Whole-Genome Analysis of a Novel Fish Reovirus (MsReV) Discloses Aquareovirus Genomic Structure Relationship with Host in Saline Environments. *Viruses* **2015**, *7*, 4282–4302. [CrossRef]
28. Sibley, S.D.; Finley, M.A.; Baker, B.B.; Puzach, C.; Armién, A.G.; Giehtbrock, D.; Goldberg, T.L. Novel Reovirus Associated with Epidemic Mortality in Wild Largemouth Bass (*Micropterus salmoides*). *J. Gen. Virol.* **2016**, *97*, 2482–2487. [CrossRef]
29. Qin, Y.; Zhang, P.; Zhang, M.; Guo, W.; Deng, S.; Liu, H.; Yao, L. Isolation and Identification of a New Strain Micropterus Salmoides Rhabdovirus (MSRV) from Largemouth Bass *Micropterus salmoides* in China. *Aquaculture* **2023**, *572*, 739538. [CrossRef]
30. Iwanowicz, L.R.; Raines, C.D.; Young, K.T.; Blazer, V.S.; Walsh, H.L.; Smith, G.; Holt, C.; Odenkirk, J.S.; Jones, T.; Hessenauer, J.-M.; et al. Novel Adomaviruses Associated with Blotchy Bass Syndrome in Black Basses. *PLoS ONE* **2025**, *20*, 657292.
31. Xu, Z.; Liao, J.; Zhang, D.; Liu, S.; Zhang, L.; Kang, S.; Xu, L.; Chen, H.; Peng, W.; Zhou, S.; et al. Isolation, Characterization, and Transcriptome Analysis of an ISKNV-like Virus from Largemouth Bass. *Viruses* **2023**, *15*, 398. [CrossRef] [PubMed]
32. Kibenge, F.; Godoy, M. Reoviruses of Aquatic Organisms. In *Aquaculture Virology*; Elsevier: Amsterdam, The Netherlands, 2016; pp. 205–236.
33. Hanson, L.A.; Petrie-Hanson, L.; Meals, K.O.; Chinchar, V.G.; Rudis, M. Persistence of Largemouth Bass Virus Infection in a Northern Mississippi Reservoir after a Die-Off. *J. Aquat. Anim. Health* **2001**, *13*, 27–34. [CrossRef]
34. Grizzle, J.M.; Brunner, C.J. Review of Largemouth Bass Virus. *Fisheries* **2003**, *28*, 10–14. [CrossRef]
35. Gao, E.-B.; Chen, G. Micropterus Salmoides Rhabdovirus (MSRV) Infection Induced Apoptosis and Activated Interferon Signaling Pathway in Largemouth Bass Skin Cells. *Fish Shellfish Immunol.* **2018**, *76*, 161–166. [CrossRef] [PubMed]

36. Ma, D.; Deng, G.; Bai, J.; Li, S.; Yu, L.; Quan, Y.; Yang, X.; Jiang, X.; Zhu, Z.; Ye, X. A Strain of *Siniperca chuatsi* Rhabdovirus Causes High Mortality among Cultured Largemouth Bass in South China. *J. Aquat. Anim. Health* **2013**, *25*, 197–204. [CrossRef] [PubMed]
37. Blazer, V.S.; Young, K.T.; Smith, G.D.; Sperry, A.J.; Iwanowicz, L.R. Hyperpigmented Melanistic Skin Lesions of Smallmouth Bass *Micropterus Dolomieu* from the Chesapeake Bay Watershed. *Dis. Aquat. Org.* **2020**, *139*, 199–212. [CrossRef]
38. Odhiambo, B.; Coxon, T.; Somers, H. Sediment and Phosphorous Fluxes Analysis in Aquia Creek, a Sub-Watershed of the Chesapeake Bay Basin, VA, USA. *Water Air Soil Pollut.* **2018**, *229*, 1–15. [CrossRef]
39. Jastram, J.D. *Streamflow, Water Quality, and Aquatic Macroinvertebrates of Selected Streams in Fairfax County, Virginia, 2007-12*; US Department of the Interior, US Geological Survey: Reston, VA, USA, 2014.
40. Froelich, A.J.; Zenone, C. *Maps Showing Geologic Terrane, Drainage Basins, Overburden, and Low Flow of Streams in Fairfax County and Vicinity, Virginia*; US Geological Survey: Reston, VA, USA, 1985.
41. Frie, S. *Little Hunting Creek Watershed Management Plan*; Fairfax County Stormwater Planning Division: Fairfax, VA, USA, 2004; p. 212.
42. Dalzell, R.F., Jr.; Dalzell, L.B. *George Washington's Mount Vernon: At Home in Revolutionary America*; Oxford University Press: Oxford, UK, 2000.
43. Li, D.; Liu, C.-M.; Luo, R.; Sadakane, K.; Lam, T.-W. MEGAHIT: An Ultra-Fast Single-Node Solution for Large and Complex Metagenomics Assembly via Succinct de Bruijn Graph. *Bioinformatics* **2015**, *31*, 1674–1676. [CrossRef] [PubMed]
44. Tisza, M.J.; Belford, A.K.; Dominguez-Huerta, G.; Bolduc, B.; Buck, C.B. Cenote-Taker 2 Democratizes Virus Discovery and Sequence Annotation. *Virus Evol.* **2021**, *7*, veaa100. [CrossRef]
45. Zimmermann, L.; Stephens, A.; Nam, S.-Z.; Rau, D.; Kübler, J.; Lozajic, M.; Gabler, F.; Söding, J.; Lupas, A.N.; Alva, V. A Completely Reimplemented MPI Bioinformatics Toolkit with a New HHpred Server at Its Core. *J. Mol. Biol.* **2018**, *430*, 2237–2243. [CrossRef]
46. Bu, C.; Zheng, X.; Zhao, X.; Xu, T.; Bai, X.; Jia, Y.; Chen, M.; Hao, L.; Xiao, J.; Zhang, Z.; et al. GenBase: A Nucleotide Sequence Database. *Genom. Proteom. Bioinform.* **2024**, *22*, qzae047. [CrossRef]
47. Lauber, C.; Seitz, S.; Mattei, S.; Suh, A.; Beck, J.; Herstein, J.; Börold, J.; Salzburger, W.; Kaderali, L.; Briggs, J.A.; et al. Deciphering the Origin and Evolution of Hepatitis B Viruses by Means of a Family of Non-Enveloped Fish Viruses. *Cell Host Microbe* **2017**, *22*, 387–399. [CrossRef]
48. Costa, V.A.; Ronco, F.; Mifsud, J.C.; Harvey, E.; Salzburger, W.; Holmes, E.C. Host Adaptive Radiation Is Associated with Rapid Virus Diversification and Cross-Species Transmission in African Cichlid Fishes. *Curr. Biol.* **2024**, *34*, 1247–1257. [CrossRef]
49. Ben, H.; Wang, X.; Yang, P.; Li, L.; Liu, P.; Gao, Y.; Wang, Y.; Liu, Y.; Huang, C.; Chen, D. Metagenomic Analysis Uncovers Novel Hepadnaviruses and Nackednaviruses. *Sci. Rep.* **2025**, *15*, 24699. [CrossRef]
50. Welch, N.L.; Yutin, N.; Dill, J.A.; Camus, A.C.; Pang, Y.-Y.S.; Schiller, J.T.; An, P.; Cantalupo, P.G.; Pipas, J.M.; Delwart, E.; et al. Adomaviruses: An Emerging Virus Family Provides Insights into DNA Virus Evolution. *BioRxiv* **2018**.
51. Koonin, E.V.; Dolja, V.V.; Krupovic, M. Origins and Evolution of Viruses of Eukaryotes: The Ultimate Modularity. *Virology* **2015**, *479*, 2–25. [CrossRef] [PubMed]
52. Edgar, R.C. MUSCLE: Multiple Sequence Alignment with High Accuracy and High Throughput. *Nucleic Acids Res.* **2004**, *32*, 1792–1797. [CrossRef]
53. Trifinopoulos, J.; Nguyen, L.-T.; von Haeseler, A.; Minh, B.Q. W-IQ-TREE: A Fast Online Phylogenetic Tool for Maximum Likelihood Analysis. *Nucleic Acids Res.* **2016**, *44*, W232–W235. [CrossRef] [PubMed]
54. Letunic, I.; Bork, P. Interactive Tree Of Life (iTOL) v5: An Online Tool for Phylogenetic Tree Display and Annotation. *Nucleic Acids Res.* **2021**, *49*, W293–W296. [CrossRef]
55. Muhire, B.M.; Varsani, A.; Martin, D.P. SDT: A Virus Classification Tool Based on Pairwise Sequence Alignment and Identity Calculation. *PLoS ONE* **2014**, *9*, e108277. [CrossRef]
56. Ford, C.E.; Dunn, C.D.; Leis, E.M.; Thiel, W.A.; Goldberg, T.L. Five Species of Wild Freshwater Sport Fish in Wisconsin, USA, Reveal Highly Diverse Viromes. *Pathogens* **2024**, *13*, 150. [CrossRef]
57. Iwanowicz, L.R.; Young, K.T.; Adams, C.R.; Blazer, V.S.; Smith, G.D.; Cornman, R.S. Draft Genome Sequence of an Adomavirus Associated with Raised Mucoid Skin Lesions on Smallmouth Bass (*Micropterus dolomieu*). *Microbiol. Resour. Announc.* **2020**, *9*, 14. [CrossRef]
58. Buck, C.B.; Welch, N.; Belford, A.K.; Varsani, A.; Pastrana, D.V.; Tisza, M.J.; Starrett, G.J. Widespread Horizontal Gene Transfer Among Animal Viruses. *bioRxiv* **2024**, preprint. [CrossRef]
59. Welch, N.L.; Tisza, M.J.; Starrett, G.J.; Belford, A.K.; Pastrana, D.V.; Pang, Y.-Y.S.; Schiller, J.T.; An, P.; Cantalupo, P.G.; Pipas, J.M.; et al. Identification of Adomavirus Virion Proteins. *bioRxiv* **2020**, 341131. [CrossRef]
60. Dill, J.A.; Camus, A.C.; Leary, J.H.; Ng, T.F.F. Microscopic and Molecular Evidence of the First Elasmobranch Adomavirus, the Cause of Skin Disease in a Giant Guitarfish, *Rhynchobatus djiddensis*. *mBio* **2018**, *9*, e00185-18. [CrossRef]

61. Kazlauskas, D.; Varsani, A.; Koonin, E.V.; Krupovic, M. Multiple Origins of Prokaryotic and Eukaryotic Single-Stranded DNA Viruses from Bacterial and Archaeal Plasmids. *Nat. Commun.* **2019**, *10*, 3425. [CrossRef]
62. Hahn, C.M.; Iwanowicz, L.R.; Cornman, R.S.; Conway, C.M.; Winton, J.R.; Blazer, V.S. Characterization of a Novel Hepadnavirus in the White Sucker (*Catostomus commersonii*) from the Great Lakes Region of the United States. *J. Virol.* **2015**, *89*, 11801–11811. [CrossRef]
63. Magnius, L.; Mason, W.S.; Taylor, J.; Kann, M.; Glebe, D.; Dény, P.; Sureau, C.; Norder, H.; Consortium, I.R. ICTV Virus Taxonomy Profile: Hepadnaviridae. *J. Gen. Virol.* **2020**, *101*, 571. [CrossRef]
64. Buigues, J.; Viñals, A.; Martínez-Recio, R.; Monrós, J.S.; Cuevas, J.M.; Sanjuán, R. Phylogenetic Evidence Supporting the Nonenveloped Nature of Hepadnavirus Ancestors. *Proc. Natl. Acad. Sci. USA* **2024**, *121*, e2415631121. [CrossRef]
65. Pfister, S.; Rabl, J.; Wiegand, T.; Mattei, S.; Malär, A.A.; Lecoq, L.; Seitz, S.; Bartenschlager, R.; Böckmann, A.; Nassal, M.; et al. Structural Conservation of HBV-like Capsid Proteins over Hundreds of Millions of Years despite the Shift from Non-Enveloped to Enveloped Life-Style. *Nat. Commun.* **2023**, *14*, 1574. [CrossRef]
66. Liu, Y.; Maya, S.; Ploss, A. Animal Models of Hepatitis B Virus Infection–Success, Challenges, and Future Directions. *Viruses* **2021**, *13*, 777. [CrossRef]
67. COX, F.E. Concomitant Infections, Parasites and Immune Responses. *Parasitology* **2001**, *122*, S23–S38. [CrossRef] [PubMed]
68. Bakaletz, L.O. Developing Animal Models for Polymicrobial Diseases. *Nat. Rev. Microbiol.* **2004**, *2*, 552–568. [CrossRef]
69. Kotob, M.H.; Menanteau-Ledouble, S.; Kumar, G.; Abdelzaher, M.; El-Matbouli, M. The Impact of Co-Infections on Fish: A Review. *Vet. Res.* **2017**, *47*, 98. [CrossRef]
70. Vonaesch, P.; Anderson, M.; Sansonetti, P.J. Pathogens, Microbiome and the Host: Emergence of the Ecological Koch's Postulates. *FEMS Microbiol. Rev.* **2018**, *42*, 273–292. [CrossRef]
71. Smith, G.; Blazer, V.; Walsh, H.; Iwanowicz, L.; Starliper, C.; Sperry, A. The Effects of Disease-Related Mortality of Young-of-Year Smallmouth Bass on the Population Characteristics in the Susquehanna River Basin, Pennsylvania and Potential Implications to Conservation of Black Bass Diversity. In Proceedings of the American Fisheries Society Symposium, Portland, OR, USA, 19 November 2015; Volume 82, pp. 319–332.
72. Arway, J.A.; Smith, G. The Susquehanna River—A Fishery in Decline. *Fisheries* **2013**, *38*, 235–236. [CrossRef]
73. Walsh, H.L.; Blazer, V.S.; Smith, G.D.; Lookenbill, M.; Alvarez, D.A.; Smalling, K.L. Risk Factors Associated with Mortality of Age-0 Smallmouth Bass in the Susquehanna River Basin, Pennsylvania. *J. Aquat. Anim. Health* **2018**, *30*, 65–80. [CrossRef] [PubMed]
74. Keplinger, B.; Rota, C.T. Effects of Poor Recruitment on Riverine Smallmouth Bass Population Dynamics. *N. Am. J. Fish. Manag.* **2024**, *44*, 247–261. [CrossRef]
75. Keplinger, B.; Hedrick, J.; Blazer, V.S. Temporal Trends in Macroscopic Indicators of Fish Health in the South Branch Potomac River. *N. Am. J. Fish. Manag.* **2022**, *42*, 277–294. [CrossRef]
76. Robertson, L.S.; Iwanowicz, L.R.; Marranca, J.M. Identification of Centrarchid Hepcidins and Evidence That 17β-Estradiol Disrupts Constitutive Expression of Hepcidin-1 and Inducible Expression of Hepcidin-2 in Largemouth Bass (*Micropterus salmoides*). *Fish Shellfish Immunol.* **2009**, *26*, 898–907. [CrossRef] [PubMed]
77. Ripley, J.; Iwanowicz, L.; Blazer, V.; Foran, C. Utilization of Protein Expression Profiles as Indicators of Environmental Impairment of Smallmouth Bass (*Micropterus dolomieu*) from the Shenandoah River, Virginia, USA. *Environ. Toxicol. Chem.* **2008**, *27*, 1756–1767. [CrossRef]
78. Willacker, J.J.; Eagles-Smith, C.A.; Blazer, V.S. Mercury Bioaccumulation in Freshwater Fishes of the Chesapeake Bay Watershed. *Ecotoxicology* **2020**, *29*, 459–484. [CrossRef] [PubMed]
79. Blazer, V.S.; Gordon, S.; Jones, D.K.; Iwanowicz, L.R.; Walsh, H.L.; Sperry, A.J.; Smalling, K.L. Retrospective Analysis of Estrogenic Endocrine Disruption and Land-Use Influences in the Chesapeake Bay Watershed. *Chemosphere* **2021**, *266*, 129009. [CrossRef]
80. Blazer, V.S.; Iwanowicz, L.R.; Starliper, C.E.; Iwanowicz, D.; Barbash, P.; Hedrick, J.; Reeser, S.; Mullican, J.; Zaugg, S.; Burkhardt, M.; et al. Mortality of Centrarchid Fishes in the Potomac Drainage: Survey Results and Overview of Potential Contributing Factors. *J. Aquat. Anim. Health* **2010**, *22*, 190–218. [CrossRef] [PubMed]
81. Rapport, D.J. Epidemiology and Ecosystem Health: Natural Bridges. *Ecosyst. Health* **1999**, *5*, 174–180. [CrossRef]
82. Vethaak, A.D.; Jol, J.G.; Pieters, J.P. Long-Term Trends in the Prevalence of Cancer and Other Major Diseases among Flatfish in the Southeastern North Sea as Indicators of Changing Ecosystem Health. *Environ. Sci. Technol.* **2009**, *43*, 2151–2158. [CrossRef]
83. Blazer, V.S.; Iwanowicz, L.R.; Iwanowicz, D.D.; Smith, D.R.; Young, J.A.; Hedrick, J.; Foster, S.; Reeser, S. Intersex (*Testicular oocytes*) in Smallmouth Bass from the Potomac River and Selected Nearby Drainages. *J. Aquat. Anim. Health* **2007**, *19*, 242–253. [CrossRef]
84. Blazer, V.S.; Iwanowicz, L.R.; Henderson, H.; Mazik, P.M.; Jenkins, J.A.; Alvarez, D.A.; Young, J.A. Reproductive Endocrine Disruption in Smallmouth Bass (*Micropterus dolomieu*) in the Potomac River Basin: Spatial and Temporal Comparisons of Biological Effects. *Environ. Monit. Assess.* **2012**, *184*, 4309–4334. [CrossRef]

85. Kolpin, D.W.; Blazer, V.S.; Gray, J.L.; Focazio, M.J.; Young, J.A.; Alvarez, D.A.; Iwanowicz, L.R.; Foreman, W.T.; Furlong, E.T.; Speiran, G.K.; et al. Chemical Contaminants in Water and Sediment near Fish Nesting Sites in the Potomac River Basin: Determining Potential Exposures to Smallmouth Bass (*Micropterus dolomieu*). *Sci. Total Environ.* **2013**, *443*, 700–716. [CrossRef] [PubMed]
86. Colin, Y.; Berthe, T.; Molbert, N.; Guigon, E.; Vivant, A.-L.; Alliot, F.; Collin, S.; Goutte, A.; Petit, F. Urbanization Constrains Skin Bacterial Phylogenetic Diversity in Wild Fish Populations and Correlates with the Proliferation of Aeromonads. *Microb. Ecol.* **2021**, *82*, 523–536. [CrossRef]
87. Stephens, W.Z.; Burns, A.R.; Stagaman, K.; Wong, S.; Rawls, J.F.; Guillemin, K.; Bohannan, B.J. The Composition of the Zebrafish Intestinal Microbial Community Varies across Development. *ISME J.* **2016**, *10*, 644–654. [CrossRef] [PubMed]
88. Sullam, K.E.; Essinger, S.D.; Lozupone, C.A.; O'CONNOR, M.P.; Rosen, G.L.; Knight, R.; Kilham, S.S.; Russell, J.A. Environmental and Ecological Factors That Shape the Gut Bacterial Communities of Fish: A Meta-Analysis. *Mol. Ecol.* **2012**, *21*, 3363–3378. [CrossRef]
89. Krotman, Y.; Yergaliyev, T.M.; Shani, R.A.; Avrahami, Y.; Szitenberg, A. Dissecting the Factors Shaping Fish Skin Microbiomes in a Heterogeneous Inland Water System. *Microbiome* **2020**, *8*, 9. [CrossRef]
90. Walsh, C.J.; Roy, A.H.; Feminella, J.W.; Cottingham, P.D.; Groffman, P.M.; Morgan, R.P. The Urban Stream Syndrome: Current Knowledge and the Search for a Cure. *J. N. Am. Benthol. Soc.* **2005**, *24*, 706–723. [CrossRef]
91. Meyer, J.L.; Paul, M.J.; Taulbee, W.K. Stream Ecosystem Function in Urbanizing Landscapes. *J. N. Am. Benthol. Soc.* **2005**, *24*, 602–612. [CrossRef]
92. Shameena, S.; Kumar, S.; Kumar, K.; Raman, R. Role of Temperature and Co-Infection in Mediating the Immune Response of Goldfish. *Microb. Pathog.* **2021**, *156*, 104896. [CrossRef] [PubMed]
93. Geoghegan, J.L.; Di Giallonardo, F.; Wille, M.; Ortiz-Baez, A.S.; Costa, V.A.; Ghaly, T.; Mifsud, J.C.; Turnbull, O.M.; Bellwood, D.R.; Williamson, J.E.; et al. Virome Composition in Marine Fish Revealed by Meta-Transcriptomics. *Virus Evol.* **2021**, *7*, veab005. [CrossRef]
94. Boonthai, T.; Loch, T.P.; Yamashita, C.J.; Smith, G.D.; Winters, A.D.; Kiupel, M.; Brenden, T.O.; Faisal, M. Laboratory Investigation into the Role of Largemouth Bass Virus (Ranavirus, Iridoviridae) in Smallmouth Bass Mortality Events in Pennsylvania Rivers. *BMC Vet. Res.* **2018**, *14*, 62. [CrossRef] [PubMed]
95. Buchmann, K. Immune Mechanisms in Fish Skin against Monogeneans—A Model. *Folia Parasitol.* **1999**, *46*, 1–9.
96. Ellis, A. Innate Host Defense Mechanisms of Fish against Viruses and Bacteria. *Dev. Comp. Immunol.* **2001**, *25*, 827–839. [CrossRef]
97. Gilliland, E. Livewell Operating Procedures to Reduce Mortality of Black Bass during Summer Tournaments. In Proceedings of the American Fisheries Society Symposium, Bethesda, MD, USA, 19 August 2002; pp. 477–488.
98. Keretz, K.R.; Dinken, C.P.; Allen, P.J.; Colvin, M.E.; Schramm, H.L., Jr. The Effect of Water Temperature, Angling Time, and Dissolved Oxygen on the Survival of Largemouth Bass Subjected to Simulated Angling and Tournament Handling Procedures. *N. Am. J. Fish. Manag.* **2018**, *38*, 606–622. [CrossRef]
99. Suski, C.D.; Killen, S.S.; Morrissey, M.B.; Lund, S.G.; Tufts, B.L. Physiological Changes in Largemouth Bass Caused by Live-Release Angling Tournaments in Southeastern Ontario. *N. Am. J. Fish. Manag.* **2003**, *23*, 760–769. [CrossRef]
100. Suski, C.D.; Killen, S.S.; Cooke, S.J.; Kieffer, J.D.; Philipp, D.P.; Tufts, B.L. Physiological Significance of the Weigh-in during Live-Release Angling Tournaments for Largemouth Bass. *Trans. Am. Fish. Soc.* **2004**, *133*, 1291–1303. [CrossRef]
101. Schramm, H.L., Jr.; Walters, A.R.; Grizzle, J.M.; Beck, B.H.; Hanson, L.A.; Rees, S.B. Effects of Live-Well Conditions on Mortality and Largemouth Bass Virus Prevalence in Largemouth Bass Caught during Summer Tournaments. *N. Am. J. Fish. Manag.* **2006**, *26*, 812–825. [CrossRef]
102. Rothlisberger, J.D.; Chadderton, W.L.; McNulty, J.; Lodge, D.M. Aquatic Invasive Species Transport via Trailered Boats: What Is Being Moved, Who Is Moving It, and What Can Be Done. *Fisheries* **2010**, *35*, 121–132. [CrossRef]
103. Kelly, N.E.; Wantola, K.; Weisz, E.; Yan, N.D. Recreational Boats as a Vector of Secondary Spread for Aquatic Invasive Species and Native Crustacean Zooplankton. *Biol. Invasions* **2013**, *15*, 509–519. [CrossRef]
104. Kemp, C.; Van Riper, C.J.; BouFajreldin, L.; Stewart, W.P.; Scheunemann, J.; van den Born, R.J. Connecting Human–Nature Relationships to Environmental Behaviors That Minimize the Spread of Aquatic Invasive Species. *Biol. Invasions* **2017**, *19*, 2059–2074. [CrossRef]
105. Murray, A.G. Using Simple Models to Review the Application and Implications of Different Approaches Used to Simulate Transmission of Pathogens among Aquatic Animals. *Prev. Vet. Med.* **2009**, *88*, 167–177. [CrossRef]
106. Crane, M.; Hyatt, A. Viruses of Fish: An Overview of Significant Pathogens. *Viruses* **2011**, *3*, 2025–2046. [CrossRef]
107. Noga, E.J. *Fish Disease: Diagnosis and Treatment*; John Wiley & Sons: Hoboken, NJ, USA, 2010.
108. Arechavala-Lopez, P.; Sanchez-Jerez, P.; Bayle-Sempere, J.T.; Uglem, I.; Mladineo, I. Reared Fish, Farmed Escapees and Wild Fish Stocks—A Triangle of Pathogen Transmission of Concern to Mediterranean Aquaculture Management. *Aquac. Environ. Interact.* **2013**, *3*, 153–161. [CrossRef]

109. Geoghegan, J.L.; Di Giallonardo, F.; Cousins, K.; Shi, M.; Williamson, J.E.; Holmes, E.C. Hidden Diversity and Evolution of Viruses in Market Fish. *Virus Evol.* **2018**, *4*, vey031. [CrossRef]
110. Dong, X.; Li, C.; Wang, Y.; Hu, T.; Zhang, F.; Meng, F.; Gao, M.; Han, X.; Wang, G.; Qin, J.; et al. Diversity and Connectedness of Brine Shrimp Viruses in Global Hypersaline Ecosystems. *Sci. China Life Sci.* **2024**, *67*, 188–203. [CrossRef]
111. Alfonso, P.; Butković, A.; Fernández, R.; Riesgo, A.; Elena, S.F. Unveiling the Hidden Viromes across the Animal Tree of Life: Insights from a Taxonomic Classification Pipeline Applied to Invertebrates of 31 Metazoan Phyla. *Msystems* **2024**, *9*, e00124-24. [CrossRef] [PubMed]
112. Zhou, K.; Zhang, T.; Chen, X.-W.; Xu, Y.; Zhang, R.; Qian, P.-Y. Viruses in Marine Invertebrate Holobionts: Complex Interactions between Phages and Bacterial Symbionts. *Annu. Rev. Mar. Sci.* **2024**, *16*, 467–485. [CrossRef] [PubMed]
113. Dong, X.; Meng, F.; Zhou, C.; Li, J.; Hu, T.; Wang, Y.; Wang, G.; Luo, J.; Li, X.; Liu, S.; et al. Enormous Diversity of RNA Viruses in Economic Crustaceans. *Msystems* **2024**, *9*, e01016-24. [CrossRef]
114. Richard, J.C.; Blevins, E.; Dunn, C.D.; Leis, E.M.; Goldberg, T.L. Viruses of Freshwater Mussels during Mass Mortality Events in Oregon and Washington, USA. *Viruses* **2023**, *15*, 1719. [CrossRef]
115. Richard, J.C.; Leis, E.M.; Dunn, C.D.; Harris, C.; Agbalog, R.E.; Campbell, L.J.; Knowles, S.; Waller, D.L.; Putnam, J.G.; Goldberg, T.L. Freshwater Mussels Show Elevated Viral Richness and Intensity during a Mortality Event. *Viruses* **2022**, *14*, 2603. [CrossRef]
116. Pao, H.-Y.; Wu, C.-Y.; Wen, C.M. Persistent Development of Adomavirus and Aquareovirus in a Novel Cell Line from Marbled Eel with Petechial Skin Haemorrhage. *J. Fish Dis.* **2019**, *42*, 345–355. [CrossRef]
117. Chinchar, V.; Logue, O.; Antao, A.; Chinchar, G. Channel Catfish Reovirus (CRV) Inhibits Replication of Channel Catfish Herpesvirus (CCV) by Two Distinct Mechanisms: Viral Interference and Induction of an Anti-Viral Factor. *Dis. Aquat. Org.* **1998**, *33*, 77–85. [CrossRef] [PubMed]
118. LaPatra, S.; Lauda, K.; Jones, G. Aquareovirus Interference Mediated Resistance to Infectious Hematopoietic Necrosis Virus. *Vet. Res.* **1995**, *26*, 455–459.
119. López-Bueno, A.; Mavian, C.; Labella, A.M.; Castro, D.; Borrego, J.J.; Alcami, A.; Alejo, A. Concurrence of Iridovirus, Polyomavirus, and a Unique Member of a New Group of Fish Papillomaviruses in Lymphocystis Disease-Affected Gilthead Sea Bream. *J. Virol.* **2016**, *90*, 8768–8779. [CrossRef] [PubMed]
120. Van Doorslaer, K.; Kraberger, S.; Austin, C.; Farkas, K.; Bergeman, M.; Paunil, E.; Davison, W.; Varsani, A. Fish Polyomaviruses Belong to Two Distinct Evolutionary Lineages. *J. Gen. Virol.* **2018**, *99*, 567. [CrossRef]
121. Buck, C.B.; Van Doorslaer, K.; Peretti, A.; Geoghegan, E.M.; Tisza, M.J.; An, P.; Katz, J.P.; Pipas, J M.; McBride, A.A.; Camus, A.C.; et al. The Ancient Evolutionary History of Polyomaviruses. *PLoS Pathog.* **2016**, *12*, e1005574. [CrossRef]
122. Monier, A.; Claverie, J.-M.; Ogata, H. Horizontal Gene Transfer and Nucleotide Compositional Anomaly in Large DNA Viruses. *Bmc Genom.* **2007**, *8*, 456. [CrossRef]
123. Salzer, J.E.; Greer, J.B.; Groner, M.; MacKenzie, A.; Gregg, J.L.; Hershberger, P. Effects of Temperature on Viral Load, Inclusion Body Formation, and Host Response in Pacific Herring with Viral Erythrocytic Necrosis (VEN). *J. Aquat. Anim. Health* **2024**, *36*, 45–56. [CrossRef]
124. Pham, P.H.; Misk, E.; Papazotos, F.; Jones, G.; Polinski, M.P.; Contador, E.; Russell, S.; Garver, K.A.; Lumsden, J.S.; Bols, N.C. Screening of Fish Cell Lines for Piscine Orthoreovirus-1 (PRV-1) Amplification: Identification of the Non-Supportive PRV-1 Invitrome. *Pathogens* **2020**, *9*, 833. [CrossRef]
125. Dhamotharan, K.; Vendramin, N.; Markussen, T.; Wessel, Ø.; Cuenca, A.; Nyman, I.B.; Olsen, A.B.; Tengs, T.; Krudtaa Dahle, M.; Rimstad, E. Molecular and Antigenic Characterization of Piscine Orthoreovirus (PRV) from Rainbow Trout (*Oncorhynchus mykiss*). *Viruses* **2018**, *10*, 170. [CrossRef] [PubMed]
126. Meyers, T.R.; Hickey, N. A Perspective: Molecular Detections of New Agents in Finfish—Interpreting Biological Significance for Fish Health Management. *J. Aquat. Anim. Health* **2022**, *34*, 47–57. [CrossRef] [PubMed]
127. Sommers, P.; Chatterjee, A.; Varsani, A.; Trubl, G. Integrating Viral Metagenomics into an Ecological Framework. *Annu. Rev. Virol.* **2021**, *8*, 133–158. [CrossRef] [PubMed]
128. Munang'andu, H.M.; Mugimba, K.K.; Byarugaba, D.K.; Mutoloki, S.; Evensen, Ø. Current Advances on Virus Discovery and Diagnostic Role of Viral Metagenomics in Aquatic Organisms. *Front. Microbiol.* **2017**, *8*, 406. [CrossRef]
129. Shi, M.; Lin, X.-D.; Chen, X.; Tian, J.-H.; Chen, L.-J.; Li, K.; Wang, W.; Eden, J.-S.; Shen, J.-J.; Liu, L.; et al. The Evolutionary History of Vertebrate RNA Viruses. *Nature* **2018**, *556*, 197–202. [CrossRef]
130. Tisza, M.J.; Pastrana, D.V.; Welch, N.L.; Stewart, B.; Peretti, A.; Starrett, G.J.; Pang, Y.-Y.S.; Krishnamurthy, S.R.; Pesavento, P.A.; McDermott, D.H.; et al. Discovery of Several Thousand Highly Diverse Circular DNA Viruses. *Elife* **2020**, *9*, e51971. [CrossRef]
131. Costa, V.A.; Holmes, E.C. Diversity, Evolution, and Emergence of Fish Viruses. *J. Virol.* **2024**, *98*, e00118-24. [CrossRef]
132. Lauber, C.; Seitz, S. Opportunities and Challenges of Data-Driven Virus Discovery. *Biomolecules* **2022**, *12*, 1073. [CrossRef] [PubMed]

133. Sharma, D.; Priyadarshini, P.; Vrati, S. Unraveling the Web of Viroinformatics: Computational Tools and Databases in Virus Research. *J. Virol.* **2015**, *89*, 1489–1501. [CrossRef] [PubMed]
134. Krishnamurthy, S.R.; Wang, D. Origins and Challenges of Viral Dark Matter. *Virus Res.* **2017**, *239*, 136–142. [CrossRef] [PubMed]

viruses

Article

Assessment and Performance of Pooled Serum Samples for Monitoring Farm-Level Immunity in Tilapia Infected with Tilapia Lake Virus

Jidapa Yamkasem [1], Puntanat Tattiyapong [1], Ian A. Gardner [2] and Win Surachetpong [1,3,*]

[1] Department of Veterinary Microbiology and Immunology, Faculty of Veterinary Medicine, Kasetsart University, Bangkok 10900, Thailand; jidapa.yam@ku.th (J.Y.); been_best@yahoo.com (P.T.)
[2] Atlantic Veterinary College, University of Prince Edward Island, 550 University Avenue, Charlottetown, PE C1A4P3, Canada; iagardner@upei.ca
[3] Laboratory of Biotechnology, Chulabhorn Research Institute, Bangkok 10210, Thailand
* Correspondence: win.s@ku.th

Abstract

Effective surveillance of viral disease in fish populations is critical for disease control and the sustainable development of global aquaculture. Here, we evaluated the application and performance of pooled serum samples using an indirect ELISA based on recombinant segment 4 protein to assess farm-level immunity in tilapia infected with Tilapia lake virus (TiLV). The TiLV-S4 ELISA was developed using a recombinant nucleoprotein (segment 4) antigen, optimized through checkerboard titration, and validated for repeatability and reproducibility, with intra- and inter-assay coefficients of variation below 10%. A pooling strategy was used to combine multiple serum samples before testing for the presence of TiLV-specific antibodies using an enzyme-linked immunosorbent assay (ELISA). Our results showed that pooling five serum samples was effective for detecting TiLV-specific antibodies, particularly when multiple seropositive individuals were presented in the pool, supporting its application for population-level surveillance. However, ELISA sensitivity may be reduced when only one seropositive sample is included in the pool, due to the dilution effects. Despite this limitation, pooled testing yielded a high proportion of positive results, suggesting similar detection performance in many cases. Overall, the pooling strategy provides a cost-effective and time-efficient approach for large-scale monitoring of immune status in tilapia populations.

Keywords: TiLV; immunity; ELISA; pooling; serum samples; tilapia; analytical sensitivity; monitoring

1. Introduction

Tilapia lake virus (TiLV) is an important pathogen affecting various species of tilapia and causing substantial economic losses in many countries [1,2]. TiLV outbreaks can lead to mortality rates as high as 90%, which poses a severe threat to the sustainability of tilapia farming [3]. Accurate and reliable diagnostic methods are crucial for mitigating these losses and controlling the spread of TiLV. Various techniques have been established to detect TiLV in infected fish samples, including quantitative polymerase chain reaction (qPCR) assays, immunohistochemistry, in situ hybridization, and loop-mediated isothermal amplification methods [4–7]. Although these assays are highly sensitive and specific, they primarily focus on detecting genomic material of the virus. These approaches make it challenging to

 https://doi.org/10.3390/v17070877

determine the TiLV-exposure status of individual fish and populations at a later stage of an outbreak or after the outbreak has ceased.

Serological methods such as enzyme-linked immunosorbent assays (ELISAs) offer a more cost-effective and practical approach for evaluating the immune response in hosts. The technique also serves as a rapid alternative for large-scale screening of samples. For TiLV, previous reports indicate that fish exposed to the virus can rapidly mount specific antibody production, with these antibodies persisting for more than four months [8]. This evidence led to the effort to develop ELISAs to assess the serum antibody concentrations using whole viral antigen [8], a recombinant protein from segment 8 [9] and segment 4 [10] of TiLV. These assays represent a significant advancement in TiLV diagnostics and can be further developed into effective tools to monitor infection status in tilapia farms. In other fish species, serological techniques such as ELISA have been applied to monitor immune responses against different viruses, such as Koi Herpes Virus, Grass Carp Reovirus, and Viral Hemorrhagic Septicemia [11–13]. Additionally, the technique can be applied to evaluate vaccine efficacy and for disease surveillance within fish farms [14–16].

However, testing individual fish serum samples is costly and time-consuming. The pooling technique, which combines multiple individual samples into a single pooled sample, presents a cost-effective and more practical approach for laboratory practices [17–19]. Therefore, studying the application of pooled serum samples to assess TiLV status within tilapia populations could significantly enhance the efficiency of monitoring the TiLV infection status of farms. While pooling techniques have been successfully used to detect TiLV genomic material in fish tissues [20], their analytical sensitivity and applicability to immunological assays have not yet been fully assessed.

The aim of this research was to develop an indirect ELISA based on a recombinant protein derived from TiLV segment 4, a structural nucleoprotein with strong antigenic properties and considered a promising target for immunological studies [8]. The assay was designed to detect TiLV-specific immunity in fish and to evaluate the sensitivity and performance of pooled serum samples. The performance of different pooling strategies was compared with individual serum samples to assess their diagnostic efficacy. This study provided important insights into the feasibility of using pooled serum samples as a diagnostic tool for TiLV in tilapia farms.

2. Materials and Methods

2.1. Serum Samples and Ethical Approval

A total of 320 fish serum samples were used in this study. All samples were stored at −20 °C until used. Based on their source and exposure to TiLV, samples were divided into two groups: laboratory challenge samples and field study samples. Sera from red hybrid tilapia (*Oreochromis* spp.) were obtained from previous experimental studies conducted in the animal facility at the Faculty of Veterinary Medicine, Kasetsart University, Thailand from 2019 to 2021. These samples included 40 serum samples from fish experimentally infected with TiLV strain KUTV08 (TiLV-seropositive) and 120 serum samples from control fish (TiLV-seronegative). The presence or absence of immunoglobulin M (IgM) serum antibodies against TiLV in these samples was previously examined using an in-house protocol of the enzyme-linked immunosorbent assay (ELISA) with whole TiLV virus as the antigen (hereafter named TiLV-WV-ELISA) [8]. In addition, 160 serum samples were collected in 2022 from four tilapia farms with a history of TiLV outbreaks in Kanchanaburi province, Thailand. Details of samples are provided in Table 1. Blood was collected from red hybrid tilapia, ranging in size from 30 to 400 g, in farms with TiLV outbreaks that occurred 28 to 90 days after first TiLV mortality.

Table 1. Details for sampling blood collection from farms with a history of a TiLV outbreak.

Farms	Total Sample	Cumulative Mortality (%) During Outbreak	Day Post Infection (dpi)	Outbreak Start Date	Blood Collection Date	Average Fish Weight (g)
A	40	33	28	10 September 2022	7 October 2022	150
B	38	23	39	28 August 2022	7 October 2022	30
C	42	40	60	6 July 2022	9 September 2022	200
D	40	40	90	6 July 2022	7 October 2022	400

All blood samples were collected in non-anticoagulation tubes for serum separation and were centrifuged at 1500× *g* for 30 min at room temperature. Serum was collected in 1.5 microcentrifuge tubes for antibody detection and stored at −20 °C until further analysis (Figure 1). This study was approved by the Animal Ethics Committee of the faculty of Veterinary Medicine, Kasetsart University, Bangkok, Thailand. All animals used in this study were handled in accordance with the Animal Ethics procedures and guidelines (protocol number ACKU65-VET-088).

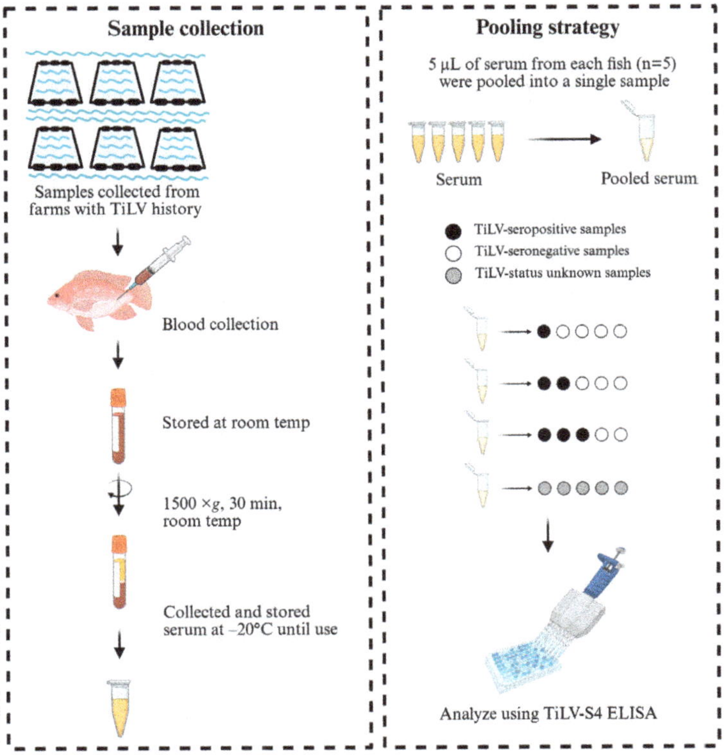

Figure 1. Sample collection and pooling strategy. (Left) Blood samples were collected from farms with a history of TiLV outbreaks. Blood was stored at room temperature and then centrifuged at 1500× *g* for 30 min at room temperature. Serum was collected and stored at −20 °C until analysis. (Right) For the pooling procedure, 5 µL of serum from each of five individual fish were combined into a single pooled sample. Pooled serum samples, which included TiLV-seropositive, TiLV-seronegative, and TiLV-status unknown samples, were then analyzed using the TiLV-S4 ELISA.

2.2. *Preparation of Recombinant TiLV Segment 4 Protein*

2.2.1. Construction of Recombinant TiLV Segment 4 Plasmid

The sequence encoding TiLV segment-4 was obtained from the NCBI GenBank database (accession number KX631924.1). To enhance protein expression and simplify purification, codon optimization was performed, and a His-tag was added to the C-terminal end. The Compute pI/MW online tool (https://web.expasy.org/compute_pi/ accessed on 25 March 2022) was used to predict the molecular weight of optimized protein. The optimized TiLV-S4 gene was synthesized by Synbio Technologies, LLC (Monmouth Junction, NJ, USA) and cloned into the pET-28a(+) expression vector using the restriction enzymes NcoI and XhoI. The ligation reaction was carried out using T4 DNA ligase.

2.2.2. Expression and Purification of Recombinant Protein

The pET-28a-TiLV-S4 was transformed into *Escherichia coli* BL21-competent cells. Transformed bacteria were incubated with isopropyl-β-D-thiogalactoside (IPTG) to stimulate expression of TiLV-S4. Briefly, positive transformants were cultured in Luria–Bertani medium containing 50 μg/mL of kanamycin at 37 °C with shaking at 225 rpm and induced with 0.2 mM IPTG when the optical density at 600 nm (OD_{600}) of culture material reached 0.5. Eighteen hours post-induction at 18 °C, bacteria were harvested and lysed using ultra-sonication (5 s on, 7 s off, total time 4 min) on ice. Following centrifugation (10,000× *g*, 30 min, 4 °C), the supernatant of bacterial lysates was collected and analyzed through SDS-PAGE followed by Coomassie blue staining. For Western blotting, proteins were transferred onto a polyvinylidene fluoride membrane (Bio-Rad, Hercules, CA, USA) using semi-dry transfer system (Bio-Rad, Hercules, CA, USA). Membranes were blocked in 3% Bovine serum albumin (Sigma, St. Louis, MO, USA) in PBS-T (PBS with 0.1% Tween-20) overnight and then incubated at 4 °C with rabbit polyclonal anti-TiLV serum (generated by immunizing rabbits with purified whole TiLV antigen) [8], diluted 1:1000. After washing, membranes were incubated with HRP-conjugated anti-rabbit IgG (SeraCare, Milford, MA, USA), followed by detection using a LumiFlash™ Infinity Chemiluminescence substrate (Visual Protein, Taipei, Taiwan) and imaging on a ChemiDoc™ system (Bio-Rad, Hercules, CA, USA). The TiLV-S4 protein was purified from supernatant using a HisTrap™ HP (Cytiva, Amersham, UK) with an ÄKTA pure™ chromatography system (Cytiva, Amersham, UK) according to manufacturer's instructions. Eluted protein solution was concentrated using a Nanosep® Centrifugal Filter with Omega membrane 10 K (Pall Corporation, Port Washington, NY, USA), and final protein concentration was determined using a Pierce BCA Protein Assay Kit (Thermo Fisher Scientific, Waltham, MA, USA).

2.3. *Optimization of TiLV-S4 ELISA*

The working concentrations of ELISA reagents, including the antigen and primary antibody were determined by checkerboard titration. The optimal concentrations for the antigen and primary antibody were selected based on the highest OD_{450} ratio of the positive sample to the negative sample (P/N ratio). Ninety-six-well ELISA plates (Thermo Fisher Scientific, Waltham, MA, USA) were coated with 100 μL of TiLV-S4 protein antigen (0, 0.5, 1, 2, and 5 μg/mL) diluted in 1X KPL coating solution (SeraCare, Milford, MA, USA) and incubated overnight at 4 °C. Plates were then blocked with 200 μL of 3% (*w*/*v*) bovine serum albumin (BSA) (Sigma-Aldrich, St. Louis, MO, USA) in PBS containing 0.05% Tween-20 (0.05% PBS-T) for 1 h at room temperature. Following the blocking step, buffer was discarded, and plates were washed three times with 200 μL of 0.05% PBS-T for 5 min each. Positive and negative serum samples, obtained from archived infection studies as described in Section 2.1, were diluted in 0.05% PBS-T containing 3% BSA to final concentrations of 1:100, 1:200, 1:400, 1:800, 1:1600, 1:3200, and 1:6400. These samples were added to wells, incubated for 1 h at room temperature, and subsequently washed three times. Next, 100 µL of secondary antibody (anti-tilapia IgM monoclonal antibody) (Aquatic Diagnostics Ltd., Stirling, UK) diluted 1:1000 in 0.05% PBS-T containing 3% BSA was added to each well and incubated for 1 h at room temperature to detect IgM against TiLV in serum samples. Plates were washed three times with 0.05% PBS-T, followed by incubation in 100 µL anti-mouse IgG HRP-conjugated antibody (SeraCare, Milford, MA, USA) at a dilution of 1:2000 in 0.05% PBS-T containing 3% BSA. Plates were incubated at room temperature for 1 h. After incubation, plates were washed three times, and 50 µL of 3,3′,5,5′-tetramethylbenzidine (TMB) substrate solution (Thermo Fisher Scientific, Waltham, MA, USA) was added to each well for the chromogenic reaction, which was carried out at room temperature for 20 min in the dark. The reaction was terminated using 50 µL of 2M sulfuric acid, and optical density at 450 nm (OD_{450}) was measured using a microplate reader and analyzed using Gen5 version 3.0 software (Biotek, Winooski, VT, USA).

2.4. Cut-Off Value for Classification of an ELISA Result as Positive or Negative

To determine the cut-off value for the TiLV-S4 ELISA, 120 TiLV-seronegative samples collected from farms with no history of TiLV infection were tested. Each sample was analyzed in triplicate, and the mean OD_{450} value was used for analysis. The percentage reactivity (PR) value for each sample was calculated using the following formula:

$$\text{PR value} = [(\text{OD value of tested sample} - \text{negative control}) \div (\text{OD value of positive control} - \text{negative control})] \times 100\%.$$

The PR cut-off value for the serum-based ELISA was determined by calculating the mean PR value ($\overline{x}$) of 120 TiLV-seronegative samples plus 3 standard deviations (SD). A sample was considered positive if its PR value was equal or greater than the cut-off value, otherwise the sample was considered negative.

2.5. Repeatability and Reproducibility of TiLV-S4 ELISA

To evaluate the repeatability (intra-assay variation) and reproducibility (inter-assay variation) of the TiLV-S4 ELISA, a total of six serum samples were selected including three TiLV-seropositive samples (with previously determined OD values of 1.07, 1.45, and 1.93 using the whole-virus ELISA) and three TiLV-seronegative samples (with OD values of 0.20, 0.25, 0.59) were selected. For repeatability assessment, each sample was tested in triplicate within a single ELISA run. For reproducibility assessment, each sample was tested once per run across three independent ELISA runs conducted on three different days. The mean, standard deviation (SD), and coefficient of variation (CV%) were calculated for each sample under both intra- and inter-assay conditions to determine the consistency and robustness of the assay.

2.6. Sensitivity and Specificity of TiLV-S4 ELISA

Point estimates of diagnostic sensitivity and specificity of the TiLV-S4 ELISA, along with 95% exact binomial confidence intervals (CI), were estimated by comparing the results to those of 40 TiLV-seropositive and 40 TiLV-seronegative serum samples obtained from controlled laboratory challenge studies. These samples had been previously tested using the TiLV-WV-ELISA, which was used as the reference method for evaluating the performance of the TiLV-S4 ELISA. The PR cut-off value of the TiLV-S4 ELISA was calculated as described in Section 2.4, and was used to classify results as TiLV-S4-positive or -negative for calculation of diagnostic sensitivity and specificity.

2.7. Viral Antigen Preparation and TiLV-WV ELISA Protocol

The protocol established by Tattiyapong (2020) [8] was applied for preparation of whole protein for the TiLV-WV ELISA assay. Briefly, TiLV strain KUTV08 was propagated by passaging in E-11 cells (ECACC, Salisbury, UK), and virus was harvested five days post-inoculation. Viral stock was then concentrated and purified. The concentration process involved passing virus through a 30% sucrose cushion at 107,000× *g* for 90 min at 4 °C using an Optima MAX-XP Ultracentrifuge (Beckman Coulter, Brea, CA, USA). Subsequently, the pellet was re-suspended in TN buffer and further purified through a sucrose gradient concentration of 30%, 40%, and 50% (*W*/*V*) at 107,000× *g* for 120 min at 4 °C. The pellet was collected from the 40% and 50% sucrose fractions, re-suspended in TN buffer, and stored at −80 °C for further analysis.

For the TiLV-WV ELISA assay, working concentrations and dilution ratios were as follows: viral antigens at 2.5 µg/mL, primary antibody (fish serum sample) diluted at 1:100, secondary antibody (anti-tilapia IgM monoclonal antibody) diluted at 1:1000, and tertiary antibody (HRP-conjugated anti-mouse IgG antibody) diluted at 1:2000 [8]. The ELISA protocol followed the same steps as described in Section 2.3.

2.8. Design of Pooling Experiments

Serum samples from individual farm fish, as described in Section 2.1, were used to evaluate the efficiency of TiLV-S4 ELISA with pooled samples. The diagnostic accuracy of pooling was assessed by selection of one, two, or three positive serum samples mixed with four, three, and two negative serum samples, respectively (Figure 1). To create a pool of five samples, 5 µL of each serum sample was aliquoted from five individuals and mixed into a 1.5 mL microcentrifuge tube. After pooling, serum samples were processed for detection of antibodies against TiLV using the new TiLV-S4 ELISA assay (Figure 1).

2.9. Data Management and Statistical Analysis

Pooled and individual TiLV-S4 ELISA results from each farm were combined. The individual TiLV-S4 ELISA result for each sample used in the pool was considered the reference standard (classified as positive if the PR value was >37.11, and negative otherwise). The corresponding pooled result was then classified as either a true positive, true negative, or false negative for each sample set. No false positives were obtained. For positive pools, the median and range of individual positive samples were calculated. Additionally, the ratio of the mean PR% of the 5 individual sample pools to the corresponding pooled PR was calculated. This value was expected to be approximately one in the absence of technical errors or antibodies aggregation/clumping. Frequencies were compared by chi-square or Fisher exact tests, and distributions of PR ratios between Farm A and Farms B, C, and D were compared with a Mann–Whitney rank sum test. p values < 0.05 were considered statistically significant.

3. Results

3.1. Production of TiLV-S4 Protein and Its Specificity

Recombinant TiLV-S4 protein was produced from *E. coli* BL21 and purified using a HisTrap™ HP column and affinity chromatography. After separation on SDS-PAGE, a distinct 38 kDa band corresponding to the TiLV-S4 protein was detected (Figure 2A). Further analysis by Western blot, using rabbit serum collected from animals exposed to TiLV, confirmed the specificity of the protein by showing a specific interaction at this band size (Figure 2B).

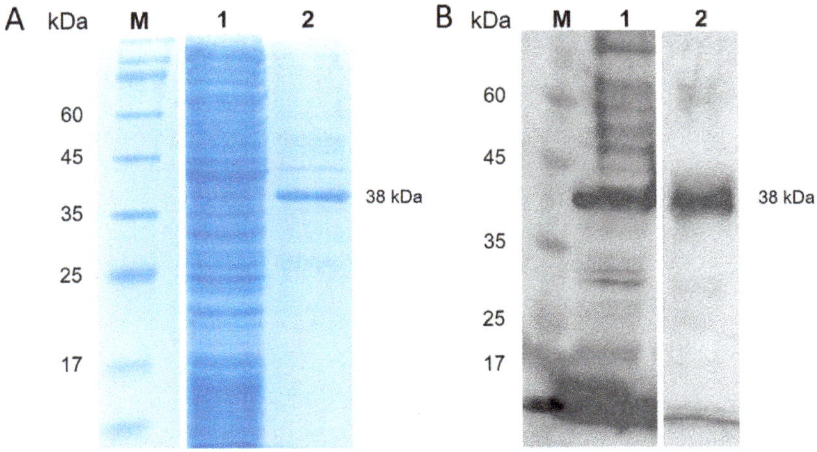

Figure 2. Expression and purification of TiLV-S4 protein. (**A**) SDS-PAGE analysis stained with Coomassie blue showing the expressed TiLV-S4 protein. (**B**) Western blot of the TiLV-S4 protein with rabbit anti-TiLV antibody. Marker (Lane M), total protein (Lane 1), and purified TiLV-S4 protein (Lane 2).

3.2. Development and Optimization of a TiLV-S4 ELISA

3.2.1. Validation of TiLV-S4 Recombinant Protein and Primary Antibody Concentrations

To develop the TiLV-S4 ELISA for detecting TiLV-IgM antibodies, a checkerboard titration was conducted to optimize assay conditions. Optimization criteria included ensuring a high OD_{450} value for positive serum and a low value for negative serum, and maximizing the positive-to-negative (P/N) ratio. The optimal concentration of TiLV-S4 protein for coating ELISA plates was determined to be 0.5 µg/mL (0.05 µg/well). Primary antibodies (fish serum samples) were used at a working dilution of 1:100, the secondary antibody (anti-tilapia IgM antibody) at 1:1000, and the tertiary antibody (HRP conjugated anti-mouse IgG antibody) at 1:2000.

3.2.2. Repeatability and Reproducibility of the TiLV-S4 ELISA

To validate the repeatability and reproducibility of the TiLV-S4 ELISA, six serum samples (three TiLV-seropositive and three TiLV-seronegative) were tested. Detailed results are presented in Table 2, which show intra-assay CVs ranging from 1.1% to 8.7%, and inter-assay CVs ranging from 1.7% to 7.7% (Table 2). These results indicate that the new TiLV-S4 ELISA demonstrated good repeatability and reproducibility.

Table 2. Estimates of coefficients of variation (CV) from six samples.

	Samples	Mean OD Value	SD	CV%
Repeatability (Intra-assay)	1	1.06	0.04	3.6
	2	0.91	0.01	1.1
	3	1.47	0.13	8.7
	4	0.08	0.00	2.7
	5	0.17	0.00	2.7
	6	0.21	0.02	7.7
Reproducibility (Inter-assay)	1	1.03	0.08	7.7
	2	0.91	0.03	3.1
	3	1.45	0.11	7.6
	4	0.08	0.01	6.8
	5	0.18	0.01	5.0
	6	0.20	0.00	1.7

Positive serum samples are indicated by shaded boxes.

3.2.3. Determination of the Cut-Off Value for TiLV-S4 ELISA for Classification of Results as Positive or Negative

To establish the cut-off value for the TiLV-S4 ELISA, 120 serum samples previously verified by TiLV-WV ELISA assays were utilized. The average PR value ($\overline{x}$) of the negative samples was calculated as $\overline{x}$ = 8.51, with a standard deviation (SD) of 9.53. The cut-off PR value was determined using the formula: PR value = $\overline{x}$ + 3SD = 37.11. Samples with a PR value ≥ 37.11 were classified as TiLV-seropositive (Figure 3).

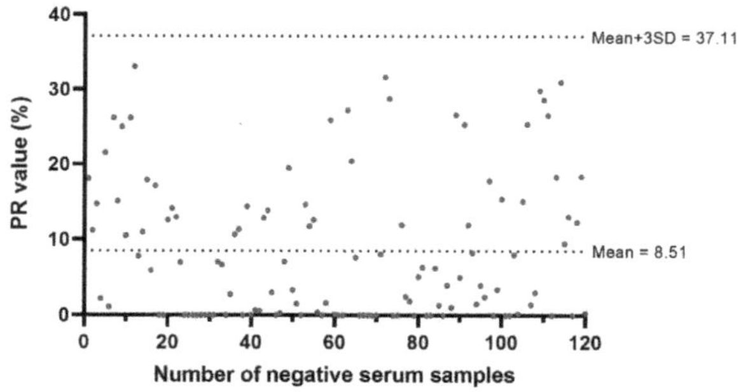

Figure 3. Dot plot of PR% values and mean and cutoff values of the TiLV-S4 ELISA for 120 fish not infected with TiLV.

3.2.4. Efficiency of TiLV-S4 ELISA Compared with the Reference Method

To evaluate the specificity and sensitivity of the TiLV S4-ELISA, a total of 80 serum samples were assessed. Prior to TiLV-S4 ELISA testing, all samples were analyzed by the TiLV WV-ELISA to confirm their TiLV serostatus. Forty serum samples were confirmed negative, and another 40 were positive, for TiLV antibodies by TiLV-WV ELISA. Using a cut-off PR value of 37.11%, the TiLV-S4 ELISA correctly identified 36 of the 40 TiLV-seronegative samples and 33 of the 40 TiLV-seropositive samples. Accordingly, the diagnostic specificity of the TiLV-S4 ELISA was 90.0% (36/40, 95% CI = 76.3 to 97.2%), and the diagnostic sensitivity was 82.5% (33/40, 95% CI =67.2 to 92.7%) compared to TiLV-WV ELISA results (Supplementary Table S1).

3.3. *Pooling Efficiency*

3.3.1. Pooling Scenarios

Of the 40 fish serum samples collected from farm A, which were taken 28 days after first mortality occurred, 32 samples (80%) were identified as TiLV-seropositive using the TiLV-S4 ELISA. The PR values for these positive samples ranged from 39.8% to 248.9%. In contrast, eight samples (20%) were below the cut-off PR value, with PR values ranging from 15.8% to 31.4%, and were classified as TiLV-seronegative (Figure 4).

To evaluate the effectiveness of using the TiLV-S4 ELISA with pooled serum samples for detecting antibodies against TiLV, various scenarios were tested using samples from farm A. When one TiLV-seropositive serum sample was added to a pool of five samples, all pooled samples tested negative (Table 3: sample no. 1–5). Similarly, adding two TiLV-seropositive serum samples resulted in only 1 of 11 pools (9.1%) testing positive (Table 3: sample no. 12), while the others remained negative. However, when three TiLV-seropositive serum samples were pooled, 11 out of 20 pools (55%) yielded positive results (Table 3: sample no. 17–36).

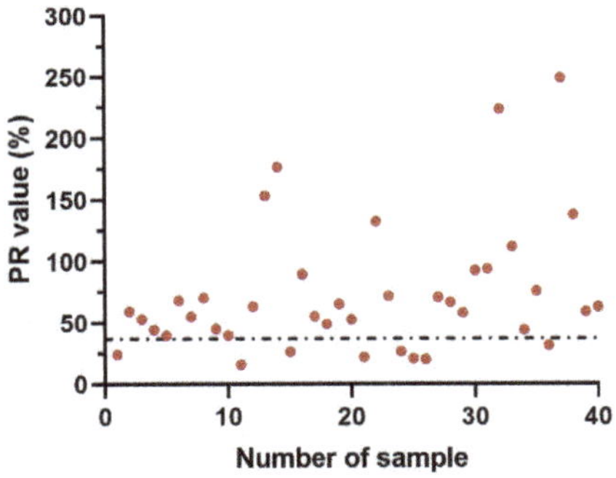

Figure 4. Detection of fish serum antibodies against TiLV in Farm A using TiLV-S4 ELISA. The dashed line indicates the assay's cut-off value for seropositivity determination.

Table 3. The pooled PR values of five serum samples, including individual component samples from farm A with one, two, or three TiLV-seropositive individuals.

Sample No.	PR Value (%) of Pooled Sample	PR Value (%) of Individual Sample				
1	14.2	223.6	26.4	21.9	26.5	20.8
2	20.8	94.0	21.9	26.5	20.8	20.1
3	14.2	55.4	15.8	26.4	21.9	26.5
4	16.5	49.2	26.4	21.9	26.5	20.8
5	12.3	44.0	21.9	26.5	20.8	20.1
6	14.9	40.1	39.7	15.8	26.4	24.6
7	29.4	176.5	44.0	20.1	31.4	24.6
8	16.2	70.9	40.1	15.8	26.4	21.9
9	26.8	67.0	44.0	20.8	20.1	26.5
10	21.1	53.4	44.0	20.1	20.8	31.4
11	26.0	45.2	44.0	20.1	31.4	24.6
12	131.2	248.9	223.6	21.9	26.5	20.8
13	35.1	153.5	67.0	15.8	26.4	21.9
14	16.2	176.5	58.5	26.5	20.8	20.1
15	20.4	70.9	63.2	26.4	21.9	26.5
16	17.1	45.2	44.5	21.9	26.5	20.8
17	16.9	44.5	39.7	40.1	15.8	26.4
18	25.7	59.7	39.7	40.1	31.4	24.6
19	13.4	52.8	39.7	40.1	24.6	15.8
20	37.7	58.5	55.4	45.2	21.9	26.5
21	15.5	59.4	49.2	39.7	31.4	24.6
22	95.8	59.7	53.4	44.5	26.4	20.8
23	22.9	63.5	45.2	40.1	15.8	26.4
24	23.9	72.0	55.4	44.0	20.8	20.1
25	54.5	76.0	58.5	45.2	24.6	15.8
26	17.5	76.0	44.0	39.7	24.6	31.4
27	72.5	89.6	68.5	55.4	20.8	20.1
28	137.5	94.0	70.9	63.5	15.8	26.4
29	60.1	92.7	70.9	63.2	21.9	26.5
30	24.6	112.0	39.7	40.1	15.8	26.4
31	133.3	153.5	58.5	49.2	31.4	24.6
32	27.4	153.5	44.5	44.0	31.4	24.6
33	80.4	176.5	89.6	72.0	26.4	21.9
34	39.0	223.6	176.5	44.5	26.5	20.8
35	87.3	248.9	39.7	40.1	24.6	15.8
36	157.5	248.9	94.0	44.0	20.1	31.4

PR values above the negative cut-off (37.11) are indicated by shaded boxes.

3.3.2. Efficiency of Pooling Technique to Detect Antibodies Against TiLV in Fish Populations

To assess the efficiency of pooling for antibody surveillance in fish populations, serum samples with unknown serostatus from farms B, C, and D were pooled into groups of five samples each. A total of ten pooled samples were created from 40, 38, and 42 individual fish serum samples in farms B, C, and D, respectively. TiLV IgM antibodies were detected in 50% (5/10), 90% (9/10), and 20% (2/10) of pooled samples from farms B, C, and D, respectively (Table 4). The PR value of the pooled samples ranged from 41.0% to 192.0% (Table 4). Notably, each pooled sample contained between two and four individual TiLV-seropositive samples, with individual PR values ranging from 37.7% to 171.9%. In contrast, pools that were entirely seronegative or contained only one seropositive sample yielded false-negative results. The latter was observed in specific cases from Farm B (samples 3 and 5) and Farm D (sample 22).

Table 4. PR value (%) of pooled samples of five tilapia and the corresponding individual serum values from farms B, C, and D.

Sample No.	Farm	PR Value (%) of Pooled Sample	PR Value (%) of Individual Sample				
1	B	41.5	60.2	63.5	43.9	11.5	13.6
2		49.2	71.6	79.7	12.5	18.1	25.2
3		26.4	60.1	29.1	15.1	29.0	8.0
4		47.5	104.2	77.0	30.5	25.5	18.7
5		19.2	40.0	24.7	13.0	31.4	7.9
6		26.2	48.6	70.4	31.9	3.6	12.6
7		84.0	57.6	75.4	47.4	41.2	31.5
8		30.2	71.6	43.9	15.1	14.3	12.9
9		40.9	104.2	48.6	57.6	29.0	7.9
10		38.1	79.7	11.5	12.9	29.1	30.5
11	C	83.5	101.1	48.4	19.2	27.7	34.7
12		146.5	63.7	160.8	128.4	27.0	27.2
13		114.5	162.9	46.7	37.7	32.7	13.5
14		16.3	18.1	2.9	12.1	16.1	24.1
15		103.9	171.9	41.9	29.4	17.4	32.6
16		192.0	60.3	68.5	142.8	27.1	13.8
17		94.9	113.1	82.2	97.5	99.7	23.7
18		66.5	101.6	15.5	24.7	35.9	23.8
19		166.7	91.3	63.7	162.9	101.1	11.5
20		89.0	113.1	60.3	18.1	32.6	23.8
21	D	14.0	8.5	8.6	19.5	24.5	9.7
22		22.3	64.8	19.5	17.4	22.5	3.7
23		27.9	10.1	16.6	18.6	29.0	35.7
24		32.7	14.1	8.0	18.4	29.9	27.5
25		12.6	6.4	2.3	1.7	7.7	7.3
26		21.1	11.5	7.4	8.4	13.9	13.6
27		100.7	95.0	123.6	32.5	16.3	30.8
28		94.5	82.6	53.3	43.2	15.0	34.4
29		14.8	8.5	3.7	10.1	2.3	6.4
30		23.4	11.5	30.8	15.0	8.6	19.5

PR values above the negative cut-off (37.11) are indicated by shaded boxes.

The prevalence of TiLV-seropositive serum varied across farms, as shown in Figure S1. Farm A, with a cumulative mortality of 33% and samples collected 28 days post first mortality, had the highest seroprevalence of 80% (32/40 samples). Farm B, with 23% cumulative mortality and samples collected 39 days post first mortality, had a prevalence of

39.5% (15/38). Farm C, with 40% cumulative mortality and samples collected 60 days post first mortality, showed a seroprevalence of 45.2% (19/42). Farm D, with 40% cumulative mortality and samples collected 90 days post first mortality, had the lowest prevalence, at 15% (6/40) (Supplementary Figure S1).

3.4. Pooled Compared with Individual Test Results in Farms A, B, C, and D

In Farm A, which had one to three positive samples in each pool by design, the proportion of TP results was significantly greater when there were three positive samples (11/20) compared with two positive results (1/11) or one positive result (0/5). Fisher exact *p* values were 0.020 and 0.046, respectively. Among the 17 TP pools in farms B, C, and D, the range was one to four positive individual samples out of five tested, with a median of 2. Among the five FN pools in the same farms, the range was one to two positives, with a median of 1 (Table 5).

Table 5. The frequency of TP, TN, and FN pools varied significantly among farms (chi-square *p* value < 0.0001).

Farms	Frequency			Total Pools
	True-Positive Pools (TP)	True-Negative Pools (TN)	False-Negative Pools (FN)	
A	12	Not applicable	24	36
B	6	0	4	10
C	9	1	0	10
D	2	7	1	10

The ratio of the mean PR% to pooled PR% values varied in Farm A ($n = 36$) from 0.39 to 4.51, with a median of 1.76 (Figure 5A). In contrast to Farm A, ratio values in the other three farms ($n = 30$) were significantly lower ($p < 0.001$), with a median of 0.67 and a range from 0.33 to 1.27 (Figure 5B).

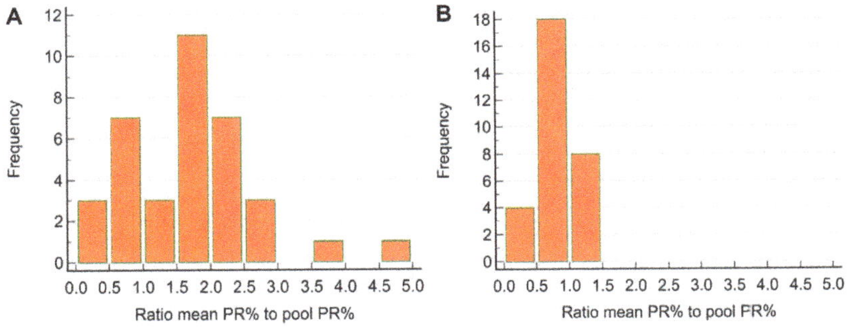

Figure 5. Ratio of the mean percentage reactivity (PR) of five individual serum samples to the pooled PR% of the component samples from (**A**) Farm A ($n = 36$) and (**B**) Farms B, C, and D ($n = 30$).

4. Discussion

Despite the relatively high prevalence of TiLV disease in tilapia world-wide [21, 22], antigen-based or serological assays to support immunological studies and disease surveillance during TiLV outbreaks remain lacking. Measuring viral-specific antibodies is essential for estimating prevalence, tracking infection dynamics, and monitoring immune responses following infection or vaccination [14–16]. Our study introduces a sensitive and specific indirect ELISA capable of quantifying antibody response against TiLV in tilapia populations. For serological analysis in fish, ELISA is cost-effective, highly specific, and suitable for large-scale detection [11–13]. In contrast, molecular assays, which are widely used to detect TiLV in infected tissues, are effective [7,23,24], but they are limited in identifying fish that have survived infection. Notably, this limitation arises as the virus may have been cleared from the tissues or reduced to levels below the detection threshold of these assays [8]. Serological tests, especially ELISA, are a low-cost and practical alternative to direct detection tests for fish viruses, including TiLV, in countries where obtaining diagnostic samples is impractical or cost-prohibitive during disease outbreaks. Because blood collection is non-lethal, it is well suited to disease surveillance programs [25].

Previous studies have demonstrated the effectiveness of ELISA tests for assessing the immune response of red tilapia infected with TiLV under laboratory conditions, highlighting that antibody levels in virus-infected fish fluctuate following exposure [8]. Specifically, antibodies against TiLV can persist for over four months, suggesting that antibody-based assessments hold practical potential for disease diagnosis and provided a valuable diagnostic window [8]. However, the application of ELISA techniques to assess the immunological status of fish in aquaculture settings has not been extensively investigated. To address this gap, we employed an indirect ELISA using a recombinant protein derived from segment 4 of TiLV to detect specific antibodies against the virus. Indeed, the recombinant proteins have been widely utilized in ELISA assays for measuring antibody responses in fish serum or tissue following pathogen infection or vaccination [13,26–28]. The use of recombinant proteins offers distinct advantages, such as eliminating the need for pathogen culture, thereby simplifying the diagnostic process and reducing the associated cost. Furthermore, previous studies have successfully developed ELISA methods for detecting TiLV-specific antibodies in fish serum, mucus, and internal organs [9,10]. For example, Hu et al. [9] demonstrated the efficacy of an ELISA based on a recombinant protein from TiLV segment 8 for detecting specific antibodies in the serum of TiLV-infected fish from both laboratory and field samples. More recently, Lalruatfela et al. [10] developed an ELISA using a recombinant protein from TiLV segment 4, which demonstrated its effectiveness in detecting TiLV-specific antibodies in fish kidney and mucus samples. Although no significant antigenic variation among TiLV strains has been confirmed to date, recent studies have reported genetic divergence among TiLV isolates from different geographic regions, such as Vietnam and Israel, which may influence antibody recognition patterns [22]. Therefore, future studies should validate the TiLV-S4 ELISA using sera from fish infected with genetically diverse TiLV strains to ensure broad diagnostic applicability.

In this study, the cut-off value for the ELISA was determined using 120 TiLV-seronegative samples from TiLV-infected fish subjected to laboratory challenges. To ensure consistency, optical density (OD) values were calculated and presented as percentage reactivity (PR) values. The PR values were chosen over OD values to minimize the variability often observed when conducting ELISA across different plates. By calculating PR values as the percentage of positive ELISA results relative to extinction of negative serum on each plate or across multiple plates, variability is reduced [11]. The use of PR values has become increasingly popular in recent ELISA development studies [29,30]. Furthermore, the repeatability and reproducibility of the TiLV-S4 ELISA were assessed, showing that intra-assay and inter-assay variations were below 10%, which demonstrates the repeatability and reproducibility of the assay [31].

The TiLV-S4 ELISA developed in this study was validated against the previously established TiLV WV-ELISA [8], demonstrating a diagnostic sensitivity of 82.5% and specificity of 90%, with minimal false positives. The WV-ELISA uses a whole-virus antigen, which presents multiple viral epitopes from both structural and non-structural proteins, allowing it to detect a broader range of antibody responses. In contrast, the TiLV-S4 ELISA detects antibodies specific to a single structural protein, which is the nucleoprotein encoded by segment 4. Therefore, if the immune response in some fish is primarily directed against other viral components, the TiLV-S4 ELISA may yield false-negative results. However, the specificity and consistency of antibody responses to immunodominant antigens of TiLV remain to be fully characterized. Similar recombinant protein-based ELISAs have shown comparable performance, such as the TiLV-S8 ELISA, with 100% sensitivity and 92.6% specificity [9], and a TiLV segment 4 ELISA, with 82.4% sensitivity and 100% specificity, against RT-PCR [10]. These results highlight the reliability of TiLV-S4 ELISA and its potential for effective disease surveillance in tilapia populations. Indeed, the TiLV-S4 ELISA was shown to be an effective tool for assessing serostatus of fish populations following a TiLV outbreak. Notably, analysis of the prevalence of TiLV-specific antibodies in surviving fish at various time points revealed that antibody titers can persist for over three months, although prevalence declines over time. The sustained immune response is consistent with previous studies that reported the presence of TiLV-specific antibodies up to four months post-infection [8]. Importantly, the prevalence of TiLV-seropositive fish was associated with higher cumulative mortality and the duration of seroreactivity following the outbreak.

Testing individual fish serum samples can be costly and time-consuming, making the pooling technique a more cost-effective and efficient alternative for laboratory workflows [17–19]. Despite its advantage, research on the effects of pooled samples on ELISA performance in aquatic animals remains limited compared to the extensive studies on terrestrial animals [32,33]. In this study, the TiLV-S4 ELISA showed no false-positive results when testing pooled seronegative samples, and the median ratio of mean PR% to pooled PR% from Farms B, C, and D was consistent and close to 1, which indicates strong agreement between pooled and individual results. We selected a pool size of five samples, as it provides a practical balance between cost-efficiency and diagnostic sensitivity, as supported by our preliminary tests. Pooling serum samples combined with the TiLV-S4 ELISA increased the detection of seropositive fish by covering a broader population at reduced cost, particularly when pools contained at least two or three seropositive samples. However, some pooled samples yielded false-negative results due to the dilution of positive sera within the pool. Similar limitations were observed in previous studies [10], where pooled kidney and mucus samples produced false negatives, which highlight the sensitivity constraints of pooling. In Farm A, a higher false-negative rate was found, likely due to the collection of samples from a population recently affected by a disease outbreak. These samples had high antibody titers and a high prevalence of seropositive sera, which may have caused antibody aggregation and clumping, which disrupt ELISA performance by impairing antibody binding and reducing assay sensitivity [34,35]. These findings emphasize the need to consider the composition of pooled samples and the potential trade-offs in analytical sensitivity when using this approach. Additionally, pooling of serum samples, as studied in the present research, is most likely to be useful for low-prevalence scenarios (e.g., one, two, or perhaps three positive samples in a pool of 5). If prevalence is high (e.g., four or five positive samples per pool), pooling would provide minimal efficiency gain over individual testing for determining population-level infection status (i.e., presence or absence of TiLV). Therefore, we did not include combinations with four or five positive sera in our pooling experiments. Future studies should also include defined serial dilutions of seropositive serum within pooled samples to better understand the detection threshold and quantitative relationship between antibody concentration and ELISA signal in pooled testing.

The application of the pooling technique with the TiLV-S4 ELISA demonstrated its potential as a cost-effective and efficient strategy for detecting TiLV antibodies at the farm level. Specifically, data from Farms B, C, and D validated the approach by testing 10 pools of serum, each comprising samples from five fish, alongside their corresponding individual samples. This method significantly reduced the number of tests required from 50 individual tests to just 10 pooled tests, with minimal additional handling and laboratory resource usage. However, the small sample size across the three farms suggests caution when extrapolating these findings to larger populations. Despite this limitation, our results highlight the practicality of pooling for large-scale surveillance of TiLV, particularly in resource-limited settings, and provides a basis for further refinement and implementation of this method in aquaculture diagnostics. Moreover, an ELISA may be applied as part of an integrated diagnostic approach, in conjunction with outbreak timing, clinical signs, and mortality patterns. Based on our observations, fish from recent and severe outbreaks (e.g., Farm A) often showed high antibody levels that made pooled ELISA results comparable to individual testing. In contrast, in farms with outbreaks occurring more than 1–3 months prior (e.g., Farms B, C, and D), declining antibody levels in surviving fish made pooling a more practical and cost-effective strategy for population-level surveillance.

5. Conclusions

Taken together, our study introduces and optimizes an indirect ELISA based on recombinant TiLV segment 4 protein, which offers a reliable and effective tool for detecting TiLV antibodies in tilapia. It also highlights the practicality of pooling serum samples as a cost-effective and efficient diagnostic strategy for large-scale surveillance. However, the sensitivity of the pooling technique depends on infection prevalence, indicating the need for careful planning and consideration during implementation. These findings present a valuable diagnostic tool and strategy for improving TiLV detection methods and highlight the potential of ELISA-based approaches in managing TiLV control and advancing tilapia health management.

Supplementary Materials: The following supporting information can be downloaded at: https://www.mdpi.com/article/10.3390/v17070877/s1, Figure S1: PR value (%) of individual serum samples in Farms A, B, C, and D using TiLV-S4 ELISA. Table S1. Optical density (OD) and percent reactivity (PR) values obtained from the TiLV-WV-ELISA and TiLV-S4-ELISA for 40 seronegative and 40 seropositive tilapia serum samples, used to evaluate the diagnostic sensitivity and specificity of the TiLV ELISA assays.

Author Contributions: Conceptualization, J.Y., I.A.G. and W.S.; methodology, J.Y., P.T. and I.A.G.; validation, I.A.G. and W.S.; formal analysis, J.Y. and I.A.G.; investigation, J.Y. and I.A.G.; data curation, J.Y., I.A.G. and W.S.; writing—original draft preparation, J.Y., I.A.G. and W.S.; writing—review and editing, I.A.G. and W.S.; visualization, J.Y., P.T., I.A.G. and W.S.; supervision, W.S.; project administration, W.S.; funding acquisition, W.S. All authors have read and agreed to the published version of the manuscript.

Funding: The project was financially supported by Kasetsart University Research and Development Institute (KURDI) under the Grant number of FF(KU)51.67. This research was also supported by the Postdoctoral Fellowship from Kasetsart University to JY.

Institutional Review Board Statement: The animal study protocol was approved by the Institutional Animal Care and Use Committee of Kasetsart University (protocol number ACKU65-VET-088) for animal handling and care.

Informed Consent Statement: Not applicable.

Data Availability Statement: The data presented in this study are available on request from the corresponding author. The data are not publicly available due to the anonymity granted to all participating parties.

Acknowledgments: We would like to acknowledge the Canada Excellence Research Chair in Aquatic Epidemiology.

Conflicts of Interest: The authors declare no conflicts of interest.

References

1. FAO. The State of World Fisheries and Aquaculture (SOFIA). In *Towards Blue Transformation [Internet]*; FAO: Rome, Italy, 2022. Available online: https://openknowledge.fao.org/handle/20.500.14283/cc0461en (accessed on 26 July 2024).
2. WOAH. Infection with Tilapia Lake Virus (TiLV)—A Novel Orthomyxo-like Virus. In *WOAH Disease Card [Internet]*; WOAH: Paris, France, 2022. Available online: https://www.woah.org/en/document/infection-with-tilapia-lake-virus-tilv/ (accessed on 26 July 2024).
3. Surachetpong, W.; Roy, S.R.K.; Nicholson, P. Tilapia lake virus: The story so far. *J. Fish Dis.* **2020**, *43*, 1115–1132. [CrossRef] [PubMed]
4. Jaemwimol, P.; Rawiwan, P.; Tattiyapong, P.; Kamlangdee, A.; Surachetpong, W. Susceptibility of important warm water fish species to tilapia lake virus (TiLV) infection. *Aquaculture* **2018**, *497*, 462–468. [CrossRef]
5. Phusantisampan, T.; Tattiyapong, P.; Mutrakulcharoen, P.; Sriariyanun, M.; Surachetpong, W. Rapid detection of tilapia lake virus using a one-step reverse transcription loop-mediated isothermal amplification assay. *Aquaculture* **2019**, *507*, 35–39. [CrossRef]
6. Piewbang, C.; Tattiyapong, P.; Techangamsuwan, S.; Surachetpong, W. Tilapia lake virus immunoglobulin G (TiLV IgG) antibody: Immunohistochemistry application reveals cellular tropism of TiLV infection. *Fish Shellfish Immunol.* **2021**, *116*, 115–123. [CrossRef]
7. Tattiyapong, P.; Sirikanchana, K.; Surachetpong, W. Development and validation of a reverse transcription quantitative polymerase chain reaction for tilapia lake virus detection in clinical samples and experimentally challenged fish. *J. Fish Dis.* **2018**, *41*, 255–261. [CrossRef]
8. Tattiyapong, P.; Dechavichitlead, W.; Waltzek, T.B.; Surachetpong, W. Tilapia develop protective immunity including a humoral response following exposure to tilapia lake virus. *Fish Shellfish Immunol.* **2020**, *106*, 666–674. [CrossRef]
9. Hu, H.; Zeng, W.; Wang, Y.; Bergmann, S.M.; Yin, J. Development and application of a recombinant protein-based indirect ELISA for detection of anti-tilapia lake virus IgM in sera from tilapia. *Aquaculture* **2020**, *520*, 734756. [CrossRef]
10. Lalruatfela; Bedekar, M.K.; Godavarikar, A.; Valsalam, A.; Gireesh Babu, P.; Rajendran, K.V. Molecular cloning and expression of codon-optimized segment 4 hypothetical protein (35 kDa) of tilapia lake virus (TiLV) in pET-28a(+) expression vector and development of indirect ELISA test. *Aquac. Int.* **2024**, *32*, 5997–6015. [CrossRef]
11. Bergmann, S.M.; Wang, Q.; Zeng, W.; Li, Y.; Wang, Y.; Matras, M. Validation of a KHV antibody enzyme-linked immunosorbent assay (ELISA). *J. Fish Dis.* **2017**, *40*, 1511–1527. [CrossRef]
12. Kim, H.J.; Park, J.S.; Kwon, S.R. Development of a stringent ELISA protocol to evaluate anti-viral hemorrhagic septicemia virus-specific antibodies in olive flounder *Paralichthys olivaceus* with improved specificity. *J. Microbiol.* **2015**, *53*, 481–485. [CrossRef]
13. Zeng, W.; Wang, Y.; Guo, Y.; Bergmann, S.M.; Yin, Y.; Li, Y. Development of a VP38 recombinant protein-based indirect ELISA for detection of antibodies against grass carp reovirus genotype II (iELISA for detection of antibodies against GCRV II). *J. Fish Dis.* **2018**, *41*, 1811–1819. [CrossRef] [PubMed]
14. Kim, H.J.; Oseka, N.; Nishizawa, T.; Yoshimizu, M. Protection of rainbow trout from infectious hematopoietic necrosis (IHN) by injection of infectious pancreatic necrosis virus (IPNV) or poly(I:C). *Dis. Aquat. Organ.* **2009**, *83*, 105–113. [CrossRef]
15. Nishizawa, T.; Takami, I.; Kokawa, Y.; Yoshimizu, M. Fish immunization using a synthetic double-stranded RNA Poly(I:C), an interferon inducer, offers protection against RGNNV, a fish nodavirus. *Dis. Aquat. Organ.* **2009**, *83*, 115–122. [CrossRef]
16. Takami, I.; Kwon, S.R.; Nishizawa, T.; Yoshimizu, M. Protection of Japanese flounder *Paralichthys olivaceus* from viral hemorrhagic septicemia (VHS) by Poly(I:C) immunization. *Dis. Aquat. Organ.* **2010**, *89*, 109–115. [CrossRef] [PubMed]
17. Fereidouni, S.R.; Harder, T.C.; Gaidet, N.; Ziller, M.; Hoffmann, B.; Hammoumi, S. Saving resources: Avian influenza surveillance using pooled swab samples and reduced reaction volumes in real-time RT-PCR. *J. Virol. Methods* **2012**, *186*, 119–125. [CrossRef] [PubMed]
18. Johnson, S.J.; Hick, P.M.; Robinson, A.P.; Rimmer, A.E.; Tweedie, A.; Becker, J.A. The impact of pooling samples on surveillance sensitivity for the megalocytivirus Infectious spleen and kidney necrosis virus. *Transbound. Emerg. Dis.* **2019**, *66*, 2318–2328. [CrossRef]
19. Nguyen, N.T.; Bish, E.K.; Bish, D.R. Optimal pooled testing design for prevalence estimation under resource constraints. *Omega* **2021**, *105*, 102504. [CrossRef]
20. Yamkasem, J.; Roy, S.R.K.; Khemthong, M.; Gardner, I.A.; Syrachetpong, W. Diagnostic sensitivity of pooled samples for the detection of tilapia lake virus and application to the estimation of within-farm prevalence. *Transbound. Emerg. Dis.* **2021**, *68*, 3519–3528. [CrossRef]
21. Kembou-Ringert, J.E.; Steinhagen, D.; Readman, J.; Daly, J.M.; Adamek, M. Tilapia Lake Virus Vaccine Development: A Review on the Recent Advances. *Vaccines* **2023**, *11*, 251. [CrossRef]
22. Tran, T.H.; Nguyen, V.T.H.; Bui, H.C.N.; Tran, Y.B.T.; Tran, H.T.T.; Le, T.T.T. Tilapia Lake Virus (TiLV) from Vietnam is genetically distantly related to TiLV strains from other countries. *J. Fish Dis.* **2022**, *45*, 1389–1401. [CrossRef]

23. Dong, H.; Siriroob, S.; Meemetta, W.; Santimanawong, W.; Gangnonngiw, W.; Pirarat, N. Emergence of tilapia lake virus in Thailand and an alternative semi-nested RT-PCR for detection. *Aquaculture* **2017**, *476*, 111–118. [CrossRef]
24. Eyngor, M.; Zamostiano, R.; Kembou Tsofack, J.E.; Berkowitz, A.; Bercovier, H.; Tinman, S. Identification of a novel RNA virus lethal to tilapia. *J. Clin. Microbiol.* **2014**, *52*, 4137–4146. [CrossRef]
25. Jaramillo, D.; Peeler, E.J.; Laurin, E.; Gardner, I.A.; Whittington, R.J. Serology in Finfish for Diagnosis, Surveillance, and Research: A Systematic Review. *J. Aquat. Anim. Health* **2017**, *29*, 1–14. [CrossRef]
26. Matsuyama, T.; Sano, N.; Takano, T.; Sakai, T.; Yasuike, M.; Fujiwara, A. Antibody profiling using a recombinant protein-based multiplex ELISA array accelerates recombinant vaccine development: Case study on red sea bream iridovirus as a reverse vaccinology model. *Vaccine* **2018**, *36*, 2643–2649. [CrossRef]
27. Rahnama, R.; Peyghan, R.; Reza Seyfi Abad Shapouri, M.; Rezaie, A.; Shahbazian, N. Designing an in-house ELISA to detect antibody against viral haemorrhagic septicaemia virus using recombinant N protein in Iranian farmed rainbow trout (*Oncorhynchus mykiss*). *Aquac. Res.* **2019**, *50*, 474–482. [CrossRef]
28. Watanabe, K.I.; Nishizawa, T.; Yoshimizu, M. Selection of brood stock candidates of barfin flounder using an ELISA system with recombinant protein of barfin flounder nervous necrosis virus. *Dis. Aquat. Organ.* **2000**, *41*, 219–223. [CrossRef] [PubMed]
29. Dong, B.; Zhang, G.; Zhang, X.; Chen, X.; Zhang, M.; Li, L. Development of an Indirect ELISA Based on Spike Protein to Detect Antibodies against Feline Coronavirus. *Viruses* **2021**, *13*, 2496. [CrossRef]
30. Lu, M.; Liu, Q.; Wang, X.; Zhang, J.; Zhang, X.; Shi, D. Development of an indirect ELISA for detecting porcine deltacoronavirus IgA antibodies. *Arch. Virol.* **2020**, *165*, 845–851. [CrossRef]
31. Jacobson, R.H. Validation of serological assays for diagnosis of infectious diseases. *Rev. Sci. Tech.* **1998**, *17*, 469–526. [CrossRef]
32. Baruch, J.; Suanes, A.; Piaggio, J.M.; Gil, A.D. Analytic Sensitivity of an ELISA Test on Pooled Sera Samples for Detection of Bovine Brucellosis in Eradication Stages in Uruguay. *Front. Vet. Sci.* **2020**, *7*, 178. [CrossRef]
33. Hernandez-Medrano, J.H.; Espinosa-Castillo, L.F.; Rodriguez, A.D.; Gutierrez, C.G.; Wapenaar, W. Use of pooled serum samples to assess herd disease status using commercially available ELISAs. *Trop. Anim. Health Prod.* **2021**, *53*, 507. [CrossRef]
34. Nezlin, R. *The Immunoglobulins: Structure and Function*; Academic Press: Cambridge, MA, USA, 1998; Volume 10.
35. Wang, W.; Nema, S.; Teagarden, D. Protein aggregation-pathways and influencing factors. *Int. J. Pharm.* **2010**, *390*, 89–99. [CrossRef]

viruses

MDPI

Article

Infectious Spleen and Kidney Necrosis Virus Triggers Ferroptosis in CPB Cells to Enhance Virus Replication

Qiushuang Zhang [1,2], Ouqin Chang [2], Qiang Lin [2], Hongru Liang [2], Yinjie Niu [2], Xia Luo [2], Baofu Ma [2], Ningqiu Li [2] and Xiaozhe Fu [2,*]

1. College of Fisheries and Life Science, Shanghai Ocean University, Shanghai 201306, China
2. Pearl River Fisheries Research Institute, Chinese Academy of Fishery Sciences, Key Laboratory of Fishery Drug Development, Ministry of Agriculture and Rural Affairs, Guangdong Provincial Key Laboratory of Aquatic Animal Immunology and Sustainable Aquaculture, Guangzhou 510380, China

* Correspondence: fuxiaozhe@prfri.ac.cn

Abstract: The role of ferroptosis—a novel iron-dependent programmed cell death pathway—in infectious spleen and kidney necrosis virus (ISKNV) infection remains poorly understood. Here, we demonstrate that ISKNV infection induces ferroptosis in CPB cells. Following ISKNV challenge, CPB cells exhibited hallmark morphological alterations including mitochondrial shrinkage, increased membrane density, and cristae reduction. Biochemical assays confirmed significant time-dependent elevations in ferroptosis markers: malondialdehyde (MDA, 1.7-fold), reactive oxygen species (ROS, 3.14-fold), and ferrous iron (Fe^{2+}, 1.42-fold) compared to controls ($p < 0.05$). Mechanistic studies revealed that ISKNV downregulated glutathione peroxidase 4 (GPx4) while upregulating acyl-CoA synthetase long-chain family member 4 (ACSL4), as validated by quantitative real-time PCR (qRT-PCR) and immunoblotting. Ferroptosis induction with erastin enhanced ISKNV replication, whereas inhibition with liproxstatin-1 suppressed viral yield. These findings establish that ISKNV exploits ferroptosis to facilitate its replication, and pharmacological blockade of this pathway significantly suppresses viral propagation, providing a new strategy and intervention approach for controlling ISKNV infection.

Keywords: ferroptosis; ISKNV; viral replication; GPx4; ACSL4

1. Introduction

The rapid expansion of *Siniperca chuatsi* (mandarin fish) aquaculture in China has been accompanied by increasing outbreaks of viral diseases, causing substantial economic losses. Among them, infectious spleen and kidney necrosis virus (ISKNV) is one of the most common outbreak diseases in *Siniperca chuatsi* [1]. ISKNV is a double-stranded DNA virus, belonging to the Megalocytivirus genus of the Iridoviridae family, with 111,362 bp DNA genome encoding 124 potential ORFs [2]. ISKNV can infect more than 50 fish species and has strong pathogenicity [3]. Under suitable temperatures, its mortality rate can reach 100% [4]. A Chinese perch brain (CPB) cell line infected with ISKNV showed characteristic cytopathic effects, including cell rounding, enlargement, and eventual detachment, consistent with megalocytivirus pathogenesis [5].

Ferroptosis has emerged as a distinct form of regulated cell death characterized by iron-dependent lipid peroxidation and reactive oxygen species (ROS) accumulation [6]. Unlike apoptosis, necrosis, or autophagy, ferroptosis exhibits unique morphological features, particularly mitochondrial shrinkage, increased membrane density, and a reduction in cristae [7–9]. Biochemically, ferroptosis induces iron overload, reactive oxygen species

https://doi.org/10.3390/v17050713

(ROS) accumulation, and lipid peroxidation through regulating iron metabolism, lipid metabolism, glutamine metabolism, and the system Xc-/glutathione peroxidase 4 (GPx4) pathway [10,11].

Growing evidence suggests that viral infections can disrupt cellular homeostasis by modulating metabolic pathways and cell death mechanisms [12–14]. Several viruses, including Newcastle disease virus (NDV) [12], human adenovirus type 7 (HAdV-7) [15], and novel duck orthoreovirus (N-DRV) [16], have been shown to induce ferroptosis through distinct mechanisms, such as iron accumulation, ROS generation, and GPx4 inhibition. However, whether ISKNV exploits ferroptosis to facilitate its replication remains unknown.

In this study, we demonstrate that ISKNV infection triggers ferroptosis in CPB cells by inducing mitochondrial dysfunction, lipid peroxidation, and iron accumulation, as well as downregulating GPx4 and upregulating ACSL4. Furthermore, pharmacological induction of ferroptosis enhanced ISKNV replication, whereas its inhibition suppressed viral propagation. These findings establish ferroptosis as a new mechanism in ISKNV infection and highlight its potential as a therapeutic target for ISKNV control in mandarin fish aquaculture.

2. Materials and Methods

2.1. Cells, Virus, and Main Reagents

The Chinese perch brain cell line (CPB) [17], ISKNV-QY strain [18], and ISKNV-MCP monoclonal antibody [19] were kept in our laboratory. Rabbit monoclonal anti-SLC7A11 was purchased from Abmart (Shanghai, China). Rabbit monoclonal anti-GPx4, FTH1, SLC40A1, ACSL4, and rabbit monoclonal anti-β-actin were purchased from Proteintech (Wuhan, China). Erastin and liproxstatin-1 were purchased from Macklin (Shanghai, China).

2.2. Cell Culture and Virus Infection

CPB cells were cultured in Leibovitz's L-15 medium supplemented with 8% fetal bovine serum (FBS) at 28 °C in a CO_2-free incubator and passaged at 1:3 ratios every 2–3 days upon reaching 80–90% confluence. For infection, semi-confluent monolayers (70% confluence) were washed with PBS and inoculated with ISKNV diluted 1:100 in serum-free L-15 medium for 2 h at 28 °C, after which the inoculum was replaced with maintenance medium (L-15 + 2% FBS). When 90% cytopathic effect was observed, the virus was harvested by freeze–thaw cycling of cell lysates and supernatants, with aliquots stored at −80 °C. Viral titers were determined by $TCID_{50}$ assay on CPB cells. Briefly, the viral suspension was serially diluted in L-15 medium at gradients ranging from 10^{-1} to 10^{-10}. Each dilution was added to a 96-well plate (100 µL/well), with eight replicates per dilution. Untreated cells served as the control group. CPE was monitored daily, and data were recorded for two weeks. The Reed–Muench two-way method was used to calculate the final results [20].

2.3. Integrated Multi-Omics Analysis of Ferroptosis During ISKNV Infection

Building upon our previous transcriptomic [21] and metabolomic [22] profiling of ISKNV-infected CPB cells, this study specifically investigated ferroptosis-related pathways through targeted multi-omics analysis. Gene expression dynamics of core ferroptosis regulators were quantified at 24 and 72 h post-infection (hpi) using normalized RNA-seq data reanalyzed with Origin 2021 (v9.8), with significance thresholds set at $|\log2FC| \geq 1$ and FDR < 0.05. Parallel LC-MS/MS-based metabolomics measured time-dependent changes in polyunsaturated fatty acids and phospholipids, critical substrates for lipid peroxidation in ferroptosis.

2.4. Cytotoxicity Assessment of Ferroptosis Modulators Using CCK-8 Assay

To establish the non-toxic working concentrations of ferroptosis inducer (erastin) and inhibitor (liproxstatin-1) for subsequent experiments, CPB cell viability was evaluated using the Cell Counting Kit-8 (CCK-8) assay. Briefly, cells were seeded in 96-well plates (2×10^5 cells/well in 100 μL of L-15 medium with 8% FBS) and allowed to adhere for 24 h at 28 °C under CO_2-free conditions. Cells were then treated with serial dilutions (1–500 μM) of erastin, liproxstatin-1, or equivalent concentrations of the vehicle control (DMSO) for 72 h. Following treatment, cells were incubated with CCK-8 reagent (10 μL) diluted in 100 μL of L-15 medium with 2% FBS for 4 h at 28 °C. Absorbance was measured at 450 nm using a microplate reader (BioTek Synergy H1, BioTek, VT, USA).

2.5. Ultrastructural Analysis by Transmission Electron Microscopy (TEM)

For TEM examination of ISKNV-induced cellular alterations, CPB cells (5×10^5 cells/mL) were seeded in T_{25} flasks and infected with ISKNV at an MOI of 100. At 24, 48, and 72 hpi, cells were fixed with 2.5% glutaraldehyde in 0.1 M phosphate buffer (pH 7.4) for 24 h at 4 °C and then post-fixed in 0.1 M phosphate buffer containing 1% osmium tetroxide for 1 h. Ultrathin sections were stained with uranyl acetate–lead citrate and examined by a Philips CM10 electron microscope (Philips, Eindhoven, The Netherlands).

2.6. Fe^{2+}, ROS, and MDA Detection

To assess ferroptosis induction during ISKNV infection, CPB cells (5×10^5 cells/mL) were seeded in 6-well plates and infected with ISKNV at an MOI of 100. At 24, 48, and 72 hpi, cells treated with erastin or liproxstatin-1 were harvested for quantification of three key ferroptosis markers. The Fe^{2+} content of the cells was detected by laser scanning confocal microscopy and a microplate reader using the fluorescent probe FerroOrange (Elabscience, E-BC-F101, Wuhan, China). The ROS content of the cells was detected by laser scanning confocal microscopy and a microplate reader using the fluorescent probe DCFH-DA (Beyotime, S0033S, Shanghai, China). MDA level was detected by a microplate reader using the Cellular Malondialdehyde Determination Kit (Jiancheng Bioengineering Institute, A003-4-1, Nanjing, China).

2.7. RNA Extraction, Reverse Transcription, and Quantitative Real-Time PCR (qRT-PCR)

To assess ferroptosis gene expression during ISKNV infection, CPB cells (5×10^5 cells/mL) were seeded in 12-well plates and infected with ISKNV at an MOI of 100. At 72 hpi, total RNA was extracted from ISKNV-infected and control CPB cells using the FastPure Complex Tissue/Cell Total RNA Isolation Kit (Vazyme, RC112-01, Nanjing, China). RNA quality and concentration were verified by spectrophotometry (A260/A280 ratio > 1.8) and agarose gel electrophoresis. First-strand cDNA was synthesized from 5 μg of total RNA using One-Step gDNA Removal and cDNA Synthesis SuperMix (TransGen Biotech, AT311-03, Beijing, China). qRT-PCR was performed in triplicate using SYBR Green Master Mix (Vazyme, Q711-02, Nanjing, China) on a QuantStudio 6 Flex Real-Time PCR System (Applied Biosystems, New York, USA) under the following conditions: 95 °C for 30 s; 40 cycles of 95 °C for 5 s; and 60 °C for 30 s. The reaction volume of SYBR Green was 20 μL, including 10 μL 2× SYBR Premix Enzyme containing ROX, 0.5 μL each of forward and reverse primer (10 μM), and 7 μL of ddH_2O and 2 μL of cDNA. The *18S rDNA* served as the endogenous control. Relative gene expression was calculated using the $2^{-\Delta\Delta Ct}$ method [23], with normalization to both *18S rDNA* and mock-infected controls. The used primers are listed in Table 1.

Table 1. Primers for detection of genes.

Primer Name	Sequence (5′-3′)	Amplicon Size (bp)	Accession Number
18S-F 18S-R	CATTCGTATTGTGCCGCTAGA CAAATGCTTTCGCTTTGGTC	120	XR_006376550.1
SLC7A11-F SLC7A11-R	GAGGAGGTAGATAACCCTGAACGG CTCCTCTGCTGACATCACAGTG	120	XM_044205657.1
GPX4-F GPX4-R	CAACAGATGATCCCAGCGTGGT CACGCACACCAATACCCTGAAG	120	XM_044218327.1
FTH1-F FTH1-R	CGCTGTGACGCTGATAATTATCC CTGCAGTTGATTGACAACTAGC	126	XM_044194482.1
SLC40A1-F SLC40A1-R	CTAACCCACTCTGAGATTGTACGG CTGGTACAGTTCATGTGGTGCTG	125	NC_058053.1
ACSL4-F ACSL4-R	GCGTAAGCCTCAGCTATTCCAG GGGAACAAACAGCGTTTCTTCAAC	132	XM_044206073.1

2.8. DNA Extraction and qPCR

To assess viral copies, CPB cells (5×10^5 cells/mL) were seeded in 12-well plates, then treated with erastin or liproxstatin-1 either before or after ISKNV infection at an MOI of 100. Viral DNA was extracted from supernatants using proteinase K lysis (160 μg/mL final concentration) with incubation at 56 °C for 2 h followed by heat inactivation (95 °C, 10 min) and centrifugation (12,000× *g*, 3 min), while intracellular DNA was isolated using the FastPure® Cell/Tissue DNA Isolation Mini Kit (Vazyme, DC102-01) according to the manufacturer's protocol. TaqMan qPCR was performed on a QuantStudio 6 Flex system. The reaction system contained 10 μL of 2×Premix, 0.4 μL of primers, 0.4 μL of probe, 0.4 μL of ROX, 6.4 μL of ddH2O, and 2 μL of DNA. The procedure was carried out at 94 °C for 1 min, 94 °C for 10 s, and 60 °C for 30 s, for 40 cycles. The primers and probe are listed in Table 2. The copy numbers of ISKNV were calculated through comparison with the standard curve ($Y = -3.314\lg X + 41.48$; Y refers to the Ct value and X refers to the copy number), as described previously.

Table 2. The primers for detection of virus copy number.

Primer Name	Sequence (5′-3′)	Amplicon Size	Reference
ISKNV-F ISKNV-R ISKNV-Probe	CGAGGCCACATCCAACATC CGCCTTTAACGTGGGATATATTG CACCAAACTGACCGCGGACTCGT	85 (bp)	Ma B. F. [24]

2.9. Western Blot

Cells were collected and lysed in RIPA buffer with 1 mM PMSF. Protein samples were separated by 12.5% SDS-PAGE (80 V for 20 min, then 120 V for 60 min) and transferred to PVDF membranes using a semi-dry transfer system. Membranes were activated in methanol for 15 s, rinsed in deionized water for 2 min, and equilibrated in transfer buffer (25 mM Tris, 192 mM glycine, 20% methanol) for 1 min. Proteins were transferred at 150 V for 55 min with the membrane positioned above the gel to ensure proper orientation. Following transfer, membranes were blocked with 3% BSA in TBST (Tris-buffered saline with 0.1% Tween-20) for 3 h at room temperature. Primary antibody (SLC7A11 (1:2000), GPx4 (1:2000), FTH1 (1:2000), SLC40A1 (1:2000), ACSL4 (1:2000), and ISKNV-MCP (1:500)) incubation was performed overnight at 4 °C in blocking buffer, followed by three 5 min TBST washes. Membranes were then incubated with HRP-conjugated secondary antibody (1:10,000 dilution) for 1 h at room temperature. After additional washing (3 × 5 min

TBST), protein bands were visualized using an enhanced chemiluminescence (ECL) substrate (Beyotime, Shanghai, China) and imaged with a ChemiDoc MP system (Bio-Rad, Shanghai, China).

2.10. Statistical Analysis

All data were analyzed using Origin 2021 (OriginLab Corporation), ImageJ-win64 (National Institutes of Health), and GraphPad Prism 8.0 (GraphPad Software). Group differences were compared using Student's *t*-tests (for two groups), one-way ANOVA with Dunnett's post hoc test (for multi-group comparisons against a single control), or two-way ANOVA with Sidak's post hoc test (for multi-factor analyses). Statistical significance thresholds were defined as * $p < 0.05$, ** $p < 0.01$, and *** $p < 0.001$, with $p > 0.05$ considered not significant (NS).

3. Results

3.1. ISKNV Induces Ferroptosis Cell Death

Although erastin is a well-characterized ferroptosis inducer in mammalian systems, its capacity to trigger ferroptosis in CPB cells remained unverified. Figure 1A demonstrates that erastin concentrations below 10 μM maintained CPB cell viability above 90%. Further comparative analysis of three erastin concentrations (5, 10, and 20 μM) administered via different treatment revealed distinct cellular responses. Pre-treatment 2 h or 24 h prior to assay showed no significant morphological alterations compared to untreated controls throughout the 24–72 h observation period (Figure 1B,C). In contrast, post-treatment, where CPB cells were continuously exposed to erastin throughout the entire duration of the experiment, the 5 μM group maintained normal morphology, and 20 μM erastin induced rapid cell detachment within 24 h, while 10 μM erastin elicited progressive cellular crumpling—a characteristic ferroptotic morphology—indicating post-treatment with 10 μM erastin was the optimal condition (Figure 1D). Upon the optimization of erastin treatment, the characteristic ferroptotic morphology, including mitochondrial cristae reduction, increased membrane density, and Fe^{2+} accumulation, was observed by TEM (Figure 1E) and confocal microscopy (Figure 1F). These results demonstrated that post-treatment with 10 μM erastin can induce ferroptosis in CPB cells.

Subsequently, ISKNV infection-induced ferroptosis in CPB cells was investigated. Firstly, ISKNV infection induces cell rounding and enlargement at 24–48 hpi followed by widespread detachment at 72 hpi, accompanied by progressive viability reduction (Figure 2). Furthermore, TEM observation revealed that mitochondrial shrinkage, increased membrane density, and cristae reduction were evident at 24 hpi; the cristae further decreased or disappeared at 48 hpi, and mitochondrial cristae loss with membrane rupture was observed at 72 hpi (Figure 3A). Confocal microscopy was used to observe the fluorescence of Fe^{2+} and ROS after ISKNV infection at 24 h, 48 h, and 72 h. The results showed ISKNV infection-dependent accumulation of Fe^{2+} (Figure 3B) and ROS (Figure 3D), and the average fluorescence intensity in the ISKNV group was higher than in the control group (Figure 3C,E). The levels of Fe^{2+}, ROS, and MDA post-ISKNV infection for 24 h, 48 h, and 72 h were significantly higher than in the control group, with Fe^{2+} increasing from 1.38-fold at 24 hpi to 1.42-fold at 72 hpi (Figure 3F), ROS elevating from 1.26-fold (24 hpi) to 3.14-fold at 72 hpi (Figure 3G), and MDA rising 1.7-fold at 72 hpi (Figure 3H). The above results suggest that ISKNV infection induces ferroptosis in CPB cells.

3.2. Ferroptosis Increases ISKNV Replication in CPB Cells

Liproxstatin-1 is a ferroptosis inhibitor, but its capacity to suppress ferroptosis in CPB cells remained unknown. Figure 4A demonstrates that liproxstatin-1 concentrations below

20 μM maintained CPB cell viability above 90%. Further comparative analysis of three liproxstatin-1 concentrations (5, 10, and 20 μM) administered via different treatments revealed distinct cellular responses. Pre-treatment for 24 h or post-treatment for 72 h induced significant cytotoxicity, such as cell aggregation and apoptotic morphology (Figure 4C,D). In contrast, pre-treatment for 2 h with 5–10 μM liproxstatin-1 maintained normal cellular morphology, while 20 μM treatment caused initial signs of cellular crumpling (Figure 4B). Therefore, 10 μM liproxstatin-1 with 2 h pre-treatment was selected for subsequent experiments. To further evaluate the inhibitory effect of liproxstatin-1 on ferroptosis, CPB cells were pre-treated with 10 μM liproxstatin-1 for 2 h prior to ISKNV infection (MOI = 100). The result showed liproxstatin-1 pre-treatment significantly decreased ISKNV-triggered ROS accumulation (Figure 4E). The above results validated the efficacy of the optimized liproxstatin-1 treatment protocol (10 μM, 2 h pre-treatment) in blocking ferroptotic signaling.

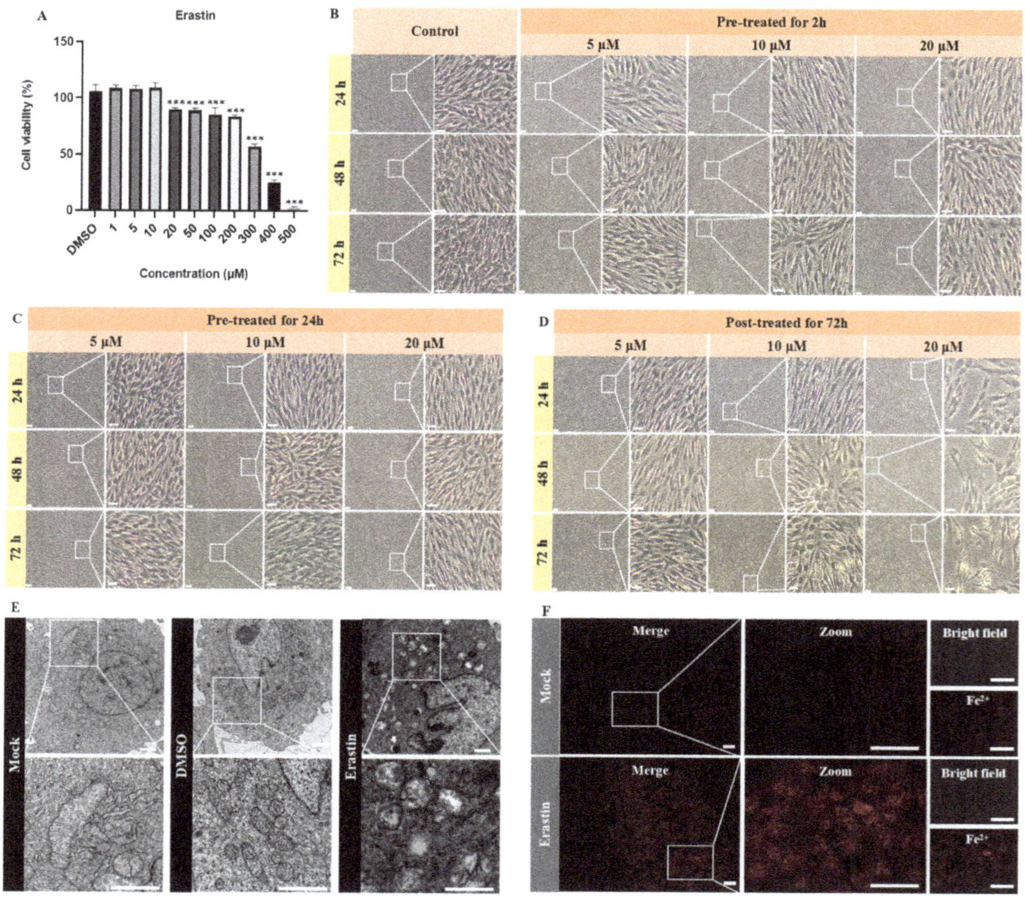

Figure 1. Ferroptosis induced with erastin in CPB cells. (**A**) Cytotoxicity of erastin in CPB cells, *** $p < 0.001$. (**B–D**) Morphological observation of CPB cells treated with erastin. CPB cells were pre-treated with erastin for 2 h and 24 h, or post-treated for 72 h with 5 μM, 10 μM, or 20 μM erastin, and then CPB morphology was observed at 24 h, 48 h, and 72 h. Scale bars = 100 μm. (**E**) Mitochondria morphological observation of CPB cells treated with erastin (10 μM, 48 h) by transmission electron microscopy. Scale bars = 1 μm. (**F**) Detection of Fe^{2+} levels in CPB cells after treatment with erastin (10 μM) for 48 h. Scale bars = 20 μm.

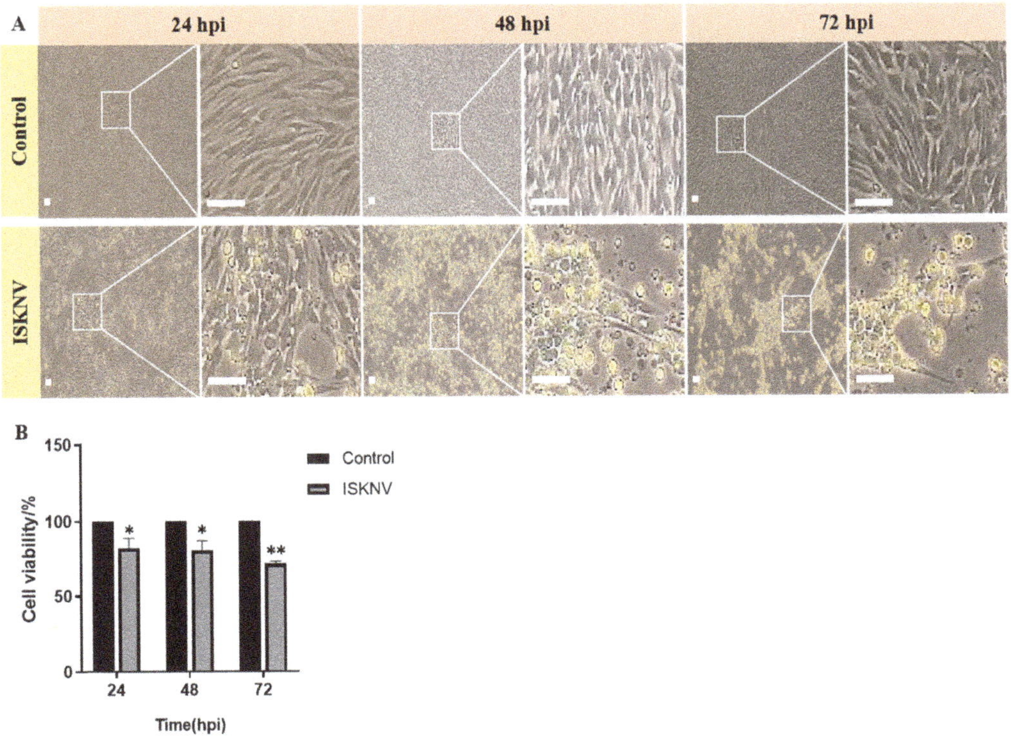

Figure 2. Changes in CPB cells in response to ISKNV. (**A**) Morphological changes in CPB cells infected with ISKNV (24–72 h). Scale bars = 100 μm. (**B**) Relative levels of cell viability post-ISKNV infection (24–72 h). * $p < 0.05$, ** $p < 0.01$.

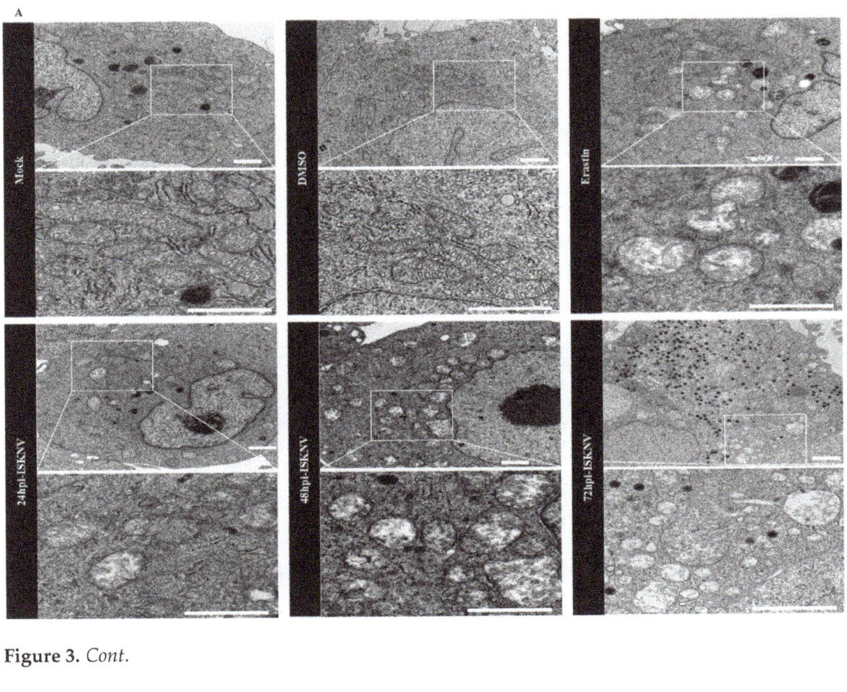

Figure 3. *Cont.*

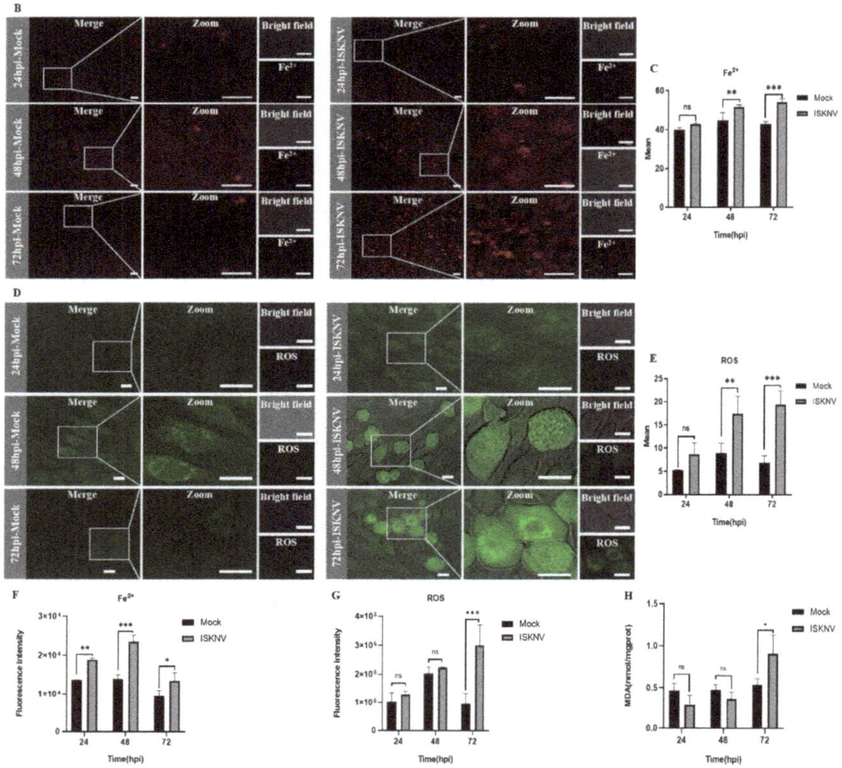

Figure 3. ISKNV infection induces ferroptosis in CPB cells. (**A**) Transmission electron microscopy of CPB cells treated with DMSO (72 h), erastin (10 μmol/L, 72 h), and ISKNV (100 MOI, 24 h, 48 h, 72 h). Scale bars = 1 μm. (**B**) Analysis of Fe^{2+} levels in CPB cells after treatment with ISKNV (100 MOI) using the fluorescent probe FerroOrange by laser scanning confocal microscopy. Scale bars = 20 μm. (**C**) Quantitative analysis of the mean fluorescence intensity of (**B**) using Image J. (**D**) Analysis of intracellular ROS levels using DCFH-DA staining, and laser scanning confocal microscopy of CPB cells treated with ISKNV (100 MOI) for 24–72 h. Scale bars = 10 μm. (**E**) Quantitative analysis of the mean fluorescence intensity of (**D**) using Image J. (**F–H**) Detection of Fe^{2+}, ROS, and MDA levels in cell lysates treated with ISKNV (100 MOI) for 24–72 h by microplate reader. * $p < 0.05$, ** $p < 0.01$, and *** $p < 0.001$, with $p > 0.05$ considered not significant (ns).

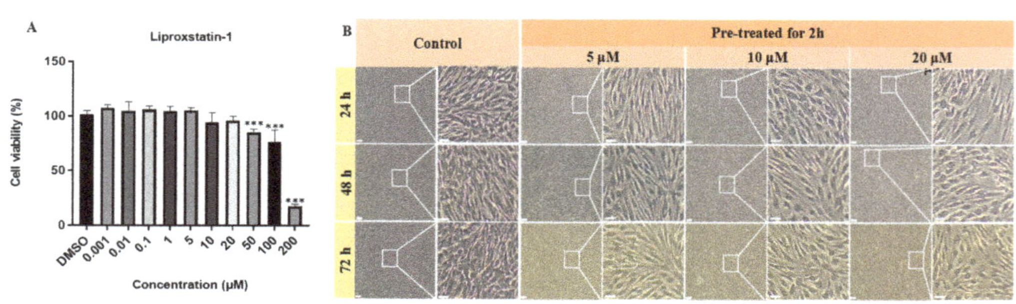

Figure 4. *Cont.*

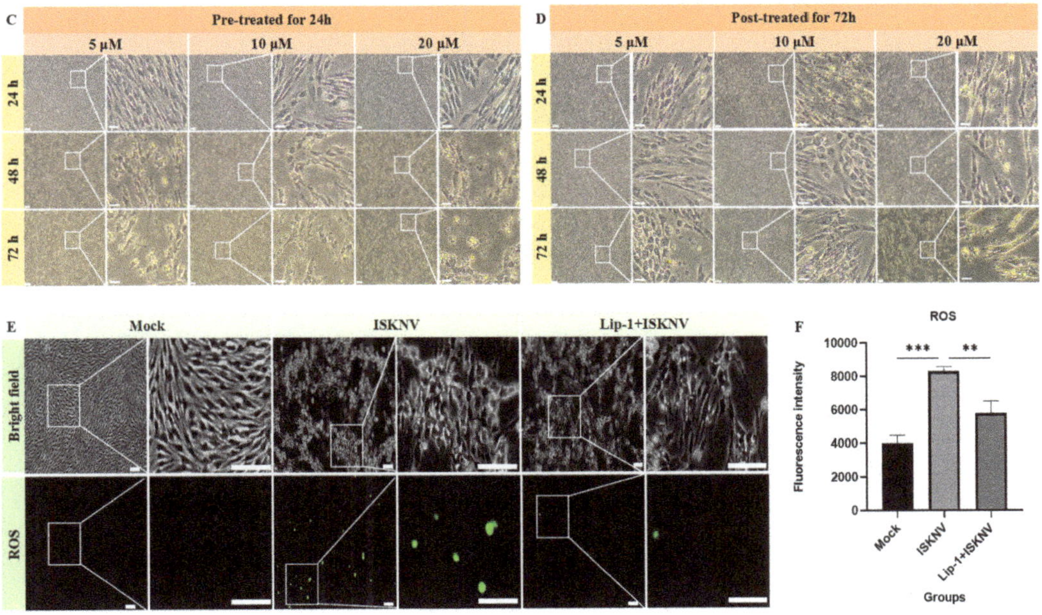

Figure 4. Liproxstatin-1 inhibited ferroptosis in CPB cells. (**A**) Cytotoxicity of liproxstatin-1 in CPB cells. (**B–D**) Morphological observation of CPB cells treated with liproxstatin-1. CPB cells were pre-treated with liproxstatin-1 for 2 h and 24 h, or post-treated with liproxstatin-1 for 72 h at 5 μm, 10 μm, and 20 μm concentration, and then CPB morphology was observed at 24 h, 48 h, and 72 h. Scale bars = 100 μm. (**E**,**F**) Detection of intracellular ROS levels treated with DMSO, ISKNV (100 MOI), and liproxstatin-1 + ISKNV for 24 h, using DCFDA staining and a microplate reader. Scale bars = 100 μm. ** $p < 0.01$, *** $p < 0.001$.

To evaluate the potential effects of erastin and liproxstatin-1 on ISKNV entry processes, three treatment methods were used, including pre-treatment (2 h prior to infection), co-treatment (with viral inoculation), and post-treatment (2 h post-adsorption) (Figure 5A). qPCR results indicated that ISKNV DNA copy numbers following intracellular or extracellular post-treatment with erastin and pre-treatment with liproxstatin-1 had no significant difference compared to untreated controls (Figure 5B,C). These results demonstrated that post-treatment with erastin and pre-treatment with liproxstatin-1 had no effect on ISKNV adsorption and invasion and were used for the subsequent experiments.

Subsequently, the relationship between ferroptosis and ISKNV replication was assessed by qRT-PCR and Western blot analyses. The results demonstrated that erastin treatment significantly enhanced ISKNV replication, as evidenced by 6.4-fold and 2-fold increases in intracellular and extracellular viral DNA copies, respectively ($p < 0.01$), accompanied by a 1.2-fold upregulation of major capsid protein (MCP) expression compared to untreated infected controls (Figure 6A). Conversely, liproxstatin-1 treatment reduced intracellular viral DNA load by 30.6% ($p < 0.01$) and extracellular virion production by 77.7% ($p < 0.01$), with a corresponding 48.1% decrease in MCP expression (Figure 6B). These results suggested that viral propagation is significantly enhanced by ferroptosis while being suppressed by ferroptosis inhibitors.

3.3. ISKNV Triggers Ferroptosis Through Suppressing GPx4 and Promoting ACSL4

Transcriptomic profiling of ISKNV-infected CPB cells revealed dynamic regulation of ferroptosis-associated pathways at 24 hpi and 72 hpi, including iron metabolism genes

(TFRC, STEAP3, ZIP8, ZIP14, DMT1, and FTH1), xCT-/GPx4 axis genes (SLC7A11, SLC3A2, and GPx4), and lipid metabolism genes (ACSL4, ALOX12, and LPCAT3), as well as ATG5/7-NCOA4 pathway components (ATG5, ATG7, and NCOA4) (Figure 7A). Metabolomic analysis identified time-dependent lipid alterations, with polyunsaturated fatty acids (ALA, AA, C9PA, EPA, LA, and TVA), phospholipid synthesis precursors (sn-G3P and GPC) and associated metabolites (G3P, betaine, pantothenate, CDP-choline, and PC) suggesting staged lipid peroxidation dynamics (Figure 7B). Subsequently, qRT-PCR validation confirmed marked downregulation of *SLC7A11*, *GPx4*, and *FTH1* at 72 hpi, while *SLC40A1* and *ACSL4* showed non-significant upward trends (Figure 7C–G). Western blot analysis revealed corresponding protein-level changes to confirm that GPx4 was significantly suppressed, ACSL4 induced, and SLC7A11 moderately upregulated, with FTH1 and SLC40A1 remaining stable (Figure 7H). These results suggest that GPx4 and ACSL4 may play a key role in ISKNV-induced ferroptosis in CPB cells.

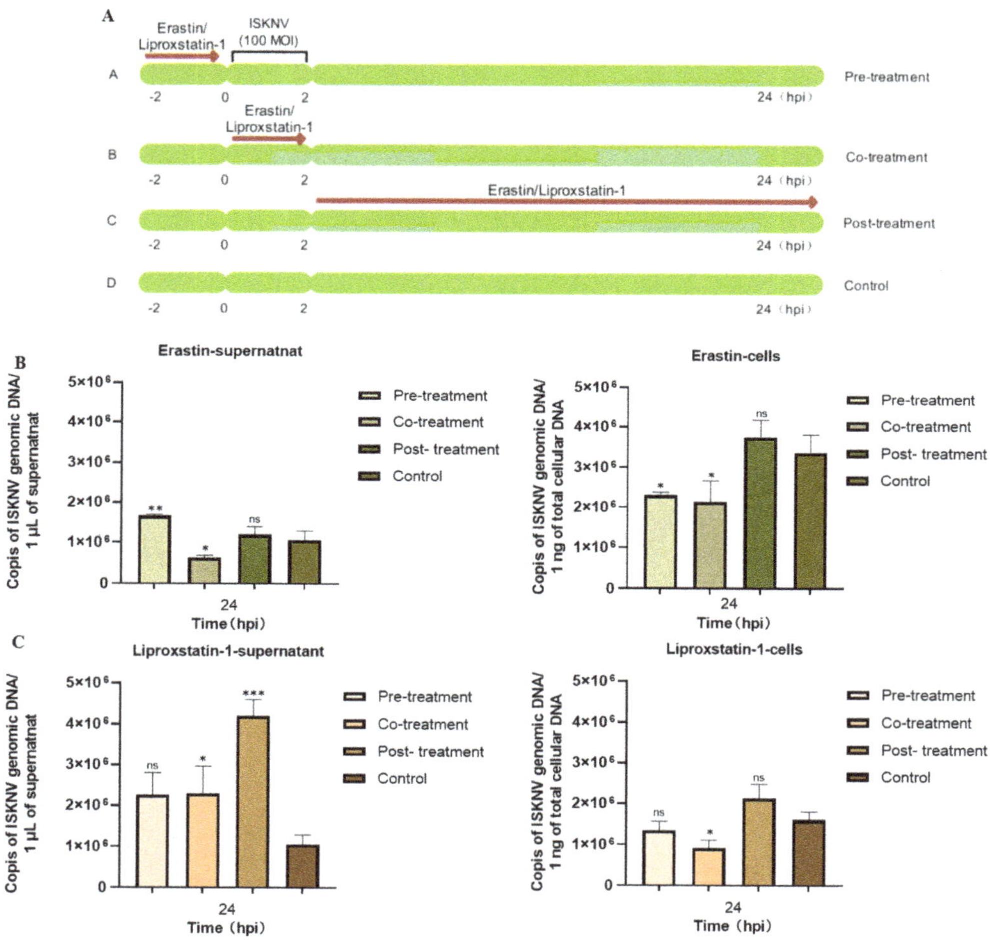

Figure 5. Detection of viral genomic DNA copies in CPB cells and supernatants after different treatments with erastin or liproxstatin-1. (**A**) Schematic diagram of erastin and liproxstatin-1 treatment methods. (**B**) CPB cells were pre-treated before ISKNV inoculation, co-treated with ISKNV inoculation, or post-treated after ISKNV inoculation with erastin (10 μmol/L). The DNA copies in the cells and supernatant

were detected by qRT-PCR at 24 hpi. (**C**) CPB cells were pre-treated before ISKNV inoculation, co-treated with ISKNV inoculation, or post-treated after ISKNV inoculation with liproxstatin-1 (10 μmol/L). The DNA copies in the cells and supernatant were detected by qRT-PCR at 24 hpi. * $p < 0.05$. ** $p < 0.01$, and *** $p < 0.001$, with $p > 0.05$ considered not significant (ns).

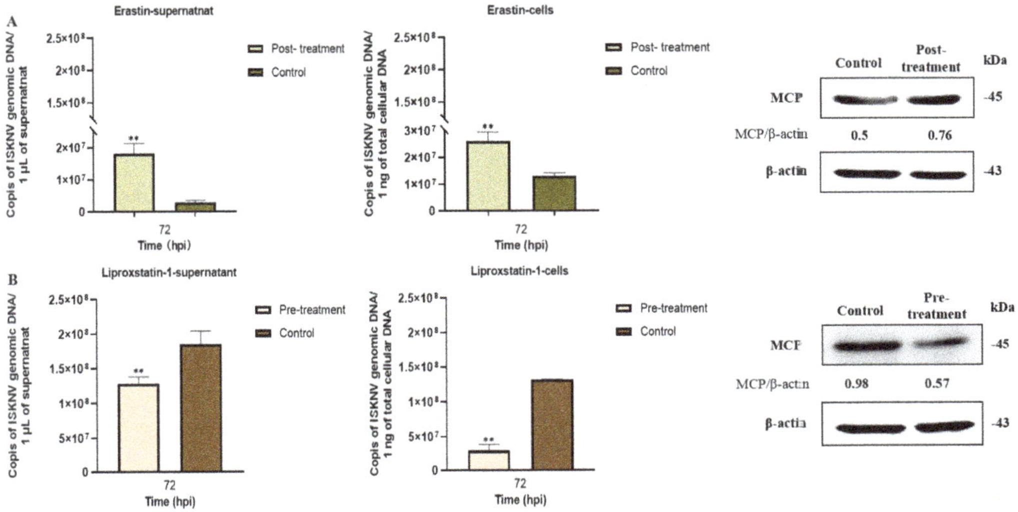

Figure 6. Effects of erastin and liproxstatin-1 on ISKNV replication. (**A**) Detection of viral genomic DNA copies and ISKNV MCP level after post-treatment with erastin (10 μM) at 72 hpi by qRT-PCR or Western blot. (**B**) Detection of viral genomic DNA copies and ISKNV MCP level after pre-treatment with liproxstatin-1 (10 μM) at 72 hpi by qRT-PCR or Western blot. ** $p < 0.01$.

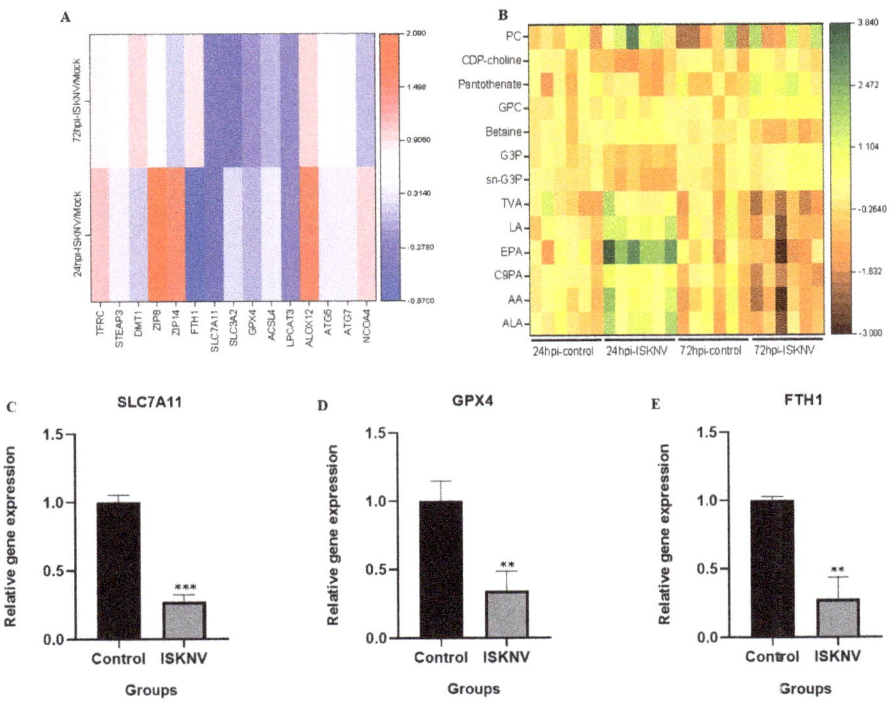

Figure 7. *Cont.*

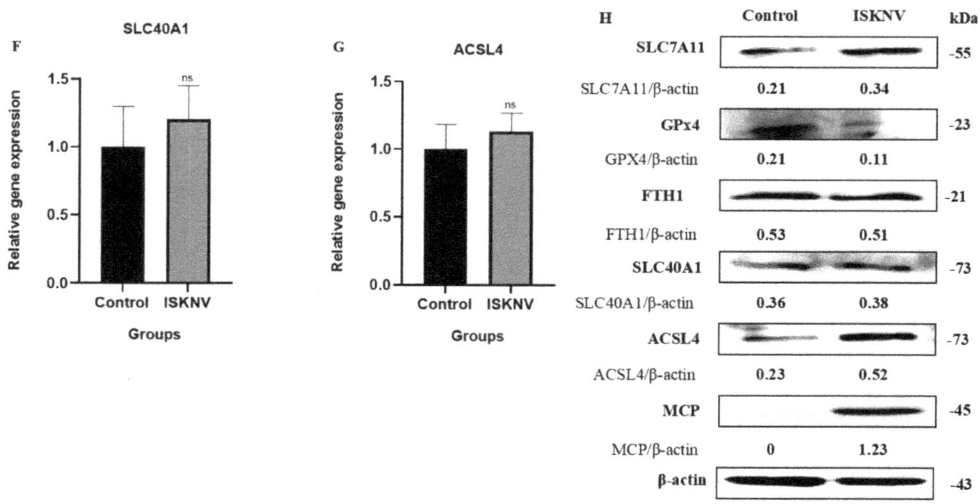

Figure 7. ISKNV induces ferroptosis by suppressing GPx4 and promoting ACSL4. (**A**) Heat map of the mRNA level involved in ferroptosis in CPB cells infected with ISKNV at 24 hpi and 72 hpi. Each column represents one sample. (Transferrin receptor (TFRC), Six-Transmembrane Epithelial Antigen of Prostate 3 (STEAP3), Zinc- and iron-related protein 8 (ZIP8), Zrt- and Irt-like protein 14 (ZIP14), Divalent metal transporter 1 (DMT1), Ferritin heavy chain (FTH1), Solute carrier family 7 member 11 (SLC7A11), Solute carrier family 3 member 2 (SLC3A2), Glutathione peroxidase 4 (GPx4), Acyl-CoA synthetase long-chain family member 4 (ACSL4), Arachidonate 12-Lipoxygenase (ALOX12), Lysophosphatidylcholine acyltransferase 3 (LPCAT3), Autophagy-related protein 5 (ATG5), Autophagy-related protein 7 (ATG7), and Nuclear receptor coactivator 4 (NCOA4).) (**B**) Heat map of differentially expressed fatty acids and phospholipids in CPB cells infected with ISKNV at 24 hpi and 72 hpi. Each column represents one sample. (Alpha-Linolenic acid (ALA), Arachidonic Acid (AA), Cis-9-Palmitoleic acid (C9PA), Eicosapentaenoic acid (EPA), Linoleic acid (LA), Trans-Vaccenic acid (TVA), sn-Glycerol 3-phosphoethanolamine (sn-G3P), Glycerophosphocholine (GPC), Glycerol 3-phosphate (G3P), Betaine, Pantothenate, Cytidine 5′-diphosphocholine (CDP-choline), and Phosphatidylcholine (PC)). (**C–G**) The mRNA expression of *SLC7A11*, *GPx4*, *FTH1*, *SLC40A1*, and *ACSL4* in CPB cells infected with ISKNV at 72 hpi, respectively. ** $p < 0.01$, and *** $p < 0.001$, with $p > 0.05$ considered not significant (ns). (**H**) The protein expression of SLC7A11, GPx4, FTH1, SLC40A1, ACSL4, and MCP in CPB cells infected with ISKNV at 72 hpi, respectively.

4. Discussion

Ferroptosis is a recently discovered form of programmed cell death, and numerous studies have shown that ferroptosis plays a role in the development and progression of various diseases, including neurological disorders and cardiovascular diseases [25]. Notably, recent advancements have revealed viruses exploit ferroptotic pathways to facilitate their replication or trigger ferroptosis as a cytopathic mechanism; for example, hepatitis B virus protein X stabilizes enhancer of zeste homolog 2 (EZH2) and promotes trimethylation of H3K27, which inhibits SLC7A11 and induces ferroptosis in acute liver failure [26], while herpes simplex virus (HSV) leads to enhanced oxidative stress, decreases GSH concentration in host cells, and induces lipid peroxidation to produce highly reactive membrane lipid hydroperoxides [27]. The hallmarks of ferroptosis include mitochondrial damage, iron homeostasis imbalance, iron-dependent ROS production, and lipid peroxidation [6,28]. Our findings demonstrated that ISKNV infection in CPB cells induced characteristic morphological and biochemical alterations consistent with ferroptotic cell death. The observed mitochondrial pathology—including marked shrinkage, increased membrane density, and progressive cristae reduction—represents the hallmark ultrastructural signature of ferroptosis. The biochemical profile of ISKNV-infected cells exhibited the accumulation of Fe^{2+}, ROS, and MDA. These morphological and biochemical alterations have been previously documented in other viral systems [29,30].

Some studies have demonstrated that viruses exploit ferroptotic pathways to facilitate their replication. Xia et al. revealed that the ferroptosis inhibitor inhibited murine hepatitis virus strain A59 (MHV-A59) propagation, inflammatory cytokine release, and cell syncytia formation in primary macrophages [31]. Similarly, Cheng et al. reported that H1N1 swine influenza virus (SIV) induced ferroptosis in host cells, and that ferroptosis inhibition using ferrostatin-1 not only reduced viral replication but also mitigated inflammatory responses [32]. In the present study, we also found ferroptosis inhibition with liproxstatin-1 suppressed ISKNV yield. Furthermore, it was interesting that ferroptosis induction with erastin enhanced ISKNV replication. Erastin induced ferroptosis through interfering with multiple sites, including voltage-dependent anion-selective channels, cystine/glutamate exchange transporters, and glutathione peroxidase 4 [33]. Cystine/glutamate exchange transporters, designated to the system *Xc-*, uptake cystine and release Glu in a 1:1 ratio [34]. Our previous study revealed ISKNV replication required glutamine [22]. Therefore, we hypothesized that the addition of an erastin-inhibiting system *Xc-* led to an increase in intracellular glutamate levels, thereby facilitating ISKNV replication.

Japanese encephalitis virus (JEV) infection led to neuronal ferroptosis by inhibiting GPx4 function and upregulating ACSL4 expression, resulting in neuronal damage and inflammation, and inhibiting ferroptosis was shown to reduce the mortality rate and alleviate neuroinflammation and brain damage in a mouse model [29]. GPx4 is the only enzyme in the cell that is capable of directly reducing lipid peroxides and plays a core role in inhibiting lipid peroxidation and preventing ferroptosis [35]. ACSL4 specifically catalyzes the conversion of polyunsaturated fatty acids such as AA and AdA into their corresponding acyl-CoA derivatives, providing substrates for phospholipid synthesis. This, in turn, promotes the accumulation of phospholipids containing polyunsaturated fatty acids (PUFA-PLs) on the cell membrane [36]. PUFA-PLs are the primary substrates for lipid peroxidation in ferroptosis, and their accumulation on the cell membrane makes the membrane more susceptible to oxidative stress [37]. Our results showed that after 72 h of ISKNV infection in CPB cells, GPx4 showed a significant reduction at both the transcription and protein levels, and ACSL4 was significantly upregulated. In the metabolomic analysis, polyunsaturated fatty acids and phospholipids were increased in ISKNV-infected CPB cells. These findings demonstrated that ISKNV infection induced ferroptosis in CPB cells through downregulation of the antioxidant enzyme GPx4 and upregulation of ACSL4.

In conclusion, it is the first study to demonstrate that ISKNV induces ferroptosis in CPB cells and ferroptosis promotes ISKNV replication. Notably, ISKNV infection induced ferroptosis through downregulation of the antioxidant enzyme GPx4 and upregulation of ACSL4. These findings significantly advance our understanding of ISKNV pathogenesis.

Author Contributions: Conceptualization, X.F.; methodology, Q.Z., O.C., Q.L., H.L. and Y.N.; software, Q.Z.; validation, X.L. and B.M. formal analysis, Q.Z. and X.F.; investigation, Q.Z. and X.F.; resources, X.F and N.L.; data curation, Q.Z.; writing—original draft preparation, Q.Z.; writing—review and editing, X.F.; visualization, X.F.; supervision, X.F.; project administration, X.F.; funding acquisition, X.F. and N.L. All authors have read and agreed to the published version of the manuscript.

Funding: This work was supported by National Key Research and Development Program of China (2023YFD2400701).

Institutional Review Board Statement: Not applicable.

Informed Consent Statement: Not applicable.

Data Availability Statement: The data that support the findings of this study are available from the corresponding author upon reasonable request.

Conflicts of Interest: The authors declare no conflict of interest.

References

1. Wu, S.Q.; Li, X.H.; Pan, H.J.; Huang, Z.B. Research on the Pathogen of the Outbreak-Ineective Disease of Siniperca Chuastsi. *J. Fish. China* **1997**, *21*, 56–60.
2. He, J.G.; Deng, M.; Weng, S.P.; Li, Z.; Zhou, S.Y.; Long, Q.X.; Wang, X.Z.; Chan, S.M. Complete Genome Analysis of the Mandarin Fish Infectious Spleen and Kidney Necrosis Iridovirus. *Virology* **2001**, *291*, 126–139. [CrossRef] [PubMed]
3. Wang, Y.Q.; Lü, L.; Weng, S.P.; Huang, J.N.; Chan, S.M.; He, J.G. Molecular epidemiology and phylogenetic analysis of a marine fish infectious spleen and kidney necrosis virus-like (ISKNV-like) virus. *Arch. Virol.* **2007**, *152*, 763–773. [CrossRef] [PubMed]
4. Zeng, K.; He, J.G.; Weng, S.P.; Huang, Z.J. Transmission, Host Range, Temperature Sensibility of InfectiousSpleen and Kidney Necrosis(ISKN)Virus from Siniperca chuatsi. *Virol. Sin.* **1999**, *14*, 353–357.
5. Fu, X.Z. Development of a Chinese Perch Cell Line Susceptible to ISKNV and Mechanism of Virus Repication Relying on Glutamine. Ph.D. Thesis, Northwest A&F University, Xianyang, China, 2017.
6. Dixon, S.J.; Lemberg, K.M.; Lamprecht, M.R.; Skouta, R.; Zaitsev, E.M.; Gleason, C.E.; Patel, D.N.; Bauer, A.J.; Cantley, A.M.; Yang, W.S.; et al. Ferroptosis: An Iron-Dependent Form of Nonapoptotic Cell Death. *Cell* **2012**, *149*, 1060–1072. [CrossRef]
7. Dixon, S.J.; Patel, D.N.; Welsch, M.; Skouta, R.; Lee, E.D.; Hayano, M.; Thomas, A.G.; Gleason, C.E.; Tatonetti, N.P.; Slusher, B.S.; et al. Pharmacological inhibition of cystine–glutamate exchange induces endoplasmic reticulum stress and ferroptosis. *eLife* **2014**, *3*, e02523. [CrossRef]
8. Yang, W.S.; SriRamaratnam, R.; Welsch, M.E.; Shimada, K.; Skouta, R.; Viswanathan, V.S.; Cheah, J.H.; Clemons, P.A.; Shamji, A.F.; Clish, C.B.; et al. Regulation of Ferroptotic Cancer Cell Death by GPX4. *Cell* **2014**, *156*, 317–331. [CrossRef]
9. Yu, H.T.; Guo, P.Y.; Xie, X.Z.; Wang, Y.; Chen, G. Ferroptosis, a new form of cell death, and its relationships with tumourous diseases. *J. Cell. Mol. Med.* **2017**, *21*, 648–657. [CrossRef]
10. Latunde-Dada, G.O. Ferroptosis: Role of lipid peroxidation, iron and ferritinophagy. *Biochim. Biophys. Acta Gen. Subj.* **2017**, *1861*, 1893–1900. [CrossRef]
11. Jiang, X.; Stockwell, B.R.; Conrad, M. Ferroptosis: Mechanisms, biology, and role in disease. *Nat. Rev. Mol. Cell Biol.* **2021**, *22*, 266–282. [CrossRef]
12. Kan, X.J.; Yin, Y.C.; Song, C.P.; Tan, L.; Qiu, X.S.; Liao, Y.; Liu, W.W.; Meng, S.S.; Sun, Y.J.; Ding, C. Newcastle-disease-virus-induced ferroptosis through nutrient deprivation and ferritinophagy in tumor cells. *iScience* **2021**, *24*, 102837. [CrossRef] [PubMed]
13. Drakesmith, H.; Prentice, A. Viral infection and iron metabolism. *Nat. Rev. Microbiol.* **2008**, *6*, 541–552. [CrossRef] [PubMed]
14. Wang, M.P.; Joshua, B.; Jin, N.Y.; Du, S.W.; Li, C. Ferroptosis in viral infection: The unexplored possibility. *Acta Pharmacol. Sin.* **2022**, *43*, 1905–1915. [CrossRef] [PubMed]
15. Yang, Z.Y.; Wei, J.H.; Ren, L.; Chen, S.Y.; Liu, E.M.; Zang, N. Ferroptosis of alveolar epithelial cells induced by HAdV-7 infection. *J. Army Med. Univ.* **2022**, *44*, 2146–2156.
16. Wang, H.Z. Pathogenicity of Novel Duck Orthoreovirus and Its Mechanism of Induction of Ferroptosis in Macrophages. Ph.D. Thesis, Shandong Agricultural University, Taian, China, 2022.
17. Fu, X.Z.; Li, N.Q.; Lai, Y.; Luo, X.; Wang, Y.; Shi, C.; Huang, Z.; Wu, S.; Su, J. A novel fish cell line derived from the brain of Chinese perch Siniperca chuatsi: Development and characterization. *J. Fish. Biol.* **2015**, *86*, 32–45. [CrossRef]
18. Fu, X.Z.; Li, N.Q.; Liu, L.H.; Lin, Q.; Wang, F.; Lai, Y.G.; Jiang, H.M.; Pan, H.J.; Shi, C.B.; Wu, S.Q. Genotype and host range analysis of infectious spleen and kidney necrosis virus (ISKNV). *Virus Genes* **2011**, *42*, 97–109. [CrossRef]
19. Fu, X.Z.; Li, N.Q.; Lin, Q.; Liu, L.H.; Wu, S.Q. Development and identification of monoclonal antibody against recombinant major capsid protein of infectious spleen and kidney necrosis virus from Sinipercachuatsi. *J. Fish. China* **2016**, *40*, 363–370.
20. Reed, L.J.; Muench, H. A simple method of estimating fifty percent endpoints. *Am. J. Hyg.* **1938**, *27*, 493–497.
21. Hu, X.Q.; Fu, X.Z.; Li, N.Q.; Dong, X.X.; Zhao, L.J.; Lan, J.F.; Ji, W.; Zhou, W.D.; Ai, T.S.; Wu, S.Q.; et al. Transcriptomic analysis of Mandarin fish brain cells infected with infectious spleen and kidney necrosis virus with an emphasis on retinoic acid-inducible gene 1-like receptors and apoptosis pathways. *Fish Shellfish Immunol.* **2015**, *45*, 619–629. [CrossRef]
22. Fu, X.Z.; Hu, X.Q.; Li, N.Q.; Zheng, F.F.; Dong, X.X.; Duan, J.; Lin, Q.; Tu, J.G.; Zhao, L.J.; Huang, Z.B.; et al. Glutamine and glutaminolysis are required for efficient replication of infectious spleen and kidney necrosis virus in Chinese perch brain cells. *Oncotarget* **2017**, *8*, 2400–2412. [CrossRef]
23. Livak, K.J.; Schmittgen, T.D. Analysis of Relative Gene Expression Data Using Real-Time Quantitative PCR and the $2-\Delta\Delta CT$ Method. *Methods* **2001**, *25*, 402–408. [CrossRef]
24. Ma, B.F.; Li, F.Y.; Fu, X.Z.; Luo, X.; Lin, Q.; Liang, H.R.; Niu, Y.J.; Li, N.Q. Asparagine Availability Is a Critical Limiting Factor for Infectious Spleen and Kidney Necrosis Virus Replication. *Viruses* **2024**, *16*, 1540. [CrossRef] [PubMed]

25. Lei, P.; Bai, T.; Sun, Y. Mechanisms of Ferroptosis and Relations with Regulated Cell Death: A Review. *Front. Physiol.* **2019**, *10*, 139. [CrossRef] [PubMed]
26. Liu, G.Z.; Xu, X.W.; Tao, S.H.; Hou, Z.H. HBx facilitates ferroptosis in acute liver failure via EZH2 mediated SLC7A11 suppression. *J. Biomed. Sci.* **2021**, *28*, 67. [CrossRef] [PubMed]
27. Palamara, A.T.; Perno, C.F.; Ciriolo, M.R.; Dini, L.; Balestra, E.; D'Agostini, C.; Francesco, P.D.; Favalli, C.; Rotilio, G.; Garaci, E. Evidence for antiviral activity of glutathione: In vitro inhibition of herpes simplex virus type 1 replication. *Antivir. Res.* **1995**, *27*, 237–253. [CrossRef]
28. Lai, L.M.; Tan, M.L.; Hu, M.M.; Yue, X.Y.; Tao, L.L.; Zhai, Y.R.; Li, Y.S. Important molecular mechanisms in ferroptosis. *Mol. Cell. Biochem.* **2025**, *480*, 639–658. [CrossRef]
29. Zhu, W.J.; Li, Q.; Yin, Y.; Chen, H.C.; Si, Y.H.; Zhu, B.B.; Cao, S.B.; Zhao, Z.K.; Ye, J. Ferroptosis contributes to JEV-induced neuronal damage and neuroinflammation. *Virol. Sin.* **2024**, *39*, 144–155. [CrossRef]
30. Gao, J.; Wang, Q.; Tang, Y.D.; Zhai, J.B.; Hu, W.; Zheng, C.F. When ferroptosis meets pathogenic infections. *Trends Microbiol.* **2023**, *31*, 468–479. [CrossRef]
31. Xia, H.; Zhang, Z.; You, F. Inhibiting ACSL1-Related Ferroptosis Restrains Murine Coronavirus Infection. *Viruses* **2021**, *13*, 2383. [CrossRef]
32. Cheng, J.; Tao, J.; Li, B.; Shi, Y.; Liu, H.L. Swine influenza virus triggers ferroptosis in A549 cells to enhance virus replication. *Virol. J.* **2022**, *19*, 104. [CrossRef]
33. Chen, L.Y.; Li, X.X.; Liu, L.B.; Yu, B.; Xue, Y.X.; Liu, Y.H. Erastin sensitizes glioblastoma cells to temozolomide by restraining xCT and cystathionine-γ-lyase function. *Oncol. Rep.* **2015**, *33*, 1465–1474. [CrossRef] [PubMed]
34. Suárez-Pozos, E.; Martínez-Lozada, Z.; Méndez-Flores, O.G.; Guillem, A.M.; Hernández-Kelly, L.C.; Castelán, F.; Olivares-Bañuelos, T.N.; Chi-Castañeda, D.; Najimi, M.; Ortega, A. Characterization of the cystine/glutamate antiporter in cultured Bergmann glia cells. *Neurochem. Int.* **2017**, *108*, 52–59. [CrossRef] [PubMed]
35. Angeli, J.P.F.; Schneider, M.; Proneth, B.; Tyurina, Y.Y.; Tyurin, V.A.; Hammond, V.J.; Herbach, N.; Aichler, M.; Walch, A.; Eggenhofer, E.; et al. Inactivation of the ferroptosis regulator Gpx4 triggers acute renal failure in mice. *Nat. Cell Biol.* **2014**, *16*, 1180–1191. [CrossRef] [PubMed]
36. Lee, H.; Gan, B. Ferroptosis execution: Is it all about ACSL4? *Cell Chem. Biol.* **2022**, *29*, 1363–1365. [CrossRef]
37. Kagan, V.E.; Mao, G.; Qu, F.; Angeli, J.P.F.; Doll, S.; Croix, C.S.; Dar, H.; Liu, B.; Tyurin, V.A.; Ritov, V.B.; et al. Oxidized Arachidonic/Adrenic Phosphatidylethanolamines Navigate Cells to Ferroptosis. *Nat. Chem. Biol.* **2017**, *13*, 81–90. [CrossRef]

viruses

MDPI

Article

Genomic and Phenotypic Characterization of a Novel Virulent Strain of *Cyvirus cyprinidallo2* Originating from an Outbreak in The Netherlands

Bo He [1], Arun Sridhar [1], Marc Thiry [2], Olga Haenen [3,†], Alain F. C. Vanderplasschen [1,4,*,‡] and Owen Donohoe [1,5,*,‡]

1 Immunology-Vaccinology, Department of Infectious and Parasitic Diseases, Fundamental and Applied Research for Animals & Health (FARAH), Faculty of Veterinary Medicine, University of Liège, B-4000 Liège, Belgium
2 GIGA Neurosciences—Cellular and Tissue Biology, Cellular Biology, Faculty of Sciences, University of Liège, B-4000 Liège, Belgium
3 National Reference Laboratory for Fish Diseases, Wageningen Bioveterinary Research, Wageningen University Research, P.O. Box 65, 8200 AB Lelystad, The Netherlands
4 WEL Research Institute, Avenue Pasteur 6, B-1300 Wavre, Belgium
5 Bioscience Research Institute, Technological University of the Shannon, N37 HD68 Athlone, Co., Westmeath, Ireland
* Correspondence: a.vdplasschen@uliege.be (A.F.C.V.); owen.donohoe@uliege.be (O.D.); Tel.: +32-4-366-42-64 (A.F.C.V.); +32-4-366-43-79 (O.D.)
† Retired.
‡ These authors contributed equally to this work.

Abstract: *Cyvirus cyprinidallo2* (CyHV-2) is the causative agent of herpesviral hematopoietic necrosis in several economically important farmed freshwater fish species of the genus *Carassius*. Despite several CyHV-2 strains being isolated and fully sequenced, there is a lack of detailed characterization and consistent information on strains that exhibit high virulence in adult goldfish through viral challenge by immersion, particularly in the context of European strains and host populations. Strains that can cause highly virulent disease via this inoculation route are much more compatible with experimental designs that are representative of natural infection; thus, their utilization provides greater biological relevance. Consequently, in this study, we isolated three novel strains of CyHV-2 (designated NL-1, NL-2, and NL-3), originating from outbreaks in The Netherlands. Full-length genome sequencing and phylogenetic analyses revealed that these newly isolated strains are distinct from known strains and from each other. Significant differences were observed between the strains, in terms of in vitro growth kinetics, with NL-2 exhibiting stable passaging and superior fitness in vitro. Importantly, the challenge of adult Shubunkin goldfish with the NL-2 strain via immersion (2000 PFU/mL) induced an average mortality of ~40%, while parallel experiments with the CyHV-2 reference strain ST-J1 resulted in no mortality. Taken together, this study represents the characterization of a new CyHV-2 in vivo infection model, much more compatible with experimental designs that are required to be representative of natural infection. This model will be extremely useful in many aspects of CyHV-2 research in the future. Importantly, the genetic and phenotypic characterization performed in this study generates hypotheses on the potential roles of CyHV-2 genes in adaptation of the virus in vitro or in vivo.

Keywords: *Cyvirus cyprinidallo2*; *Cyprinid herpesvirus* 2; CyHV-2; cyprinivirus; *Alloherpesviridae*; *Carassius auratus*; aquaculture

1. Introduction

Among the viruses that infect widely cultivated freshwater fish, viral species from the family *Alloherpesviridae* can be particularly problematic. Disease caused by these viruses can lead to high mortality and large economic loss within the freshwater aquaculture industry [1,2], which is an extremely important sector in the context of global food security [3]. Within the family *Alloherpesviridae*, viruses belonging to the recently renamed genus *Cyvirus* [4] (formerly referred to as *Cyprinivirus* and collectively referred to as cypriniviruses [5,6]), can be among the most pathogenic, leading to high mortality outbreaks when water temperatures are optimum for viral replication [1,2]. Some of these cyvirus species are relatively well studied, including *Cyvirus cyprinidallo2* [4] (or CyHV-2, formerly known as *Cyprinid herpesvirus* 2). CyHV-2 is the causative agent of a disease referred to as herpesviral hematopoietic necrosis disease (HVHND), which is associated with elevated mortality rates in goldfish (*Carassius auratus*) and related species such as crucian carp (*Carassius carassius*) and gibel carp (*Carassius gibelio*), as well as other *Carassius* spp. [7,8]. The disease caused by CyHV-2 was initially documented in 1995, and since then, its global prevalence has increased substantially [9,10]. It represents a significant threat to the sustainability of goldfish aquaculture and the culture of other economically important species such as gibel carp and crucian carp, which are farmed for consumption and thus represent an important food production sector [10].

As per other members of the genus *Cyvirus* [11,12], live-attenuated vaccines against CyHV-2 have been demonstrated to be effective [13,14]. Notably, previous CyHV-2 vaccine development studies involving goldfish, utilized the Wakin-variety goldfish as models, with viral challenge trials conducted via immersion in water contaminated with CyHV-2 [13]. However, recently, we observed notable resistance to CyHV-2 challenge by immersion among adult goldfish populations sourced from European suppliers, including the popular Shubunkin goldfish breed. This was observed with the CyHV-2 ST-J1 strain, which is the designated CyHV-2 reference strain in GenBank (NC_019495.1). Notably, this was despite the fact that this strain was initially isolated from goldfish undergoing clinical disease [7]. We also observed a similar outcome using the CyHV-2 YC-01 strain using the same populations [15].

The Shubunkin goldfish breed is heavily associated with international trade [16] and thus represents a highly epidemiologically and economically relevant model in terms of vaccine development. Furthermore, Shubunkin goldfish are relatively easy to maintain and reproduce, which is extremely useful in terms of sustaining a supply of hosts for experimental infection trials. However, the lack of a broadly virulent and lethal CyHV-2 strain that can be used with this model, particularly when long-term, stable culture in vitro and infection by immersion in vivo is required, limits the study of CyHV-2 pathogenesis in the most biologically relevant contexts.

Despite the clear resistance to some CyHV-2 strains (including the ST-J1 reference strain) exhibited by some goldfish breeds (or at least some populations thereof), novel uncharacterized or recently emerged strains of CyHV-2 still remain a problem for both domesticated and wild *Carassius* spp. [16,17]. Consequently, in the present study, we sought to identify novel CyHV-2 strains from ongoing outbreaks that were (i) compatible with long-term stable culture in vitro and (ii) capable of causing sufficient and consistent virulence and lethality in Shubunkin goldfish when challenged by immersion. If successful, this would facilitate the establishment of a stronger CyHV-2 in vivo infection model to use in advancing our understanding of this virus and the development of effective disease mitigation strategies.

As part of this process, we successfully isolated and cultured three novel CyHV-2 isolates from different internal organ samples originating from CyHV-2 outbreaks among goldfish and gibel carp in The Netherlands, which we designated NL-1, NL-2, and NL-3. We then compared their genetic characteristics to those of other sequenced strains. Subsequently, in order to identify the optimum strain to use as part of a future CyHV-2 in vivo infection model, we compared their biological properties in vitro and in vivo. Through this process, we selected the most suitable isolate for further comparison to the CyHV-2 ST-J1 reference strain. This revealed that the NL-2 isolate was (*i*) highly compatible with long-term stable culture in vitro and (*ii*) much more virulent in a CyHV-2 Shubunkin goldfish model compared to the CyHV-2 ST-J1 reference strain. Importantly, in contrast to the reference strain, the NL-2 strain was able to induce mortality after inoculation by immersion in infectious water. These properties make the NL-2 isolate more valuable for incorporation into CyHV-2 Shubunkin goldfish studies. Finally, the full-length genome sequencing of the new isolates generated data on CyHV-2 genetic diversity and created new hypothesis for the identification of key CyHV-2 virulence factors.

2. Materials and Methods

2.1. *Sample Collection and Handling*

Three independent pools of internal organs (spleen, heart, kidney, pooled from up to ten fish per pool) were collected between 2016 and 2018 in The Netherlands from diseased goldfish and gibel carp. Samples were designated by the names NL-1 (unknown goldfish breed; WBVR 16009271), NL-2 (Shubunkin goldfish; WUR-NL-18015159-3), and NL-3 (wild gibel carp; WUR-NL-16010689). Each sample was processed separately. Samples were first homogenized with pestle and mortar and resuspended in 2 mL of sterile M199 medium (HEPES buffered; Sigma, Kawasaki, Japan); the suspension was clarified by centrifugation at 3000× *g*, 4 °C for 15 min (Allegra X-15R, Beckman, Brea, CA, USA). After centrifugation, the supernatant was aseptically separated from the tissue pellet. The supernatant was filtered twice using a sterile cell strainer with a diameter of 100 μm, then filtered with a sterile disposable syringe filter with a pore size of 0.22 μm. Purified supernatant was then aliquoted and stored at −80 °C (designated as passage #0 or p#0). DNA extraction was performed with the remaining tissue pellet using a commercial nucleic acid extraction kit (QIAamp DNA Mini Kit, QIAGEN, Hilden, Germany). DNA samples were stored at −20 °C for further analysis.

2.2. *In Vitro Methods*

2.2.1. CyHV-2 PCR Detection

To confirm the presence of CyHV-2 in the samples, the DNA extracted from the tissue pellets was analyzed by PCR. Briefly, this was performed using the High-Fidelity PCR Mix Buffer (New England Biolabs, Ipswich, MA, USA) per manufacturer instructions, with a total reaction volume of 25 μL and 2 μL template. The forward and revers primers were 5′-GGACTTGCGAAGAGTTTGATTTCTAC-3′, and 5′-CCATAGTCACCATCGTCTCATC-3′, respectively (as described by [18]). The PCR cycling conditions included an initial denaturation at 98 °C for 2 min, followed by 35 cycles at 98 °C for 30 s, 60 °C for 45 s, 72 °C for 45 s, and a final extension at 72 °C for 10 min. The PCR products were loaded into electrophoresis gels and imaged using an ImageQuant 800 CCD imager (Cytiva, Marlborough, MA, USA).

2.2.2. Cells and Virus

The RyuF-2 cell line kindly provided by Prof. Sano (Tokyo University of Marine Science and Technology, Tokyo, Japan) was used to propagate all viral isolates. Cells were cultured at 25 °C using Medium 199 (HEPES buffered; Sigma, Kawasaki, Japan), supplemented with 10% fetal bovine serum (FBS; Gibco, Life Technologies, Carlsbad, CA,

USA), penicillin (100 U/mL), streptomycin (100 μg/mL), and Amphotericin B (0.25 μg/mL). The CyHV-2 ST-J1 strain (NC_019495.1) was used as the reference strain in this study. All isolates were cultured in RyuF-2 cells at 25 °C with no CO_2 using the cell culture media described above, supplemented with goldfish kidney extract (final concentration 0.2 v/v%, prepared as Shibata et al. described utilizing Shubunkin goldfish kidney) [19].

2.2.3. Initial In Vitro Culture and Characterization of the CyHV-2 Isolates

Serial dilutions (10-fold) of the purified organ supernatant (NL-1 #0; NL-2 #0, and NL-3 #0) were prepared. Briefly, 100 μL of supernatant was mixed with 900 μL serum-free sterile medium, and this was repeated to create a dilution series ranging from 10^{-1} to 10^{-3}. Six-well plates with RyuF-2 fresh cells (100% confluent, approximately 24 h post-seeding) were inoculated with 800 μL of diluted supernatant per well, and the cells were incubated for 2 h at 25 °C. After this initial incubation, the initial inoculum was removed from each well and replaced with 2 mL sterile complete cell culture medium (0.2% kidney extract added) containing 2% w/v carboxymethylcellulose (CMC) [19]. These plates were then incubated for 10 days at 25 °C. Cytopathic effect (CPE) was visually inspected and imaged using an optical microscope (Olympus CKX41, Tokyo, Japan), followed by indirect immunofluorescence staining as described previously [15]. The stained samples were imaged using an epifluorescence microscope (Leica DM2000, Leica Microsystems, Wetzlar, Germany).

2.2.4. Plaque Purification and Amplification

The earlier experimental procedure describing the initial in vitro culture and characterization of the isolates was replicated. At 10 dpi (days post-inoculation), using an optical microscope, isolated viral plaques were identified in inoculated wells. Infectious material was collected by submerging a sterile pipette tip in the 2% CMC complete media, allowing it to come into physical contact with the center of the plaque, and aspirating 10 μL of the culture medium. This 10 μL media was then transferred to 90 μL of sterile serum-free culture medium, as a means of collecting plaque-purified infectious material for later use. These 100 μL samples of plaque-purified isolates were immediately stored at −80 °C. Three independent viral plaques were purified for each isolate and designated as the first passage #1 (or p#1). In order to amplify these plaque-purified sub-cultures, they were serially passaged in cell culture flasks. As part of this process, the 100 μL of each isolate was thawed, diluted in 900 μL sterile complete medium, and used to inoculate confluent cell culture flasks (175 cm^2). After two hours, kidney extract was aseptically added to the culture medium to reach a final concentration of 0.2%. Once CPE became widespread, viral culture medium was collected and centrifuged at 300× g for 5 min at 4 °C, and the viral supernatant was split into 1 mL aliquots. These aliquots were either immediately stored at −80 °C or used to inoculate new cell culture flasks as part of serial viral passaging, with viral culture proceeding as described above.

2.2.5. Viral DNA Extraction and Restriction Fragment Length Polymorphism (RFLP) Analysis

Plaque-purified sub-cultures were diluted in 1 mL complete culture medium (NL-1 #1, NL-2 #1, and NL-3 #1) and used to inoculate confluent cell culture flasks (175 cm^2), as described earlier. At 10 dpi, viral culture medium was collected and centrifuged at 3000× g, 4 °C for 15 min (Allegra X-15R Centrifuge, Beckman, Brea, CA, USA) to remove cell debris. Viral supernatant (35 mL from each 175 cm^2 flask) was transferred to ultracentrifuge tubes and centrifuged at 100,000× g for 1 h at 4 °C. The supernatant was discarded after centrifugation, and residual liquid drops on the tube borders were removed with a swab. The viral pellet was suspended in 1 mL sterile TE buffer (0.1% NP40) and incubated in a 37 °C water bath for 20 min. It was then incubated at 56 °C for an additional 10 min to accelerate pellet dissolution. 30 mL of cold TE buffer (4 °C) was added to the tube after complete pellet dissolution. Then, 5 mL of cold 30% sucrose was added to the bottom of the tube using a glass Pasteur pipette. The tube was centrifuged at 100,000× g for 2 h at 4 °C. After this, the supernatant was discarded, and the pellet was resuspended in 450 µL of TE buffer. A total of 50 µL of filtered 10% SDS and 20 µL of proteinase K (25 mg/mL) was added and mixed thoroughly. The tube was covered with a lid and incubated at 56 °C in a water bath for 2 h, shaking occasionally. After this, the sample was transferred to a 2 mL tube and mixed with 510 µL phenol chloroform (4 °C). The mixture was agitated and then centrifuged at 18,000× g for 15 min at 20 °C. After centrifugation, the aqueous phase was carefully removed and transferred to a new 2 mL tube, and the volume was topped up to 500 µL with ddH_2O. Then, 1 mL of absolute ethanol and 50 µL of 3M sodium acetate were added to the mixture. The sample was submerged in liquid nitrogen until frozen, then thawed by warming in a 37 °C water bath until it formed a jelly-like consistency. At this point, it was centrifuged at 18,000× g for 15 min at 4 °C. The supernatant was discarded, and 600 µL of 70% ethanol was added, followed by centrifugation again at 18,000× g for 15 min at 4 °C. The pellet was air-dried at room temperature (RT) until it became transparent. It was then resuspended in 50 µL ddH_2O by incubating overnight at 4 °C. After this, the extracted and purified viral DNA was aliquoted and stored at −20 °C for future use. The genomic diversity of the isolated strains was then investigated by SacI restriction RFLP analysis. Briefly, 3 µg of genomic DNA was digested using SacI (New England Biolabs, Ipswich, MA, USA). After 6 h of digestion, fully digested DNA productions were separated in 0.8% electrophoresis agarose gel at 60 V for 18 h and imaged.

2.2.6. Genome Sequencing, Assembly, Phylogenetic and Mutation Analysis

DNA was extracted from 3 independent plaque-purified subcultures from NL-1, NL-2, and NL-3 (9 samples in total) as described earlier, and full-length genome sequencing was performed. DNA samples were submitted for Illumina paired-end sequencing (MiSeq Kit v3 600 cycles). Raw reads (in fastq format) were processed using BBduk (v38.26) [20] facilitating adaptor sequence removal and quality trimming. This was followed by an assessment of processed fastq files using FastQC (v0.11.8). Processed reads were used as input for reference guided assembly using an in-house modified version of the popular de Bruijn graph-based assembly tool "spades" v3.15.2 [21].

The modifications to the publicly available version of spades specifically involved increasing the maximum kmer size from a default of 127 to 251, thus utilizing larger portions of individual reads during the assembly process, with the aim of improving the performance of spades, specifically the assembly of low-complexity regions, where ambiguous assembly (regarding the number of repeat units) reduced assembly scaffold contiguity. To achieve this, the following modifications were made to the publicly available source code: (*i*) Line 224 in the spades_compile.sh script was replaced with "cmake -G "Unix Makefiles" -DCMAKE_INSTALL_PREFIX = "$PREFIX" -DSPADES_MAX_K = 251 $* "$BASEDIR/src", (*ii*) CMake (v3.15.3) was loaded in local Linux environment, and the modified compilation script was run per standard installation instructions. (*iii*) After compilation and the creation of the python script "options_storage.py" in /share/spades/spades_pipeline/, this new script was edited at line 59 to "MAX_K = 251" to ensure longer kmers would be tolerated when running spades, and line 74 was changed to "K_MERS_250 = [21, 33, 55, 77, 99, 127, 249]" to ensure that longer kmers were automatically included when spades generated combined assembly from different kmer sizes (note: for memory reasons, a value of 249 is used here as it is just under the new maximum limit of 251). (*iv*) As these modifications to utilize longer kmers caused run failure in initial testing due to memory issues, lines 69 and 70 of the python script "options_storage.py" were also changed to "THREADS = 32" and "MEMORY = 800" respectively, thus allowing assembly processes to successfully run to completion. The assembly quality was assessed via QUAST [22] (v5.2.0), which revealed improved N50 metrics (measurement of contiguity of assemblies) using the in-house version of the spades tool compared to the publicly available version.

Assembly graphs were generated using the in-house modified version of spades. The resulting scaffold(s), in .gfa format, were opened up using Bandage (v0.8.1) [23] for inspection and alignment to a local BLAST database representing the CyHV-2 reference strain, ST-J1 (NC_019495.1). Individual scaffolds mapping to reference sequences, and thus collectively representing the full-length, assembled genome, were identified from inspection of the BLAST output in Bandage. Where multiple scaffolds matched the reference genome, full-length genomes were derived from concatenating these scaffolds based on overlaps at their terminals, which was performed using SnapGene (v6.2.1) [24]. For isolates NL-1, NL-2, and NL-3, there were three reassembled genomes (each representing an independent plaque-purified subculture). Taking each isolate separately, genome reassemblies of the three plaque-purified subcultures were aligned to each other using MAFFT (v7) [25], and the output alignment was exported as a fasta file. This revealed little or no difference between plaque-purified subcultures derived from the same isolates. The alignments of genomes from these subcultures were imported into SnapGene to generate consensus sequences for each isolate (>50% agreement).

Once the consensus sequences for NL-1, NL-2, and NL-3 were established as described earlier, these three sequences were aligned to other complete CyHV-2 genome sequences. Complete CyHV-2 genome sequences for this analysis were compiled based on the accession numbers of fully sequenced CyHV-2 genomes listed in the NCBI Viral Genome Browser [26,27]. This consisted of fully sequenced genomes from the following CyHV-2 strains: CNDF-TB2015 (MN201961.1), YZ-01 (MK260012.1), SY (KT387800.1), SY-C1 (KM200722.1), ST-J1 (NC_019495.1), and YC-01 Unverified (MN593216.1). This analysis also included an additional full length sequence of the YC-01 isolate that was also sequenced and assembled in-house as described earlier (this specific YC-01 isolate was available to us from a previous study we conducted [15]). These full-length CyHV-2 genomes were aligned to the CyHV-2 ST-J1 strain using MAFFT, and the resulting alignment was exported as a fasta file and then imported into MEGA (v11) [28] as an alignment. This alignment was then saved as a MEGA format data file (.meg). This MEGA format file was opened up in MEGA and subsequently used to generate a phylogenetic tree using the UPGMA method.

In order to understand the consequences of the differences between the newly assembled genomes and the CyHV-2 ST-J1 reference genome, a list of differences relative to the reference genome were compiled. This was followed by inspection of their genomic context (i.e., coding/noncoding region etc.) and consequences (synonymous, non-synonymous amino acid change, frameshifts, ORFs impacted etc.). To achieve this, the consensus sequence for each fully assembled isolate was aligned to the CyHV-2 ST-J1 strain using MAFFT, and the resulting alignment was exported as a fasta file. All differences in the consensus relative to ST-J1 were summarized in variant call format (VCF) by converting the fasta alignment to a VCF file using the msa2vcf.jar tool [29,30]. The VCF file was opened in Excel, and each variant coordinate, which was in the context of the alignment only, was transformed to the corresponding coordinates in the ST-J1 reference strain. This was achieved by using an Excel formula to take into account the cumulative gaps in the ST-J1 and consensus sequence that had occurred prior to each variant call. The updated coordinates of these variants were verified by manually inspecting the raw read mapping data. This was done by building an index of the ST-J1 reference sequence and mapping processed reads to from each new isolate to the index using HISAT2 (v2.1.0) [31,32]. The mapping output was then converted from SAM to compressed BAM format using SAMTools (v1.9) [33], and finally, raw mapping data (BAM format) was inspected using Integrative Genomics Viewer (IGV) (v2.8.0) [34]. Inspection involved sampling random variants described in the updated VCF file (coordinates and expected differences relative to ST-J1 reference) and using IGV to locate and verify the presence of these differences in reads from new isolates that were mapped to the ST-J1 reference by HISAT2.

The updated VCF files (with variant coordinates corresponding to the ST-J1 reference) were utilized to identify the consequence of each mutation in the consensus sequence relative to the ST-J1 reference strain using VEP (v107.0) [35]. To achieve this, the updated VCF, the ST-J1 sequence (fasta format) and ST-J1 feature coordinates (GFF format) were used as input for VEP. Using a combination of grep, sort and bgzip tools in Linux, the first line of the GFF file was removed, contents were sorted by coordinates, and finally, it was compressed in .gz format before being used as input for VEP.

Maps of each fully assembled genome were generated using Geneious Prime (v11.0.4+11) [36]. First, using SnapGene, the sequence of each assembled genome was annotated with the location of CyHV-2 ORFs and terminal repeats by importing features (≥99% similarity) from a separate SnapGene file of the ST-J1 genome (initially downloaded as .gb file from GenBank, opened in SnapGene and saved as SnapGene format .dna file). Using SnapGene, the annotated versions of each newly assembled genome were saved as GenBank (.gb) files and imported into Geneious Prime. These maps were converted into GFF files and exported from Geneious Prime in GFF format. These GFF files were opened in Excel and modified to remove excess annotation. The fasta files for each assembly were then imported into Geneious Prime, along with the modified GFF files defining ORF annotations, and together, they were used to generate a map of each fully assembled genome with ORFs. The coordinates of all sequence variants relative to the ST-J1 reference strain were also added these maps by importing another GFF file defining the locations of each variant. This GFF file was generated by converting VCF files (in this case, the earlier VCF files used as input for VEP) into GFF format using the python script vcf_gff.py [37,38]. After this, the GFF was opened in Excel, and the locations of each variant (initially provided as ST-J1 coordinates) were converted to their corresponding coordinates in each newly assembled isolate using an Excel formula (described earlier). The lengths of all variants in the GFF file were calculated by utilizing the variant sequence information, which was stored in the "attributes" field (column 9) as a consequence of earlier conversion from a VCF file. Using this length information, Excel formulas were used to classify each variant as either a single nucleotide variant (SNV) or multiple-base-change/indel. This information was added to the "type" field (column 3) for later utilization by Geneious Prime when generating maps. Feature names were also deleted from the "attributes" field of the GFF using Excel. This modified GFF was then imported into Geneious Prime and used to update the maps with the location of each variant relative to the ST-J1 strain. Maps were then exported as PNG images.

All Linux-based work on this project was conducted by our team using NIC5 computing cluster, which is part of the Consortium des Équipements de Calcul Intensif (CECI) and maintained by ULiège.

2.2.7. Viral Growth Assay

Viral growth assays were conducted using the same inoculation method. CyHV-2 supernatant was diluted in serum-free media to achieve a specific multiplicity of infection (MOI) in each well. Triplicate cultures of RyuF-2 cells grown in 6-well plates were infected with 1 mL viral supernatant. After a 2 h incubation period, the cells were washed with sterile, serum-free media and overlaid with complete, sterile cell culture medium supplemented with 0.2% kidney extract. At selected timepoints post-infection, both viral supernatant and infected cells were collected from each well, and the mixture was separated by centrifugation and stored at −80 °C. These samples were later thawed and analyzed by viral titration using triplicate plaque assays in RyuF-2 cells. Different inoculum doses and sampling time points were used across various experiments (*i*) An MOI = 0.0001 was used for the comparison between NL-1 #10, NL-2 #10, and NL-2 #2, with time points at 8, 12, and 16 dpi. (*ii*) An MOI = 0.01 was used for the comparison between ST-J1 #10 and NL-2 #10, with time points at 4, 6, 8, and 10 dpi.

2.2.8. Viral Plaque Size Assay

RyuF-2 cells were cultured in six-well plates and inoculated with 100 PFU/well of virus, followed by a 2 h incubation period. After incubation, the cells were washed with serum-free medium and then overlaid with culture medium supplemented with 1% (*w*/*v*) CMC and 0.2% kidney extract. At different time points, viral plaques were visualized by indirect immunofluorescent staining as described in the CPE Imaging section. After a final wash with PBS, 20 randomly selected individual plaques were imaged using the Incucyte live cell analysis system (Sartorius, Goettingen, Germany), and their areas were measured manually using ImageJ software (Version 2.14).

2.3. *In Vivo*

2.3.1. Fish

Two different sizes of Shubunkin goldfish (*Carassius auratus*) were utilized in this study, both at the adult developmental stage. The larger fish had an average weight of 12 g (12 ± 1.4 g), while the smaller fish group averaged 6 g per fish (5.84 ± 0.6 g). Live, mature Shubunkin goldfish were obtained from an accredited commercial company (Ruinemans Aquarium, Montfoort, The Netherlands). Microbiological, parasitic, and clinical examinations were conducted immediately upon arrival at the animal facility and then monthly to monitor for fish health. PCR analysis of a kidney homogenate confirmed that the adult goldfish from the colony were free of CyHV-2. Fish were maintained in 60 L freshwater recirculation tanks at 25 °C until they were transferred to A2 facilities for infection experiments.

2.3.2. Infection of Fish with CyHV-2

Different modes of inoculation were used depending on the different test subjects and viral strains. (*i*) Basic evaluation of the inherent pathogenicity of NL-1, NL-2, and NL-3: The original supernatant (p#0) of the three new isolates (NL-1 #0, NL-2 #0, and NL-3 #0) was diluted 10-fold using serum-free medium and kept on ice. For this preliminary and exploratory experiment, due to the low amounts of p#0 material available, the viral titers in NL-1 #0, NL-2 #0, and NL-3 #0 were not established prior to use. However, viral titers were established, and viral doses were normalized, for later experiments (see later). Shubunkin goldfish (big: average weight 12 ± 1.4 g; small: 5.84 ± 0.6 g) were anesthetized by immersion in water containing benzocaine (25 mg/L). Once the fish were anesthetized, the virus was injected intraperitoneally using a 300 µL syringe (BD Micro-Fine), with each fish receiving a volume of 100 µL (big) or 50 µL (small). After injection, the fish were placed in an aerated recovery bath for 10 min and then returned to the tank. Similarly, a mock group was injected with the same dose of sterile culture medium. The experiment included 10 fish per group. Survival status and clinical manifestations were recorded daily, and the survival rate was calculated at the end of the experiment. At 5 dpi, 10 sentinel goldfish were added to each tank for cohabitation experiments. (*ii*) Comparison of in vivo pathogenicity of NL-2 strain and ST-J1 reference strain after 10 in vitro passages: The concentrations of NL-2 #10 and ST-J1 #10 were adjusted using culture medium to a final concentration of

2×10^4 PFU/mL. After the administration of anesthesia as described above, inoculation was performed by intraperitoneal injection. Smaller Shubunkin goldfish (5.84 ± 0.6 g) were used, with three replicates of 10 fish for each virus strain. Each fish received a dose of 1000 PFU in 50 μL. The mock group was treated the same way, except sterile complete culture medium was injected instead of virus. Survival status and clinical manifestations were recorded daily, and the survival rate was calculated after 30 days post-injection. At 5 dpi, two replicate groups for each virus strain were selected and placed with the same number of sentinel goldfish, and fish status was monitored as described earlier for 30 days. (*iii*) Comparison of virus pathogenicity after inoculation by immersion after 10 passages: The fish (5.84 ± 0.6 g) were inoculated by immersion in water containing the virus, thus simulating the natural infection route. ST-J1 #10 and NL-2 #10 were diluted ten-fold using water at 25 °C to reach a final concentration of 2000 PFU/mL. After a 2 h immersion (10 fish/L), the fish were returned to the 60 L tank. The mock group underwent the same process using sterile complete viral culture medium. Three replicates were included for each virus strain, with 10 fish in each replicate. Survival status and clinical symptoms were recorded daily, and the survival rate was calculated after 30 days post-injection.

2.3.3. Transmission Electron Microscopy

Kidney tissue was sampled from a moribund goldfish which was inoculated by IP injection using 100 μL of NL-2 (p#0, 10-fold dilution). It was fixed in 2.5% glutaraldehyde in 0.1 M Sörensen's buffer (pH 7.4) for 2 h at room temperature. After several washes in the same buffer, the samples were post-fixed for 60 min with 2% osmium tetroxide in Sörensen's buffer, washed in the same buffer, dehydrated at room temperature through a graded ethanol series (70, 90, and 100%), and embedded in Epon for 72 h at 60 °C. Ultrathin sections (70 nm thick) were cut using an ultramicrotome (Ultracut S, Leica, Wetzlar, Germany) equipped with a diamond knife (Diatome, Sierre, Switzerland), mounted on copper grids, contrasted in the dark for 15 min in uranyl acetate solution and for 15 min in lead citrate solution, and examined under a Jeol TEM JEM-1400 transmission electron microscope at 80 kV. Random fields were photographed using an 11-megapixel camera system (Quemesa, Olympus, Tokyo, Japan).

2.4. *Statistical Analysis*

Two-way omnibus tests on data from the viral growth curve and plaque size experiments were conducted. First, a two-way ANOVA model was generated using the anova function in R (v4.2.2) (part of the R core package) [39]. The residuals of this model were checked for normality using the Shapiro–Wilk test, implemented using the shapiro.test function in R (part of the R core package). If normal distribution was observed, the significance of each variable on outcome was taken from the two-way ANOVA model, and multiple comparisons were conducted using a Tukey post hoc test, implemented using the TukeyHSD function in R (part of the R core package). In the cases where normal distribution was not observed, analysis was performed using a generalized linear model (GLM) implemented using the glm function in R (part of the R core package), with Family option: "gamma" and Link option: "log". To determine the significance of each variable on outcome, the model was analyzed using type III sum of squares test, implemented using the Anova tool from the R "car" package (v3.0-6) in R [40], with the default contrast coding in R adjusted appropriately prior to the use of the glm function to facilitate the use of type III sum of squares test. In such cases, post hoc multiple comparison tests were conducted on the same model using a least squares means test, implemented using lsmeans tool from the "lsmeans" package (v2.3.0) in R [41], with BH correction. Survival curves were compared using the log-rank test, implemented in GraphPad prism 8.0. All graphs were also generated using GraphPad 8.0. Only significant p-values (<0.05) are reported in Section 3. For the purposes of visual clarity, only results from post hoc multiple comparisons are indicated in each corresponding figure and are represented using the following symbols: ns = not significant, * = $p < 0.05$; ** = $p < 0.01$; *** = $p < 0.001$; **** = $p < 0.0001$

2.5. Ethics Statement

The experiments, maintenance, and care of fish complied with the guidelines of the European Convention for the Protection of Vertebrate Animals used for Experimental and other Scientific Purposes (CETS No. 123). The animal studies were approved by the local ethics committee of the University of Liège, Belgium (Laboratory accreditation No. 1610008). All efforts were made to minimize suffering of the fish.

3. Results

3.1. Confirmation of CyHV-2 Presence in Pooled Organ Samples

Three pooled organ samples from diseased fish initially suspected of being infected with CyHV-2 were generously provided by WBVR (Wageningen Bioveterinary Research), part of Wageningen University in The Netherlands. We assigned them the names NL-1, NL-2, and NL-3 (see Section 2 for details). Upon arrival in our lab, these samples were screened for the presence of CyHV-2 using a PCR assay targeting the CyHV-2 helicase gene. Through this process, we confirmed the presence of the virus, with the expected 366 bp PCR product forming in all three samples (Figure 1A).

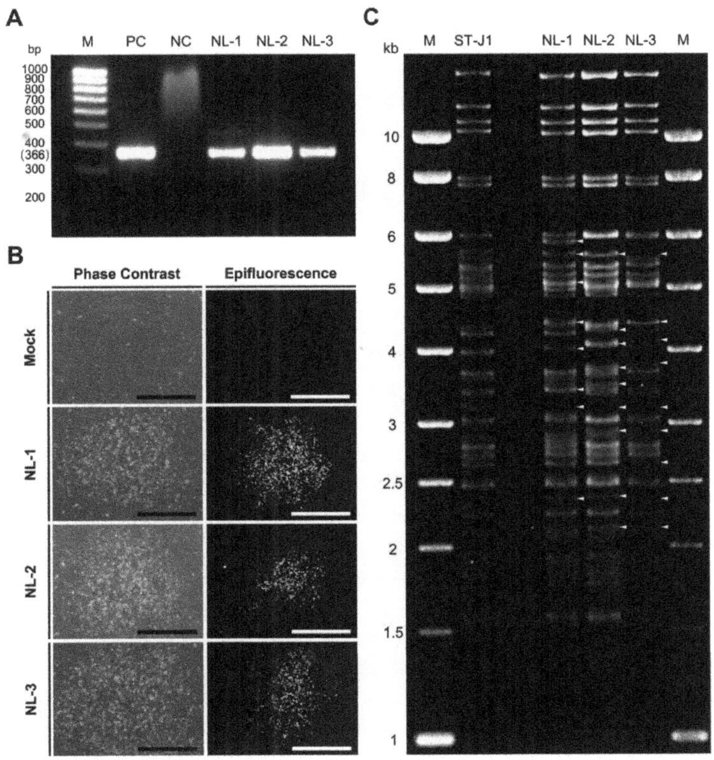

Figure 1. (**A**) PCR screening for CyHV-2 in three samples from fish from The Netherlands. M: molecular weight marker (bp: base pair); PC: positive control (from ST-J1); NC: negative control;

NL-1: NL-1 tissue DNA; NL-2: NL-2 tissue DNA; NL-3: NL-3 tissue DNA. (**B**) CPE imaging via microscopy at 10 dpi. The left and right panels represent the brightfield and epifluorescence channels, respectively, with different representative plaques shown in each channel. CPE was only observed in RyuF-2 cells inoculated with purified tissue supernatant from NL-1 #0, NL-2 #0, and NL-3 #0, and was entirely absent from the mock control. Black scale bar = 600 μm; white scale bar = 500 μm. (**C**) Genomic analysis of CyHV-2 strains. Viral DNA from plaque-purified sub-cultures from ST-J1, NL-1, NL-2, and NL-3 (one from each) were compared by RFLP analysis after digestion using the SacI restriction enzyme. The white arrowhead indicates some of the most notable differences relative to the ST-J1 reference strain. M: molecular weight maker (kb: kilobase).

In parallel, RyuF-2 fresh cells were inoculated with dilutions of purified supernatant prepared from the NL-1, NL-2, and NL-3 pooled tissue samples (passage #0 or p#0; see Section 2) and incubated at 25 °C. At 10 dpi, the characteristic morphology of CyHV-2 associated CPE became visible via brightfield phase contrast microscopy and eventually formed plaques. The observed characteristics included focal areas of granulation, cell vacuolization, and the emergence of rounded phase-bright cells (Figure 1B left panel). By approximately 14 dpi, infected cells in these plaques began to detach, with this process starting from the central regions and radiating outwards as time progressed. The same CPE-like and plaque morphology was not observed in the mock group. Notably, no syncytial plaque formation was observed in cells infected with any of the three isolates.

These regions exhibiting signs of CyHV-2-induced CPE and plaque formation via brightfield phase contrast microscopy also stained positive for CyHV-2 via indirect immunofluorescent staining and epifluorescence microscopy (Figure 1B, right panel). This indicated (i) that the p#0 supernatant samples prepared from the CyHV-2 PCR-positive samples contained infectious CyHV-2 virus particles and (ii) that these three CyHV-2 isolates, NL-1, NL-2, and NL-3 could be cultured in vitro.

3.2. *Virus Isolation and Amplification*

While the initial in vitro culture of the p#0 supernatant led to the formation of CPE and plaques, for any of these new isolates to be useful as part of future CyHV-2 in vivo infection models, it was crucial that they could be stably passaged in cell culture. Subsequent in vitro passaging of these three isolates revealed significant differences in the in vitro propagation of NL-1, NL-2, and NL-3 in RyuF-2 cells. Only NL-1 and NL-2 isolates could be consistently passaged successfully, reaching up to ten passages (#10). Conversely, the NL-3 isolate could not be passaged beyond p#3 in RyuF-2 cells, with cytopathic effects disappearing in the subsequent passages.

3.3. *Genomic and Phylogenetic Analysis*

At p#1, three independent plaques were sub-cultured from each isolate, amplified for one passage in cell culture flasks, generating p#2. At 10 dpi, viral culture media was used to prepare viral supernatant for subsequent virus purification and for extraction of pure viral DNA (i.e., free of host cell DNA), which would be suitable for genomic analysis. In addition, the CyHV-2 reference strain ST-J1 was used as a reference for comparative analysis.

Since the samples originated from distinct geographic regions and hosts, we hypothesized that the NL-1, NL-2, and NL-3 CyHV-2 isolates may represent genetically distinct CyHV-2 lineages. To investigate this further, we used RFLP to conduct a quick preliminary analysis on one plaque-purified sub-culture from each isolate. The results are presented in Figure 1C. This indicated that the NL-1, NL-2, and NL-3 exhibited major genetic differences relative to the ST-J1 strain. While some of these differences were common to two or more of these isolates, indicating that they were more closely related to each other than to the

ST-J1 reference, the results also indicated notable differences between the three isolates, indicating that they were also somewhat genetically distinct from each other.

To determine the exact nature of the differences, viral DNA prepared from purified particles of each NL-1, NL-2, and NL-3 were also subject to genomic sequencing. As part of this process, three independent plaque-purified sub-cultures (or replicates) were sequenced from each isolate. Genome assembly revealed little or no differences between the plaque-purified sub-cultures of individual isolates. For each isolate, the sequences of the three purified sub-cultures were used to generate a single representative consensus genome sequence, which was used in further analysis. The representative genome sequences for the NL-1, NL-2, and NL-3 isolates are available on GenBank under the accession numbers PQ738159.1, PQ723683.1, and PQ738160.1, respectively. The results of genome sequencing were in agreement with the RFLP analysis, indicating that each isolate exhibited many differences (each exhibiting > 600 differences) relative to the ST-J1 reference strain. These are presented in Figure 2, described in detail in Tables S1–S6, and summarized in Table 1. Some of these differences to the reference strain included major changes, such as frameshifts in several ORFs of unknown function; however, as expected, most of the changes consisted of single nucleotide variants (SNVs) (Figure 3A left and Table 1) and either occurred outside of protein coding regions or were synonymous mutations, therefore having no impact on the amino acid sequence (Figure 3A Right).

Table 1. Summary of genomic variants identified in each of the newly sequenced isolates relative to the CyHV-2 ST-J1 reference strain. Variant classes are defined as per https://www.ensembl.org/info/genome/variation/prediction/classification.html (accessed on 1 April 2025). Consequence types are as defined by the Sequence Ontology (SO) project and are summarized here: https://www.ensembl.org/info/genome/variation/prediction/predicted_data.html (accessed on 1 April 2025).

		NL-1		NL-2		NL-3	
		Number	Proportion	Number	Proportion	Number	Proportion
Variants Class	Indel	13	2.2%	27	3.7%	23	3.1%
	Substitution	40	6.6%	45	6.1%	31	4.2%
	Insertion	117	19.4%	130	17.7%	153	20.5%
	Deletion	138	22.9%	179	24.4%	165	22.1%
	SNV	295	48.9%	353	48.1%	374	50.1%
Consequences	Stop gained	0	0%	0	0%	1	0.1%
	Frameshift *	5	0.8%	7	1%	8	1.1%
	Inframe insertion	29	4.8%	33	4.5%	41	5.5%
	Inframe deletion	48	8%	61	8.3%	55	7.4%
	Protein altering	2	0.3%	5	0.7%	5	0.7%
	Missenses	93	15.4%	118	16.1%	102	13.7
	Start retained	1	0.2%	0	0%	2	0.3%
	Stop retained	0	0%	1	0.1%	1	0.1%
	Synonymous	106	17.6%	125	17%	124	16.6%
	Coding sequence	1	0.2%	1	0.1%	2	0.3%
	Upstream gene	306	50.7%	358	48.8%	390	52.3%
	Downstream gene	9	1.5%	22	3%	11	1.5%
	Intergenic	3	0.5%	3	0.4%	4	0.5%

* This includes multiple frameshifts in the same ORF(s).

The high number of differences to the ST-J1 reference (Table 1, Figures 2 and 3A) and the commonalities among the isolates observed in the RFLP analysis (Figure 1C) indicated that these newly sequenced isolates were more related to each other than to the ST-J1 strain. To examine this further, a phylogenetic tree was constructed based on sequences of all currently available CyHV-2 genomes and these newly sequenced isolates (Figure 3B). This analysis indicated that, indeed, the new NL-1, NL-2, and NL-3 isolates were more closely related to each other than to the ST-J1 reference strain, the latter of which represents the J genotype [42]. Notably, the three new isolates are more related to the C genotype strains [42,43]. Within this large C genotype clade, NL-1, NL-2, and NL-3 occupy separate sub-clades, indicating that they are genetically distinct from each other. As a positive control for our genome assembly workflow, we also assembled the YC-01 genome. Our YC-01 assembly was almost identical to the publicly available YC-01 genome sequence, with notable genomic rearrangements and inversions relative to other strains [44], which is currently described as "Unverified" in GenBank. Indeed, the YC-01 strain is more distantly related to all other fully sequenced strains (~96.7% vs. >97% identity to the CyHV-2 reference sequence) and may represent a new distinct CyHV-2 genotype clade, which we refer to as "Undefined" (Figure 3B).

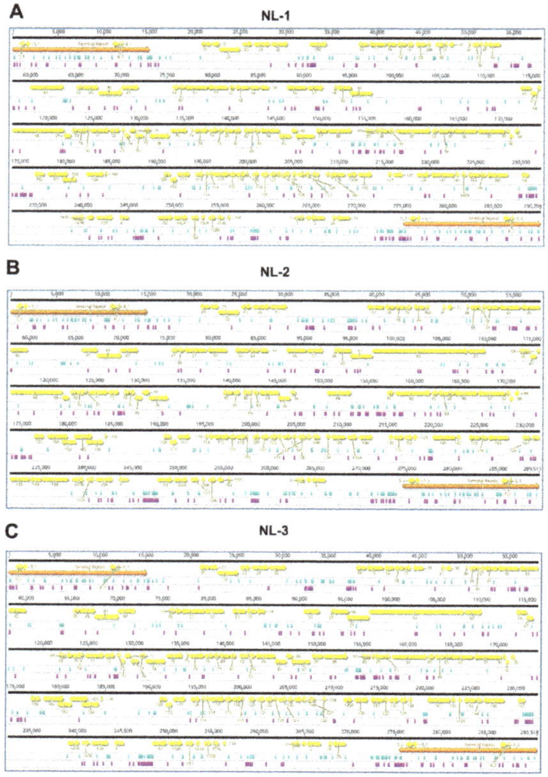

Figure 2. Schematic of (**A**) NL-1, (**B**) NL-2, and (**C**) NL-3 assembled genomes. Direct terminal repeats are indicated in orange. ORFs with 99% sequence match to ORFs in CyHV-2 ST-J1 reference genome (NC_019495.1) are indicated in yellow and labelled by their corresponding ORF number in the reference genome. Substitutions involving more than one nucleotide, insertions, or deletions are highlighted in turquoise. SNVs are indicated in purple. The locations of all variants are based on their starting point in each of their respective genomes, with feature lengths enlarged to ensure visualization on the map (coordinates correspond to the data in Tables S1–S3).

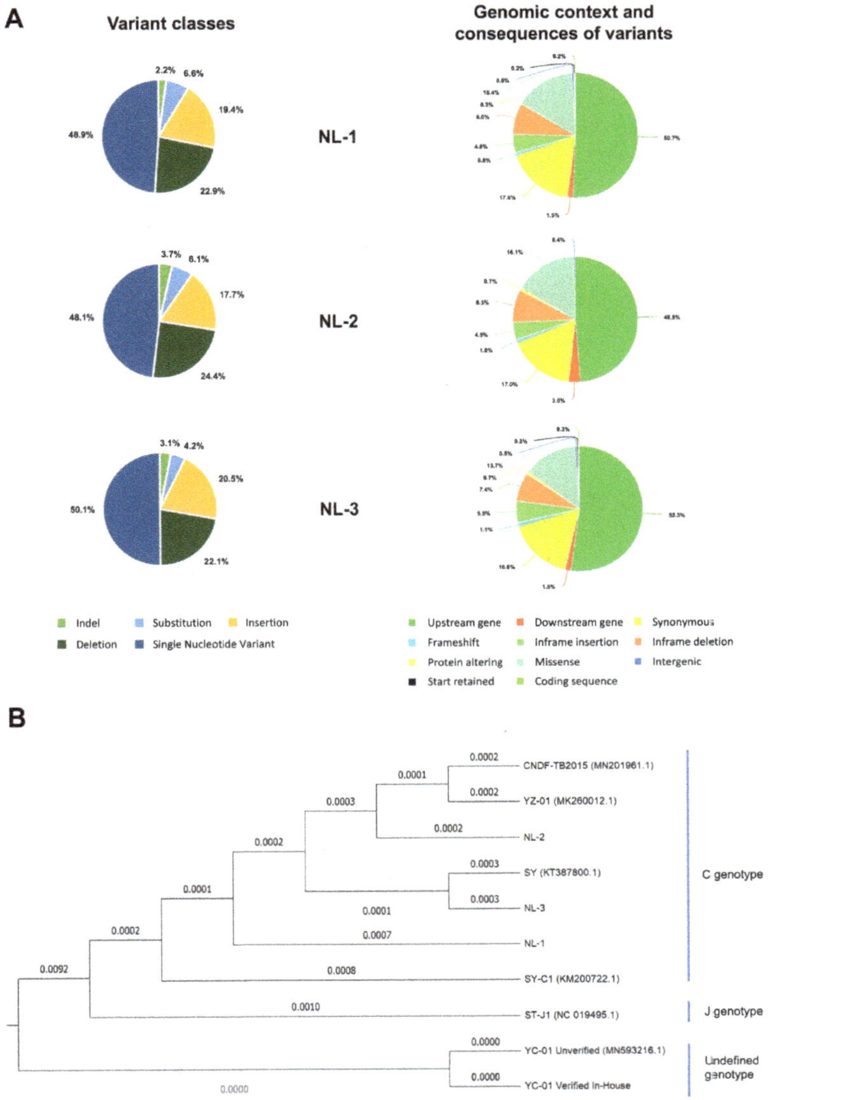

Figure 3. Genomic comparison of isolates to other CyHV-2 strains (**A**) (Left) Pie charts indicating the breakdown of different variant classes that were identified in each of the newly sequenced isolates relative to the CyHV-2 ST-J1 reference strain. Variant classes are defined as per https://www.ensembl.org/info/genome/variation/prediction/classification.html (accessed on 1 April 2025). (Right) Pie charts indicating the breakdown of corresponding genomic context and consequences of all variants identified. Consequence types are as defined by the Sequence Ontology (SO) project and are summarized here: https://www.ensembl.org/info/genome/variation/prediction/predicted_data.html (accessed on 1 April 2025). (**B**) Phylogenetic tree based on complete genome sequences of CyHV-2 strains. Values above or below the branches represent the number of substitutions per site. The tree is rooted using the undefined genotype clade as an outgroup.

3.4. Pathogenicity of the Three Isolates

In order for any of the newly sequenced isolates to be useful in terms of utilization as a robust CyHV-2 in vivo infection model, it was important to first determine if the p#0 inoculums were actually capable of causing clinical disease and mass mortality in commercially and economically relevant goldfish breeds. As part of this preliminary assessment of the virulence of these new isolates, we conducted in vivo challenge experiments using an in-house batch of adult Shubunkin goldfish. In this trial, two sizes of adult Shubunkin goldfish were used. These were referred to as "small" and "big". Fish in the "big" group were twice the mean weight of those in the small group and thus received twice the volume of the non-titered p#0 inoculum relative to the small group. Subjects were inoculated either by intraperitoneal (IP) injection or by cohabitation with IP injected fish (Figure 4). This initial exploratory experiment was limited to 10 fish per group due to the low amounts of p#0 infectious material that was available.

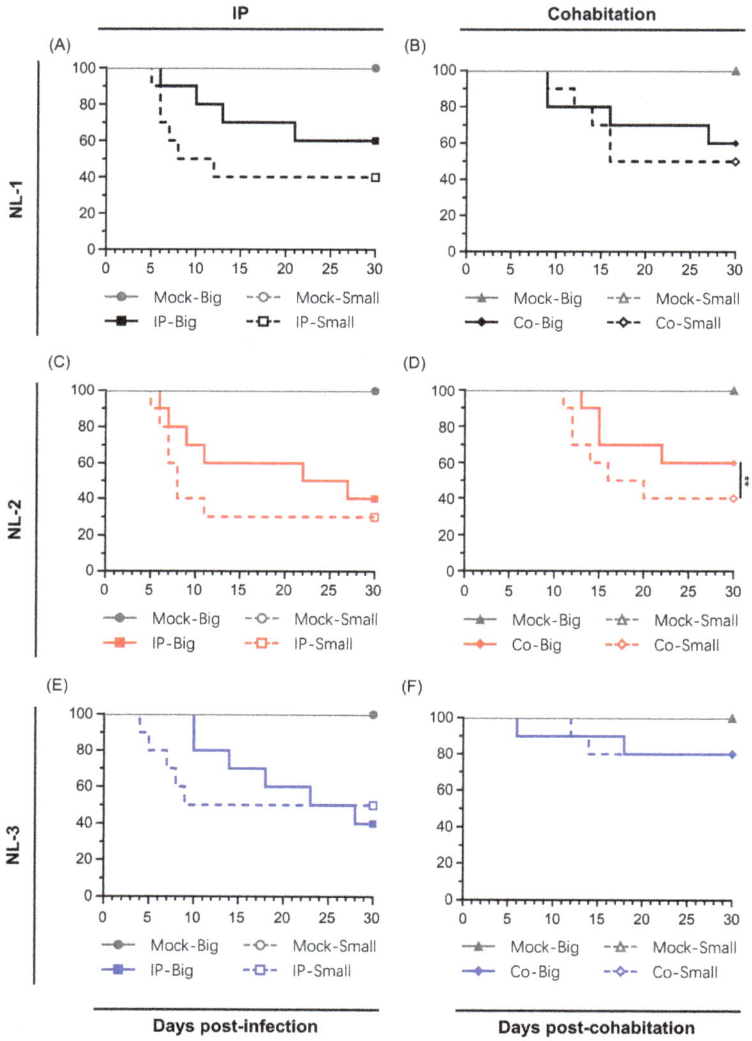

Figure 4. Primary comparison of inherent in vivo pathogenicity of the three isolates (**A**,**B**) NL-1, (**C**,**D**) NL-2 and (**E**,**F**) NL-3. The pathogenicity of the indicated strains was tested using adult Shubunkin goldfish (big, average weight 12 ± 1.4 g) and the young adult Shubunkin goldfish (small,

average weight 5.84 ± 0.6 g). Each group consists of 10 subjects. Fish were either mock-infected or infected via IP injection with the respective strains (100 μL in big fish, 50 μL in small fish). For the cohabitation trial, at 5 days post-injection, 10 more fish were added to each tank. The fish were monitored daily for clinical signs of CyHV-2 disease, and fish reaching the endpoints were euthanized. In all graphs, the y-axis represents percentage survival.

Mortality rates among the NL-1 and NL-2 groups were similar, with both the IP and cohabitation groups showing high mortality, particularly in small fish. The IP-injected fish exhibited a more rapid onset of illness and clinical symptoms, with mortality in small fish primarily occurring between 5 and 12 dpi (60% for NL-1, 70% for NL-2). Big fish had a longer disease course, higher survival rates and delayed onset of death, relative to that of small fish (Figure 4A,C). Sentinel fish cohabitating with IP-injected fish exhibited lower mortality rates of 50% for NL-1 and 60% for NL-2 in small fish, both higher than the 40% seen in big fish, with NL-1 deaths concentrated between 9 and 16 dpi and NL-2 between 11 and 20 dpi (Figure 4B,D). Despite the consistent differences in mortality between big and small fish inoculated with NL-1 and NL-2, due to the limited sample size in this preliminary experiment, these differences were only statistically significant for groups inoculated with NL-2 by cohabitation (Figure 4D).

The patterns observed in the NL-3 group were different than the NL-1 and NL-2 groups. In big fish, the NL-3 inoculum induced slightly more mortality than the NL-1 and NL-2 inoculums. However, it induced similar mortality in small fish (Figure 4E). Furthermore, in contrast to NL-1 and NL-2, in cohabitation experiments with NL-3, only a limited mortality of 20% was observed among both small and big sentinel fish (Figure 4F). No statistically significant difference was observed between big and small fish inoculated with NL-3 via the routes tested. Moreover, even though one early mortality was recorded among the big sentinel fish, this mortality was not found to be attributed to CyHV-2 infection.

However, regardless of the isolate that was used, in all IP infected groups, the fish exhibited classical clinical symptoms, including diminished appetite, lethargic swimming velocity, protruding eyeballs, distended abdomen, drooping of dorsal fin, and loss of balance, with only very few minor instances of hemorrhaging on the skin. Post-mortem examinations on subjects undergoing mortality during the experiment revealed severe ascites, enlarged spleen and kidneys, and white dot-like nodules within the kidneys, while the intestines were swollen with no food material remaining, and no abnormalities were found in the gills (Figure S1). Sentinel fish from the cohabitation experiment exhibited similar but less severe clinical symptoms. No clinical symptoms, deaths, or CyHV-2 PCR positive results were observed in the mock group.

Kidneys from fish that underwent mortality were promptly sampled and stored at -80 °C, whereas kidneys from surviving fish were collected at the end of the experiment. DNA was extracted from all kidneys, and subsequent PCR testing confirmed the presence of CyHV-2 in all infected subjects (data not shown). Separately, some kidney samples from fish infected with the NL-2 isolate (IP injection) were immediately fixed after sampling for later analysis using high-magnification transmission electron microscopy. This provided clear evidence of herpesvirus replication and assembly in kidney tissue (Figure 5). Infected kidney cells exhibited nuclear changes characterized by central nucleoplasmic transparency and marginal chromatin aggregation. The nucleus contained aggregates of characteristic naked herpes-like virus nucleocapsids (90–100 nm), including incomplete particles with empty or partially complete cores and capsids with a central electron-dense core. Potential organelle-enveloped virus particles were also observed. These ultrastructural characteristics of the virus particles are consistent with those of CyHV-2 virions described in other reports [45–47]. Furthermore, the virus was successfully re-isolated from the kidneys in cohabitating goldfish (data not shown).

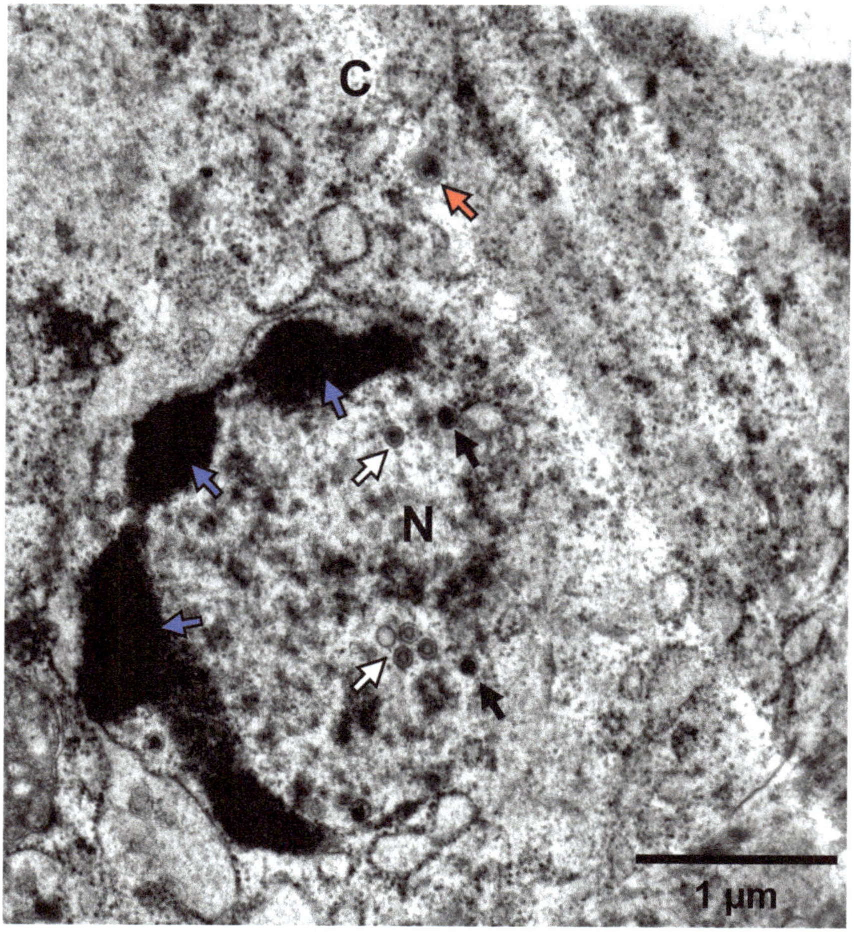

Figure 5. Transmission electron microscopy of the CyHV-2 NL-2 isolate. This image represents the ultrastructure of CyHV-2-infected kidney cells. Nuclear chromatin margination consistent with herpesvirus infection is visible (blue arrows). Virus particles at different stages of assembly are also visible, including some prime examples of incomplete nuclear particles with empty or partially complete cores (white arrows) and capsids with a central electron-dense core (black arrows). Cytoplasmic virions, which appear to be larger than nuclear viral particles and thus potentially organelle-enveloped as part of the process of egress from the cell, are also visible (red arrows). N = nucleus; C = cytoplasm.

3.5. In Vitro Growth Kinetics Comparison

All of the p#0 samples were shown to contain infectious CyHV-2 capable of causing clinical disease in the Shubunkin goldfish model. However, to be useful in future studies, it was important that these isolates could also be easily cultured in vitro in order to be easily titrated and to facilitate the sufficient expansion of viral stock for in vivo studies. In general, the ability of a viral isolate to replicate in vivo is not indicative of its ability to replicate in vitro or adapt quickly to this specific biological niche, and there are many factors that influence this. For example, while causing clinical disease in vivo, as described earlier, the passaging of the NL-3 isolate could not be sustained beyond p#3 in vitro, indicating that NL-3 was not practical for future use as a robust CyHV-2 in vivo infection model. Therefore, we proceeded with further characterization of NL-1 and NL-2 only, which, in contrast to NL-3, grew well in vitro, with viral titers increasing over several passages, thus providing sufficient viral stock for further characterization. Notably, the NL-2 reached higher titers faster, indicating that it might be better adapted to in vitro culture. To investigate this further, we compared the in vitro growth kinetics of NL-1 and NL-2. Importantly, to maximize the validity of the comparison, in addition to using the same dose of each isolate (MOI = 0.0001) for viral growth curves, we also compared them at the same passage number, in this case p#10, (referred to as NL-1 #10 and NL-2 #10, respectively). The results indicated that viral titers remained low initially, from 8 to 12 dpi. However, from 12 dpi to 16 dpi, the titers of both isolates increased rapidly. At the end of the experiment, at 16 dpi, NL-2 CyHV-2 reached titers close to 300 PFU/mL, nearly twice that of NL-1 #10, and we observed a significant difference in the growth kinetics of these two isolates at p#10, indicating that, indeed, NL-2 is better suited to in vitro culture compared to NL-1 (Figure 6A).

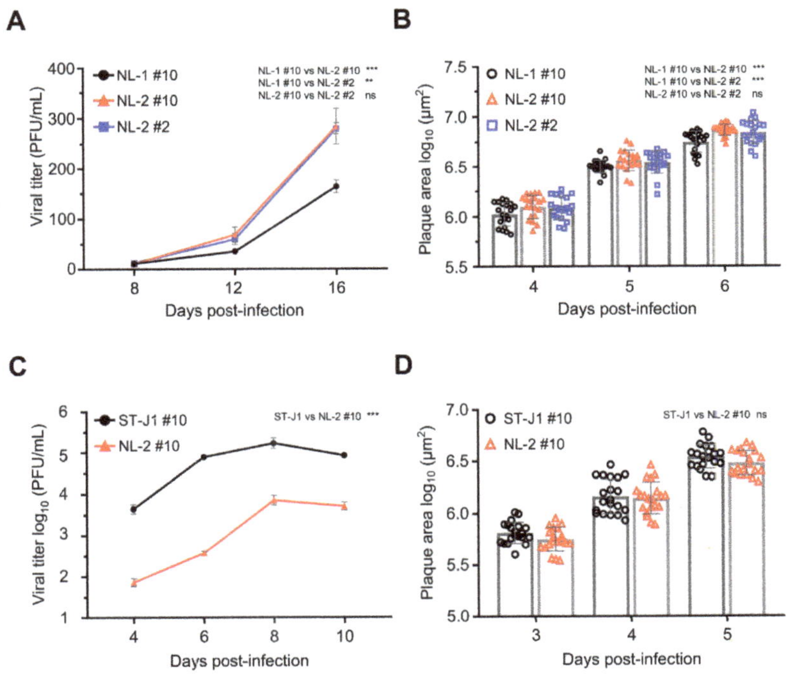

Figure 6. Comparison of in vitro growth kinetics and plaque sizes between the NL-1 and NL-2 isolates. (**A**) Viral growth curve comparison between NL-1 #10, NL-2 #10, and NL-2 #2. RyuF-2 cells were infected with the indicated strains (MOI = 0.0001), and the $\log_{10}$ value of the titer (PFU/mL) in the supernatant was determined at 8, 12, and 16 dpi. Data represent the mean ± SEM of triplicate measurements. (**B**) Viral plaque size comparison between NL-1 #10, NL-2 #10, and NL-2 #2. RyuF-2 cells were infected with the respective strains, and plaque areas were measured at 4, 5, and 6 dpi. Data represent the mean ± SEM of 20 individual plaques. (**C**) Viral growth curve comparison between ST-J1 and NL-2. RyuF-2 cells were infected with the indicated strains (MOI = 0.01), and the $\log_{10}$ value of the titer (PFU/mL) in the supernatant was determined at 4, 6, 8, and 10 dpi. Data represent the mean ± SEM of triplicate measurements. (**D**) Viral plaque size comparison in ST-J1 and NL-2. RyuF-2 cells were infected with the respective strains, and plaque areas were measured at 3, 4, and 5 dpi. Data represent the mean ± SEM of 20 individual plaques.

To explore the possibility that the superior growth kinetics of NL-2 may have been simply due to incremental adaptation to in vitro culture during the previous 10 passages, in parallel, we included a growth curve for NL-2 at passage #2. This indicated that there was no significant change growth kinetics of NL-2 between early (p#2) and later (p#10) passages (Figure 6A). Similar patterns were also observed when comparing plaque size between the same isolates, with plaques from NL-2 #10 being slightly but significantly larger than NL-1 #10, with no difference between NL-2 #2 and NL-2 #10 in terms of plaque size. (Figure 6B). Taken together, these experiments indicated that out of the three new isolates, the biological properties of NL-2 made it intrinsically better suited to stable and efficient culture in vitro and therefore potentially represented a more useful candidate for utilization in a more robust CyHV-2 in vivo infection model.

Next, we compared the in vitro growth kinetics NL-2 #10 with the CyHV-2 reference strain ST-J1, also at p#10, using an MOI of 0.01 for both. At this MOI, the progression of the viral growth curve over time was similar for both strains, with viral titers increasing over the first three timepoints, peaking at 8 dpi, and decreasing by 10 dpi (Figure 6C). However, the viral titer of ST-J1 was higher than NL-2 at each timepoint, resulting in a significant difference between the two strains overall, demonstrating that relative to NL-2, the ST-J1 strain exhibits superior replication kinetics in vitro. Despite this, the comparison of viral plaque size at early timepoints in the infection (3, 4, and 5 dpi) revealed no consistent difference between the two strains, with no statistically significant difference overall (Figure 6D).

Taken together, these experiments indicated that NL-2 was better suited to in vitro culture compared to NL-1 and NL-3, and furthermore, its properties remained stable over multiple passages. While the ST-J1 reference strain exhibited superior in vitro replication properties, the replication kinetics of the NL-2 isolate was sufficient to build up enough viral stock to facilitate its use in subsequent in vivo experiments.

3.6. Comparison of In Vivo Pathogenicity of CyHV-2 NL-2 Strain and Reference Strain ST-J1

After examining the in vitro growth kinetics, we compared the virulence of the NL-2 strain to that of the CyHV-2 reference strain ST-J1 in vivo. In this experiment, young adult fish were inoculated with the same dose of each CyHV-2 strain via IP injection, allowing a valid comparison between both strains. Analysis of the survival curves indicated a significant difference between the strains in terms of mortality rates (Figure 7A). NL-2 induced more intense and faster-developing clinical symptoms associated with CyHV-2 disease compared to ST-J1. Symptoms in the NL-2 group began at 4 dpi, with mortality starting at 5 dpi, rapidly increasing after, and finally ceasing approximately two weeks post-infection, with an average mortality rate of 76.6%. In contrast, the ST-J1 group exhibited only mild symptoms and a short recovery period, with mortality events limited to between 6 and 9 dpi and a significantly lower average mortality rate of 20%. No mortality was observed in the mock infected groups.

The cohabitation experiments also revealed a big difference between the two CyHV-2 strains, with only sentinel fish in the NL-2 group exhibiting mortality after cohabitation with IP-injected fish. Mortality among sentinel fish occurred mainly between 2 and 3 weeks post-cohabitation (Figure 7B), leading to an overall mortality rate of approximately 35%, which was lower than that of the IP-inoculated group. Notably, this pattern, including the mortality window, was consistent with the earlier NL-2 cohabitation experiment (Figure 4D), but the mortality rates were slightly lower in this later cohabitation experiment. In stark contrast to this, no mortality or clinical disease was observed in the ST-J1 cohabitation group or the mock group (Figure 7B).

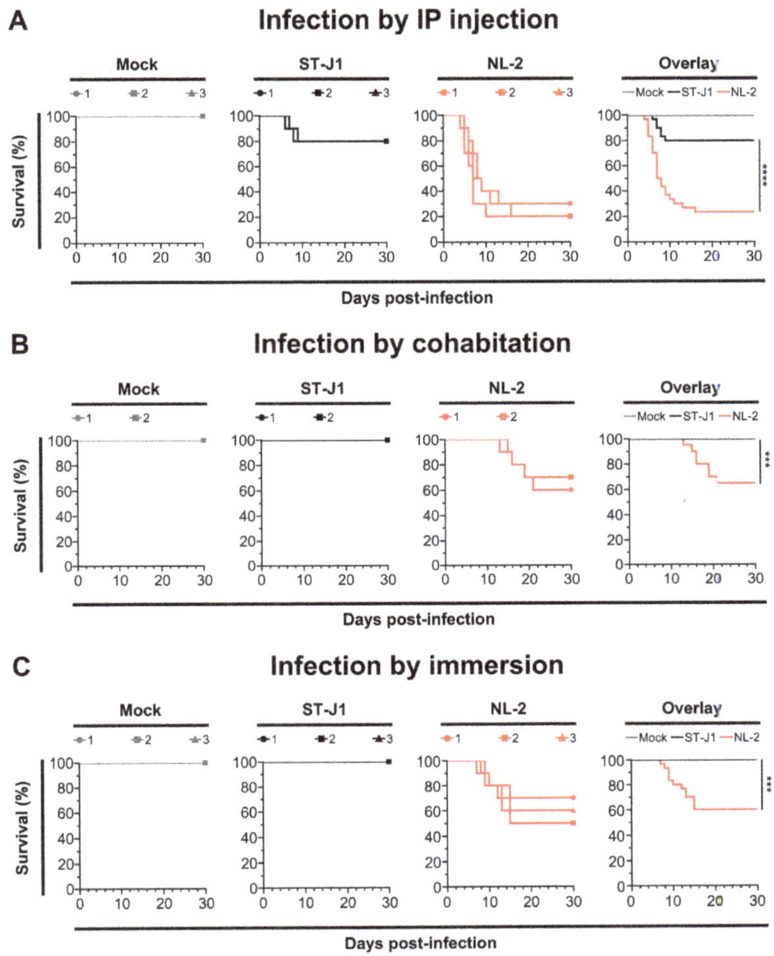

Figure 7. Comparison of NL-2 and ST-J1 in young adult Shubunkin goldfish in vivo. The pathogenicity of the two strains (both at passage #10) was tested in young adult Shubunkin goldfish (average weight 5.84 ± 0.6 g). (**A**) Comparison of in vivo pathogenicity of the ST-J1 and NL-2 strains after infection by IP injection. Fish were mock-infected or infected by IP injection (triplicate groups each consisting of 10 subjects) with the indicated strains. Each subject received a 50 μL injection of inoculum (1000 PFU/fish). The fish were monitored daily for clinical signs of CyHV-2 disease, and fish reaching the endpoints were euthanized. In the overlay graph, the mean survival results were compared. (**B**) Comparison of in vivo pathogenicity of the ST-J1 and NL-2 strains after infection by cohabitation. The same number of goldfish were placed as sentinel fish in the first two replicates of each group in the experiment described in (**A**) (5 days after infection by IP injection). Fish were examined daily, and those reaching the endpoints were euthanized. In the overlay graph, the mean survival results were compared. (**C**) Comparison of in vivo pathogenicity of the ST-J1 and NL-2 strains following infection by immersion. Fish were mock-infected or infected by immersion in water containing each virus strain at a final concentration of 2000 PFU/mL (triplicate groups each consisting of 10 subjects). Fish were examined daily, and those reaching the endpoints were euthanized. In the overlay graph, the mean survival results were compared.

We observed a similar contrast between the two strains in an additional experiment, which involved inoculation by immersion in water containing virus. This involved the immersion of 10 subjects in 1L of water containing 2000 PFU/mL of each strain, for a period of 2 h, before being returned to 60 L tanks. Unlike IP injection, this mode of inoculation is more representative of the natural waterborne infection route. Using this inoculation strategy, the NL-2-inoculated fish exhibited an average mortality rate of approximately 40%, which occurred during one to two weeks post-infection (Figure 7C). As per earlier experiments, the typical clinical symptoms such as abdominal swelling were observed, but more severe skin lesions were observed in fish from the NL-2 infection group. Moribund fish exhibited multiple skin hemorrhages and enlarged kidneys (Figure S2A). In contrast to the earlier experiments, a white necrosis appeared on the tail fins of all goldfish three days after infection, with some fish experiencing tail fin structure destruction and impaired swimming posture in the later stages of infection (Figure S2B). In contrast, no mortality or clinical disease was observed within the ST-J1 group of fish. Taken together, these experiments indicated that the NL-2 isolate was much more suitable for in vivo infection experiments in European-sourced Shubunkin goldfish populations, compared to the ST-J1 CyHV-2 reference.

4. Discussion

A crucial element in the study of any virus is the identification of a suitable virus–host model for in vivo experimental infection. Being able to study a virus in its natural host facilitates more biologically relevant insights into many aspects of virus–host interaction. Like all known members of the genus *Cyvirus*, CyHV-2 can be easily studied in its natural hosts, including goldfish, making them interesting models to use in the study of herpesvirus infections, with strong relevance to other vertebrate species. Indeed, this process is greatly facilitated by the availability of a host that can be easily bred and maintained under controlled conditions to allow a consistent and sufficient supply of subjects for experiments.

With regard to CyHV-2, there are many breeds of goldfish that meet these criteria, including Shubunkin goldfish. In addition to being susceptible to CyHV-2, Shubunkin goldfish are heavily associated with international trade [16], which is likely to be a key epidemiological factor in the global spread of CyHV-2, making them relevant host models in the context of research into disease control and mitigation. However, while the CyHV-2 ST-J1 strain can induce mortality in European-sourced adult Shubunkin goldfish populations by IP injection, these populations exhibit no mortality when challenged with this virus strain by immersion (Figure 7), with similar observations after YC-01 CyHV-2 challenge [15]. However, it also remains possible that ORF disruptions in these CyHV-2 strains—which may or may not have emerged during passaging [7,44]—may also be responsible for this lack of virulence, which we will discuss later. Regardless of whether this is due to host or viral factors, this renders these CyHV-2 strains sub-optimal to use, and with these, a wide range of economically and epidemiologically relevant cohorts. At the same time, outbreaks of CyHV-2 among European populations of Shubunkin goldfish indicate that they remain susceptible to new and uncharacterized CyHV-2 variants [16]. We therefore sought to identify and characterize some of these new CyHV-2 isolates, in order to assess their utility in future CyHV-2 studies utilizing the Shubunkin goldfish infection model.

In Europe, where CyHV-2 has a wide geographical distribution, CyHV-2 outbreaks are frequently described in goldfish and gibel carp populations in both aquaculture facilities and in the wild [45,48–52]. Like other regions within Europe, The Netherlands has experienced outbreaks as early as 2011, with repeated occurrences since then [53]. As per patterns across the globe, these represent a mixture of cases in both domesticated and wild fish [16,17,54]. Given the frequency of CyHV-2 detection in The Netherlands, it represented an ideal source from which to obtain high quality tissue samples from diseased fish for further investigation, ultimately leading to the isolation of three new isolates, NL-1, NL-2, and NL-3. Further in vitro and in vivo characterization established that the NL-2 isolate (*i*) could be stably cultured in cells, leading to the production of a viral stock, and (*ii*) exhibits much greater virulence in adult Shubunkin goldfish relative to the CyHV-2 ST-J1 reference strain. Given their properties, the use of this CyHV-2 NL-2 isolate in combination with Shubunkin goldfish represents a more optimal in vivo infection model compared to the use of the CyHV-2 ST-J1 reference strain. Thus, this newly characterized in vivo infection model represents an ideal tool to use in future research into CyHV-2 disease mitigation, control, and viral host interaction.

Whole genome sequencing of the NL-1, NL-2, and NL-3 isolates and genetic comparison of these strains to existing strains indicated that they are distantly related to the CyHV-2 ST-J1 reference strain and the CyHV-2 YC-01 strain, both of which cause limited or no mortality in adult Shubunkin goldfish when exposed to virus by cohabitation and/or immersion (see [15] and Figure 7). Instead, these new isolates are more related to other CyHV-2 strains that occupy the C genotype clade (Figure 3B). Furthermore, these three new isolates are also genetically distinct from each other and thus occupy separate sub-clades within the C genotype clade. However, the genetic differences between these new isolates and the ST-J1 reference strain may potentially provide some interesting insights into phenotypic differences observed in further characterization (discussed in more detail later).

The initial in vivo characterization of these new CyHV-2 isolates revealed that they could all cause clinical disease associated with CyHV-2 in adult Shubunkin goldfish subjected to IP inoculation. In all cases, this also led to the subsequent transmission of virus to cohabiting fish (Figure 4). While this experiment was conducted using unknown titers of each isolate, making it difficult to make a valid comparison of virulence between the strains, importantly, it did indicate that they were capable of inducing mortality in adult Shubunkin goldfish. Notably, in this experiment, the results with NL-1 and NL-2 were consistent with our previous observations with the CyHV-2 Shubunkin goldfish model [15], which indicated that earlier developmental stages are more permissive of CyHV-2 replication and exhibit higher mortality rates. However, again, NL-3 was slightly different in this regard; while mortality certainly occurred earlier and faster in smaller fish in this group, it ceased very early. Taken together these experiments demonstrated that the p#0 inoculums prepared from NL-1, NL-2, and NL-3 tissue homogenates contained infectious CyHV-2, capable of causing clinical disease and mass mortality in the Shubunkin goldfish model, and therefore, we concluded that they may represent ideal candidates to replace CyHV-2 ST-J1 or YC-01 strains in this in vivo infection model, as the latter cause relatively little disease after infection by immersion.

However, in addition to causing virulence in vivo, for these new viral isolates to be useful, it was equally important that they could be stably and efficiently passaged in vitro. The initial in vitro culture of the three CyHV-2 isolates revealed slow CPE development in RyuF-2 cells post-inoculation (0 to #1), with virus plaques identified after approximately 10 days (Figure 1B). This was similar across the three isolates and consistent with the previous isolation of other CyHV-2 strains [45,55–57]. However, differences between the isolates emerged during continuous passage. Most notably, we observed that CPE caused by NL-3 could not be observed beyond the third passage. While this does not mean that the virus is totally absent after the third passage, it does indicate that the initial NL-3 phenotype may be intrinsically unstable in vitro. It also suggests that high levels of productive lytic infection may be limited to early passages only when using the NL-3 isolate. Similar instances of instability of CyHV-2 isolates in vitro have been observed using KF-1 cell lines [45,57] and the goldfish-derived cell lines GFSe and GFKf [58]. This suggests that the NL-3 isolate is intrinsically less fit in vitro compared to the other isolates. Alternatively, while it could not be passaged continuously in RyuF-2 cells, it remains possible that it exhibits greater fitness in other cells that are permissive to CyHV-2 replication. For example, in previous studies, it was observed that some CyHV-2 isolates were capable of causing CPE in EPC cells, while others were not [10]. However, it is also important to consider experimental differences between these studies that may also contribute to these differences. In contrast to NL-3, the NL-1 and NL-2 isolates could be passaged more stably in RyuF-2 cells, with NL-2 exhibiting superior replication kinetics relative to NL-1 in this cell line. However, the CyHV-2 ST-J1 reference strain exhibited much higher replication kinetics, compared to NL-2. Taken together, these observations allow us to infer a simple ranking system in terms of replication kinetics or fitness in RyuF-2 cells, i.e., "ST-J1 > NL-2 > NL-1 > NL-3".

The genetic determinants of these different phenotypes are unclear as each isolate exhibits hundreds of mutations relative to the ST-J1 reference strain (Figure 2, Tables S1–S3). Most of these are SNVs that occur either outside of ORFs or result in synonymous changes within ORFs, with no impact on protein sequences (Figure 3). However, other mutations, such as insertions and deletions, can cause frameshifts within ORFs, and hence, these may be much more likely to result in phenotypic impacts (Tables S4–S6). Notably, all of the new isolates have frameshifts in ORF138, ORF151A, ORF25B, and ORF156, relative to ST-J1. The frameshifts in ORF138 and ORF25B are identical in all of the new isolates and are also present in other CyHV-2 strains. The frameshifts in ORF156 are identical in NL-1 and NL-3, and a very similar frameshift is present in NL-2 ORF156, leading to a similar ORF156 truncation. These common mutations point towards a monophyletic origin of these three new isolates separate from the J genotype. Conversely, the ORF151A frameshifts in the NL-1, NL-2, and NL-3 isolates are not identical to each other, and only the NL-2 version of ORF151A is found in other CyHV-2 strains. It is possible that the ST-J1 proteome is more optimum for CyHV-2 fitness in vitro. Outside of these four major differences, which are common to all three isolates, additional changes relative to this "optimum proteome" may further reduce fitness in vitro. Consistent with this, it is notable that NL-3, which exhibited the least fitness in vitro, has more frameshifts (a total of eight, including frameshifts in ORF28A, ORF30, ORF89, and ORF90) relative to ST-J1 than the other two isolates. Thus, the NL-3 proteome, which is predicted to be more divergent from that of ST-J1 than those from the other two isolates, may be less optimal for CyHV-2 fitness in vitro. However, NL-2 proteome, exhibiting seven frameshifts relative to ST-J1, is also quite divergent from the ST-J1 proteome, yet can be stably passaged in vitro. This raises the possibility that the specific ORFs disrupted in NL-3 may be important for facilitating stable passaging of CyHV-2 in vitro, and it may be interesting to investigate this further in future studies. However, it is important to note that outside of highly disruptive changes such as frameshifts, it remains possible that the smaller protein changes, such as single nucleotide changes leading to non-synonymous amino acid substitutions, may also impact fitness in vitro, as we described recently in the closely related virus, *Cyvirus cyprinidallo3* [59] (or CyHV-3 formerly known as *Cyprinid herpesvirus 3* [4]).

The direct comparisons between the CyHV-2 NL-2 isolate and the CyHV-2 reference strain ST-J1 in vitro and in vivo, also yielded some interesting results. Despite the ST-J1 being more fit in vitro, it was less fit in vivo, and vice versa for NL-2. This is similar to our previous observations with different CyHV-3 strains [59,60]. This may occur as a consequence of continuous passaging in vitro, which introduces a selective pressure that favors variants that are adapted to in vitro conditions. In the absence of any selective pressure to retain in vivo fitness, these in vitro adapted strains may develop an attenuated phenotype in vivo, as we described previously with CyHV-3 variants [59]. It is unclear if the same has occurred with ST-J1, leading to an attenuated phenotype in some goldfish breeds (or populations thereof). Notably, there were no genomic differences between the original ST-J1 stocks used to generate a viral stock for these experiments and the ST-J1 reference sequence in GenBank (data not shown).

Interestingly, a total of seven frameshifted ORFs exist in NL-2 relative to the ST-J1 strain. These include ORF1, ORF16, ORF25B, ORF52, ORF138, ORF151A, and ORF156 (Table S5). In the case of NL-2 ORF1, while a frameshift variant is present at the start of the coding sequence, as the start codon is still retained, there is actually no impact on the protein sequence. Given that these represent the biggest differences in their respective proteomes, it may be interesting to determine if ST-J1 virulence in vivo may be improved through replacement of one or more of these with the equivalent NL-2 allele(s). With the exception of ORF156, the functions of the proteins encoded by these CyHV-2 ORFs have been predicted based on homology with other proteins in related viruses such as CyHV-3 [7], which may allow us to hypothesize which of these changes are more likely to be responsible for these differences in phenotype between ST-J1 and NL-2 in vivo.

Some of these frameshifts result in the truncation of predicted proteins in NL-2 relative to ST-J1 and other CyHV-2 strains, some of which may be responsible for the reduced fitness of NL-2 relative to ST-J1 in vitro. For example, ORF52 is predicted to encode a signal peptide that is truncated in NL-2 relative to ST-J1 and most other fully sequenced strains. However, the potential functional impact of this truncation on NL-2 fitness in vitro relative to ST-J1 is unclear. In the case of other frameshift mutations causing truncated predicted protein products in NL-2, the potential impact in vitro may be more apparent. For example, in NL-2, ORF16 and ORF138 are both predicted to encode for truncated proteins relative to ST-J1 and other fully sequenced CyHV-2 strains. ORF16 and ORF138 are predicted to encode for type III and type I viral membrane proteins, respectively, both of which are important for herpesvirus viral entry and egress [61,62]. Although potentially important for viral entry and egress, these viral membrane protein truncations in the NL-2 isolate do not totally prevent host cell entry or egress in vitro, but it remains possible that they may reduce fitness relative to ST-J1 in this environment (Figure 6C). However, it should be noted that the same truncated membrane proteins did not prevent NL-2 from exhibiting a much higher virulence and transmission in vivo relative to ST-J1 (Figure 7), although smaller differences elsewhere in either strains may also contribute to these phenotypic observations.

Conversely, other frameshifts in NL-2 relative to ST-J1, such as those in ORF25B and ORF151A, represent truncations of these protein products in ST-J1 relative to NL-2 and other fully sequenced CyHV-2 strains. Notably, ORF25B is a type I viral membrane protein and may therefore be important for viral entry and egress. However, this ORF25B truncation in ST-J1 does not reduce its fitness in vitro relative to NL-2 (Figure 6C). While we observe a lack of virulence and transmission when ST-J1 is delivered by immersion in vivo (Figure 7), it is unclear if this is directly connected to this ORF25B truncation. Notably, in CyHV-3, we previously provided evidence that mutations in type I membrane proteins can impact virion stability in extracellular environments such as mucosal surfaces [59], and it would be interesting to investigate this possibility with CyHV-2 ORF25B in the future in a similar manner. In contrast to other major gene disruptions between NL-2 and ST-J1, the in vivo impact of the truncation of ORF151A in ST-J1 may be more readily apparent. ORF151A encodes for a protein that is predicted to be a viral receptor for Tumor Necrosis Factor (TNF) and is a member of the TNF receptor (TNFR) family of proteins that are encoded by CyHV-1, CyHV-2, and CyHV-3 [60]. TNF is an important cytokine of the innate immune response and is produced in fish in the early stages of viral infection in response to the nuclear factor-κB (NF-κB) inflammatory pathway. Through interaction with its natural cellular receptors (cellular TNFRs), TNF stimulates signaling that leads to an inflammatory response in order to help recruitment of fish immune cells to the site of infection, stimulate their activation, and also cause behavioral fever [63–66], all of which are important for rapid clearance of viral infections. However, virally encoded TNFRs can act as "decoy" TNFRs, preventing interaction with the cellular TNFRs and thus reducing the antiviral effects of cellular TNF [63,66]. Indeed, we previously demonstrated that a soluble decoy TNFR expressed by CyHV-3 can interfere with behavioral fever response during infection [67]. However, ST-J1 encodes a truncated form of ORF151A (129AA) in which the TNFR domain is missing. Conversely, unlike ST-J1, NL-2 and other CyHV-2 strains encode a full-length protein that includes the TNFR domain (324 AA) (InterProScan analysis, data not shown). Therefore, it is possible that the ability of the ST-J1 strain to counteract host cell TNF during infection is compromised due to this missing decoy TNF receptor encoded by the full-length ORF151A. Given the roles that TNF plays during viral infection, this deficiency may have a greater impact in vivo than in vitro, leading for ST-J1 to exhibit an attenuated phenotype in vivo, relative to NL-2. This may partially account for the opposite patterns we observed between ST-J1 and NL-2 in terms of in vitro and in vivo fitness. Notably, ST-J1 is the only fully sequenced CyHV-2 strain that has been observed to encode truncations in both ORF25B and ORF151A, and it is possible that the reduced virulence in vivo relative to that of NL-2 may be caused by the combined impact of both these frameshifts. Taken together, the genomic and biological comparisons of CyHV-2 from this study have allowed the formation of some interesting hypotheses on some potential determinants of CyHV-2 virulence in vivo, which provide scope for further investigation in future studies.

According to the literature, laboratory challenges with CyHV-2 are primarily conducted via intraperitoneal (IP) injection of infected tissue homogenates or cell culture fluid containing the virus, which can rapidly induce clinical symptoms and high mortality [56,68–71]. However, this highly artificial challenge method has significant drawbacks. It amplifies the pathogenicity and virulence of the virus, fails to represent the series of events that occur during a natural infection, and does not reflect the natural immune response of the host during a CyHV-2 infection. In fish, mucosal surfaces, including the skin, gills, and gastrointestinal mucosa, serve as the primary entry routes for viruses [72], and previously, we identified the skin as a major entry portal for CyHV-2 [15]. Therefore, identifying an in vivo infection model that will allow us to simulate this natural infection process via the mucosal surfaces is crucial in order to have biologically relevant infection experiment designs. In this study, the inoculation of adult goldfish with NL-2 by immersion using a final concentration of 2000 PFU/mL led to 40% mortality in goldfish (Figure 7). Notably, this dose is 50 times lower than the dose we previously used in equivalent experiments with the YC-01 strain (1×10^5 PFU/mL) [15], which, despite the higher dose, led to little or no mortality in similar adult Shubunkin goldfish. This would indicate that the NL-2 Shubunkin goldfish model is ideal for future studies of CyHV-2, maximizing the biological relevance of observations, whereby it also represents a useful tool to the wider CyHV-2 research community.

5. Conclusions

Three isolates of the Dutch CyHV-2 isolates from goldfish or wild gibel carp, NL-1, NL-2, and NL-3, were characterized in this study. They were found to be genetically distinct from the ST-J1 CyHV-2 reference strain and from each other. They were all found to cause disease associated with CyHV-2 in experiments that involved infection of adult Shubunkin goldfish by either IP injection or cohabitation. Among these, NL-2 could be stably passaged in vitro, and it also exhibited the most efficient replication in vitro. NL-2 was also capable of inducing CyHV-2-related mortality in adult Shubunkin goldfish through inoculation by immersion. These two properties make the NL-2 Shubunkin model an optimum in vivo infection model to use in future studies, as it facilitates both stable maintenance and generation of a viral stock in vitro and the infection of hosts in a way that is more representative of a natural infection with CyHV-2. Importantly, the genetic and phenotypic characterization performed in this study generates hypotheses on the potential roles of CyHV-2 genes in adaptation of the virus in vitro or in vivo.

Supplementary Materials: The following supporting information can be downloaded at https://www.mdpi.com/article/10.3390/v17050658/s1: Figure S1. Representative examples of autopsy observations in Shubunkin goldfish undergoing mortality after IP infection with p#0 inoculums of novel CyHV-2 isolates. Figure S2. Representative examples of clinical symptoms in Shubunkin goldfish after infection with CyHV-2 NL-2 isolate by immersion. Table S1. Summary of NL-1 variants. Table S2. Summary of NL-2 variants. Table S3. Summary of NL-3 variants. Table S4. Summary of NL-1 variant effects. Table S5. Summary of NL-2 variant effects. Table S6. Summary of NL-3 variant effects.

Author Contributions: Conceptualization, B.H., A.F.C.V. and O.D.; data curation, B.H. and O.D.; formal analysis, B.H. and O.D.; funding acquisition, A.F.C.V. and O.D.; investigation, B.H., A.F.C.V. and O.D.; methodology, B.H., A.S., M.T., O.H., A.F.C.V. and O.D.; project administration, A.F.C.V.; supervision, A.F.C.V. and O.D.; validation, B.H. and O.D.; visualization, B.H. and O.D.; writing—original draft, B.H. and O.D.; writing—review and editing, B.H., A.S., A.F.C.V. and O.D. All authors have read and agreed to the published version of the manuscript.

Funding: Bo He is a research fellow of the Chinese Scholarship Council. Arun Sridhar is a post-doctoral research fellow of the Wel Research Institute (Welbio). This work was supported by the University of Liège (FSR-SS-2022/64, FSR-SS-2023/49), the FNRS (PDR T.0241.19), the Welbio (WELBIO-CR-2022 A-14 12), and the Walloon Region (ERANets NucNanoFish 2010052). Computational resources have been provided by the Consortium des Équipements de Calcul Intensif (CÉCI), funded by the F.R.S.-FNRS under Grant No. 2.5020.11 and by the Walloon Region.

Institutional Review Board Statement: Not applicable.

Informed Consent Statement: Not applicable.

Data Availability Statement: The data presented in this study are available on request from the corresponding authors.

Acknowledgments: The authors are grateful to Lorène Dams, Caroline Deketelaere, Aurélie Vanderlinden (Faculty of Veterinary Medicine, ULiège), David Colignon (CECI, ULiège), and Patricia Piscicelli (Faculty of Sciences, ULiège) for their excellent technical assistance and to Natacha Delrez for transporting the tissue samples from The Netherlands.

Conflicts of Interest: The authors declare no conflicts of interest.

References

1. Hanson, L.A.; Doszpoly, A.; van Beurden, S.; de Oliveira Viadanna, P.H.; Waltzek, T. Chapter 8—Alloherpesviruses of Fish. In *Aquaculture Virology*, 2nd ed.; Kibenge, F.S.B., Godoy, M.G., Eds.; Academic Press: Cambridge, MA, USA, 2024; pp. 165–189. ISBN 978-0-323-91169-6.
2. Boutier, M.; Morvan, L.; Delrez, N.; Origgi, F.; Doszpoly, A.; Vanderplasschen, A. Fish and Amphibian Alloherpesviruses (Herpesviridae). In *Encyclopedia of Virology*, 4th ed.; Bamford, D.H., Zuckerman, M., Eds.; Academic Press: Oxford, UK, 2021; pp. 306–315. ISBN 978-0-12-814516-6.
3. Pradeepkiran, J.A. Aquaculture Role in Global Food Security with Nutritional Value: A Review. *Transl. Anim. Sci.* **2019**, *3*, 903–910. [CrossRef] [PubMed]
4. ICTV. Available online: https://ictv.global/taxonomy (accessed on 5 September 2024).
5. Donohoe, O.; Zhang, H.; Delrez, N.; Gao, Y.; Suárez, N.M.; Davison, A.J.; Vanderplasschen, A. Genomes of Anguillid Herpesvirus 1 Strains Reveal Evolutionary Disparities and Low Genetic Diversity in the Genus Cyprinivirus. *Microorganisms* **2021**, *9*, 998. [CrossRef] [PubMed]

6. Streiff, C.; He, B.; Morvan, L.; Zhang, H.; Delrez, N.; Fourrier, M.; Manfroid, I.; Suárez, N.M.; Betoulle, S.; Davison, A.J.; et al. Susceptibility and Permissivity of Zebrafish (*Danio rerio*) Larvae to Cypriniviruses. *Viruses* **2023**, *15*, 768. [CrossRef] [PubMed]
7. Davison, A.J.; Kurobe, T.; Gatherer, D.; Cunningham, C.; Korf, I.; Fukuda, H.; Hedrick, R.P.; Waltzek, T.B. Comparative Genomics of Carp Herpesviruses. *J. Virol.* **2013**, *87*, 2908–2922. [CrossRef]
8. Padhiary, S.; Paul, A.; Tripathy, D.K.; Nayak, D.; Mohanty, J.; Pillai, D.; Rejish Kumar, V.J.; Krishnan, R.; Haridas, L.; Abdul Majeed, S.; et al. Current understanding of Cyprinid herpesvirus 2, and prospects in its management. *Indian J. Anim. Health* **2023**, *62*, 49–60. [CrossRef]
9. Jung, S.J.; Miyazaki, T. Herpesviral Haematopoietic Necrosis of Goldfish, *Carassius auratus* (L.). *J. Fish Dis.* **1995**, *18*, 211–220. [CrossRef]
10. Thangaraj, R.S.; Nithianantham, S.R.; Dharmaratnam, A.; Kumar, R.; Pradhan, P.K.; Thangalazhy Gopakumar, S.; Sood, N. Cyprinid Herpesvirus-2 (CyHV-2): A Comprehensive Review. *Rev. Aquac.* **2021**, *13*, 796–821. [CrossRef]
11. Zhang, H.; Sridhar, A.; Delrez, N.; He, B.; Fourny, S.; Gao, Y.; Donohoe, O.; Vanderplasschen, A.F.C. Development Using Bioluminescence Imaging of a Recombinant Anguillid Herpesvirus 1 Vaccine Candidate Associated with Normal Replication In Vitro but Abortive Infection In Vivo. *Vaccines* **2024**, *12*, 1423. [CrossRef]
12. Boutier, M.; Ronsmans, M.; Ouyang, P.; Fournier, G.; Reschner, A.; Rakus, K.; Wilkie, G.S.; Farnir, F.; Bayrou, C.; Lieffrig, F.; et al. Rational Development of an Attenuated Recombinant Cyprinid Herpesvirus 3 Vaccine Using Prokaryotic Mutagenesis and In Vivo Bioluminescent Imaging. *PLoS Pathog.* **2015**, *11*, e1004690. [CrossRef]
13. Saito, H.; Okamura, T.; Shibata, T.; Kato, G.; Sano, M. Development of a Live Attenuated Vaccine Candidate against Herpesviral Hematopoietic Necrosis of Goldfish. *Aquaculture* **2022**, *552*, 737974. [CrossRef]
14. Sun, Y.; Xu, C.; Wang, H.; Qiao, G.; Wang, Z.; Li, Z.; Li, Q.; Wei, C. An Attenuated Strain of Cyprinid Herpesvirus 2 as a Vaccine Candidate against Herpesviral Hematopoietic Necrosis Disease in Gibel Carp, *Carassius auratus gibelio*. *Fish Shellfish Immunol.* **2023**, *138*, 108826. [CrossRef] [PubMed]
15. He, B.; Sridhar, A.; Streiff, C.; Deketelaere, C.; Zhang, H.; Gao, Y.; Hu, Y.; Pirotte, S.; Delrez, N.; Davison, A.J.; et al. In Vivo Imaging Sheds Light on the Susceptibility and Permissivity of *Carassius auratus* to Cyprinid Herpesvirus 2 According to Developmental Stage. *Viruses* **2023**, *15*, 1746. [CrossRef] [PubMed]
16. Ito, T.; Kurita, J.; Haenen, O.L.M. Importation of CyHV-2-Infected Goldfish into the Netherlands. *Dis. Aquat. Org.* **2017**, *126*, 51–62. [CrossRef] [PubMed]
17. Kurobe, T.; Kurita, J.; Haenen, O.; Voorbergen-Laarman, M.; Ito, T. Mass Mortality Events Associated with Cyprinid Herpesvirus 2 (CyHV-2) Infection in Wild Prussian Carp *Carassius gibelio* in the Netherlands, and Molecular Biology of Virus Strains. *J. Fish Dis.* **2024**, *47*, e13868. [CrossRef]
18. Waltzek, T.B.; Kurobe, T.; Goodwin, A.E.; Hedrick, R.P. Development of a Polymerase Chain Reaction Assay to Detect Cyprinid Herpesvirus 2 in Goldfish. *J. Aquat. Anim. Health* **2009**, *21*, 60–67. [CrossRef]
19. Shibata, T.; Nanjo, A.; Saito, M.; Yoshii, K.; Ito, T.; Nakanishi, T.; Sakamoto, T.; Sano, M. In Vitro Characteristics of Cyprinid Herpesvirus 2: Effect of Kidney Extract Supplementation on Growth. *Dis. Aquat. Org.* **2015**, *115*, 223–232. [CrossRef]
20. BBTools. Available online: https://jgi.doe.gov/data-and-tools/software-tools/bbtools/ (accessed on 5 October 2022).
21. Antipov, D.; Korobeynikov, A.; McLean, J.S.; Pevzner, P.A. hybridSPAdes: An Algorithm for Hybrid Assembly of Short and Long Reads. *Bioinformatics* **2016**, *32*, 1009–1015. [CrossRef]
22. Mikheenko, A.; Prjibelski, A.; Saveliev, V.; Antipov, D.; Gurevich, A. Versatile Genome Assembly Evaluation with QUAST-LG. *Bioinformatics* **2018**, *34*, i142–i150. [CrossRef]
23. Wick, R.R.; Schultz, M.B.; Zobel, J.; Holt, K.E. Bandage: Interactive Visualization of de Novo Genome Assemblies. *Bioinformatics* **2015**, *31*, 3350–3352. [CrossRef]
24. SnapGene LLC. SnapGene | Software for Everyday Molecular Biology. Available online: https://www.snapgene.com/ (accessed on 7 May 2024).
25. Katoh, K.; Rozewicki, J.; Yamada, K.D. MAFFT Online Service: Multiple Sequence Alignment, Interactive Sequence Choice and Visualization. *Brief. Bioinform.* **2019**, *20*, 1160–1166. [CrossRef]
26. Complete Genomes: Alloherpesviridae. Available online: https://www.ncbi.nlm.nih.gov/genomes/GenomesGroup.cgi?taxid=548682 (accessed on 7 August 2024).
27. Viral Genomes. Available online: https://www.ncbi.nlm.nih.gov/genome/viruses/ (accessed on 7 August 2024).
28. Kumar, S.; Stecher, G.; Li, M.; Knyaz, C.; Tamura, K. MEGA X: Molecular Evolutionary Genetics Analysis across Computing Platforms. *Mol. Biol. Evol.* **2018**, *35*, 1547–1549. [CrossRef] [PubMed]
29. Lindenbaum, P. JVarkit: Java-Based Utilities for Bioinformatics. 2015. Available online: https://github.com/lindenb/jvarkit (accessed on 1 April 2025).
30. MsaToVcf. Available online: http://lindenb.github.io/jvarkit/MsaToVcf.html (accessed on 27 June 2024).
31. Kim, D.; Paggi, J.M.; Park, C.; Bennett, C.; Salzberg, S.L. Graph-Based Genome Alignment and Genotyping with HISAT2 and HISAT-Genotype. *Nat. Biotechnol.* **2019**, *37*, 907–915. [CrossRef] [PubMed]

32. Pertea, M.; Kim, D.; Pertea, G.M.; Leek, J.T.; Salzberg, S.L. Transcript-Level Expression Analysis of RNA-Seq Experiments with HISAT, StringTie and Ballgown. *Nat. Protoc.* **2016**, *11*, 1650–1667. [CrossRef] [PubMed]
33. Danecek, P.; Bonfield, J.K.; Liddle, J.; Marshall, J.; Ohan, V.; Pollard, M.O.; Whitwham, A.; Keane, T.; McCarthy, S.A.; Davies, R.M.; et al. Twelve Years of SAMtools and BCFtools. *Gigascience* **2021**, *10*, giab008. [CrossRef]
34. Robinson, J.T.; Thorvaldsdóttir, H.; Winckler, W.; Guttman, M.; Lander, E.S.; Getz, G.; Mesirov, J.P. Integrative Genomics Viewer. *Nat. Biotechnol.* **2011**, *29*, 24–26. [CrossRef]
35. McLaren, W.; Gil, L.; Hunt, S.E.; Riat, H.S.; Ritchie, G.R.S.; Thormann, A.; Flicek, P.; Cunningham, F. The Ensembl Variant Effect Predictor. *Genome Biol.* **2016**, *17*, 122. [CrossRef]
36. Geneious | Bioinformatics Software. Available online: https://www.geneious.com/ (accessed on 4 July 2024).
37. Baldwin, S.; Revanna, R.; Thomson, S.; Pither-Joyce, M.; Wright, K.; Crowhurst, R.; Fiers, M.; Chen, L.; Macknight, R.; McCallum, J.A. A Toolkit for Bulk PCR-Based Marker Design from next-Generation Sequence Data: Application for Development of a Framework Linkage Map in Bulb Onion (*Allium cepa* L.). *BMC Genom.* **2012**, *13*, 637. [CrossRef]
38. McCallum, J. Cfljam/Galaxy-Pcr-Markers 2022. Available online: https://github.com/cfljam/galaxy-pcr-markers (accessed on 1 April 2025).
39. R Core Team. *R: A Language and Environment for Statistical Computing*; R Foundation for Statistical Computing: Vienna, Austria, 2022.
40. John, F.; Sanford, W. *An R Companion to Applied Regression*; SAGE Publications, Inc.: Los Angeles, CA, USA, 2018; ISBN 978-1-5443-3647-3.
41. Lenth, R.V. Least-Squares Means: The R Package Lsmeans. *J. Stat. Softw.* **2016**, *69*, 1–33. [CrossRef]
42. Li, L.; Luo, Y.; Gao, Z.; Huang, J.; Zheng, X.; Nie, H.; Zhang, J.; Lin, L.; Yuan, J. Molecular Characterisation and Prevalence of a New Genotype of Cyprinid Herpesvirus 2 in Mainland China. *Can. J. Microbiol.* **2015**, *61*, 381–387. [CrossRef]
43. Wang, F.; Xu, Y.; Zhou, Y.; Ding, C.; Duan, H. Isolation and Characterization of a Cyprinid Herpesvirus Strain YZ01 from Apparently Healthy Goldfish after Rising Water Temperature. *bioRxiv* **2021**. [CrossRef]
44. Yang, J.; Xiao, S.; Lu, L.; Wang, H.; Jiang, Y. Genomic and Molecular Characterization of a Cyprinid Herpesvirus 2 YC-01 Strain Isolated from Gibel Carp. *Heliyon* **2024**, *10*, e32811. [CrossRef] [PubMed]
45. Jeffery, K.R.; Bateman, K.; Bayley, A.; Feist, S.W.; Hulland, J.; Longshaw, C.; Stone, D.; Woolford, G.; Way, K. Isolation of a Cyprinid Herpesvirus 2 from Goldfish, *Carassius auratus* (L.), in the UK. *J. Fish Dis.* **2007**, *30*, 649–656. [CrossRef] [PubMed]
46. Luo, Y.Z.; Lin, L.; Liu, Y.; Wu, Z.X.; Gu, Z.M.; Li, L.J.; Yuan, J.F. Haematopoietic Necrosis of Cultured Prussian Carp, Carassius Gibelio (Bloch), Associated with Cyprinid Herpesvirus 2. *J. Fish Dis.* **2013**, *36*, 1035–1039. [CrossRef]
47. Stephens, F.J.; Raidal, S.R.; Jones, B. Haematopoietic Necrosis in a Goldfish (*Carassius auratus*) Associated with an Agent Morphologically Similar to Herpesvirus. *Aust. Vet. J.* **2004**, *82*, 167–169. [CrossRef]
48. Adamek, M.; Hellmann, J.; Jung-Schroers, V.; Teitge, F.; Steinhagen, D. CyHV-2 Transmission in Traded Goldfish Stocks in Germany—A Case Study. *J. Fish Dis.* **2018**, *41*, 401–404. [CrossRef]
49. Boitard, P.-M.; Baud, M.; Labrut, S.; de Boisséson, C.; Jamin, M.; Bigarré, L. First Detection of Cyprinid Herpesvirus 2 (CyHV-2) in Goldfish (*Carassius auratus*) in France. *J. Fish Dis.* **2016**, *39*, 673–680. [CrossRef]
50. Daněk, T.; Kalous, L.; Vesel, T.; Krásová, E.; Reschová, S.; Rylková, K.; Kulich, P.; L, M.P.; Pokorová, D.; Knytl, M. Massive Mortality of Prussian Carp *Carassius gibelio* in the Upper Elbe Basin Associated with Herpesviral Hematopoietic Necrosis (CyHV-2). *Dis. Aquat. Organ.* **2012**, *102*, 87–95. [CrossRef]
51. Doszpoly, A.; Benkő, M.; Csaba, G.; Dán, Á.; Lang, M.; Harrach, B. Introduction of the Family Alloherpesviridae: The First Molecular Detection of Herpesviruses of Cyprinid Fish in Hungary. *Magy. Allatorvosok Lapja* **2011**, *133*, 174–181.
52. Giovannini, S.; Bergmann, S.M.; Keeling, C.; Lany, C.; Schütze, H.; Schmidt-Posthaus, H. Herpesviral Hematopoietic Necrosis in Goldfish in Switzerland: Early Lesions in Clinically Normal Goldfish (*Carassius auratus*). *Vet. Pathol.* **2016**, *53*, 847–852. [CrossRef]
53. Haenen, O.; Way, K.; Gorgoglione, B.; Ito, T.; Paley, R.; Bigarré, L.; Waltzek, T. Novel Viral Infections Threatening Cyprinid Fish. *Bull.-Eur. Assoc. Fish Pathol.* **2016**, *36*, 11–23.
54. Schiphouwer, M.; Kessel, N.; Matthews, J.; Leuven, R.S.E.W.; Koppel, S.; Kranenbarg, J.; Haenen, O.; Lenders, H.; Nagelkerke, L.; Van der Velde, G.; et al. *Risk Analysis of Exotic Fish Species Included in the Dutch Fisheries Act and Their Hybrids*; Nederlands Expertise Centrum Exoten (NEC-E): Nijmegen, The Netherlands, 2014; Available online: https://edepot.wur.nl/295342 (accessed on 1 April 2025).
55. Ito, T.; Kurita, J.; Ozaki, A.; Sano, M.; Fukuda, H.; Ototake, M. Growth of Cyprinid Herpesvirus 2 (CyHV-2) in Cell Culture and Experimental Infection of Goldfish *Carassius auratus*. *Dis. Aquat. Org.* **2013**, *105*, 193–202. [CrossRef] [PubMed]
56. Xiao, Z.; Xue, M.; Xu, C.; Jiang, N.; Luo, X.; Li, Y.; Fan, Y.; Meng, Y.; Liu, W.; Zeng, L.; et al. First Report of Cyprinid Herpesvirus 2 Isolated from the Golden Crucian Carp in China. *Aquaculture* **2022**, *558*, 738361. [CrossRef]
57. Xu, J.; Zeng, L.; Zhang, H.; Zhou, Y.; Ma, J.; Fan, Y. Cyprinid Herpesvirus 2 Infection Emerged in Cultured Gibel Carp, *Carassius auratus* Gibelio in China. *Vet. Microbiol.* **2013**, *166*, 138–144. [CrossRef]

58. Jing, H.; Gao, L.; Zhang, M.; Wang, N.; Lin, X.; Zhang, L.; Wu, S. Establishment from the Snout and Kidney of Goldfish, *Carassius auratus*, of Two New Cell Lines and Their Susceptibility to Infectious Pancreatic Necrosis Virus. *Fish Physiol. Biochem.* **2016**, *42*, 303–311. [CrossRef]
59. Gao, Y.; Sridhar, A.; Bernard, N.; He, B.; Zhang, H.; Pirotte, S.; Desmecht, S.; Vancsok, C.; Boutier, M.; Suárez, N.M.; et al. Virus-Induced Interference as a Means for Accelerating Fitness-Based Selection of Cyprinid Herpesvirus 3 Single Nucleotide Variants in Vitro and in Vivo. *Virus Evol.* **2023**, *9*, vead003. [CrossRef]
60. Gao, Y.; Suárez, N.M.; Wilkie, G.S.; Dong, C.; Bergmann, S.; Lee, P.-Y.A.; Davison, A.J.; Vanderplasschen, A.F.C.; Boutier, M. Genomic and Biologic Comparisons of Cyprinid Herpesvirus 3 Strains. *Vet. Res.* **2018**, *49*, 40. [CrossRef]
61. Ahmad, I.; Wilson, D.W. HSV-1 Cytoplasmic Envelopment and Egress. *Int. J. Mol. Sci.* **2020**, *21*, 5969. [CrossRef]
62. Jambunathan, N.; Clark, C.M.; Musarrat, F.; Chouljenko, V.N.; Rudd, J.; Kousoulas, K.G. Two Sides to Every Story: Herpes Simplex Type-1 Viral Glycoproteins gB, gD, gH/gL, gK, and Cellular Receptors Function as Key Players in Membrane Fusion. *Viruses* **2021**, *13*, 1849. [CrossRef]
63. Al Rumaih, Z.; Tuazon Kels, M.J.; Ng, E.; Pandey, P.; Pontejo, S.M.; Alejo, A.; Alcamí, A.; Chaudhri, G.; Karupiah, G. Poxvirus-Encoded TNF Receptor Homolog Dampens Inflammation and Protects from Uncontrolled Lung Pathology during Respiratory Infection. *Proc. Natl. Acad. Sci. USA* **2020**, *117*, 26885–26894. [CrossRef]
64. Janeway, C.A., Jr.; Travers, P.; Walport, M.; Shlomchik, M.J. Induced Innate Responses to Infection. In *Immunobiology: The Immune System in Health and Disease*, 5th ed.; Garland Science: New York, NY, USA, 2001.
65. Netea, M.G.; Kullberg, B.J.; Van der Meer, J.W.M. Circulating Cytokines as Mediators of Fever. *Clin. Infect. Dis.* **2000**, *31*, S178–S184. [CrossRef]
66. Rahman, M.M.; Lucas, A.R.; McFadden, G. Viral TNF Inhibitors as Potential Therapeutics. In *Madame Curie Bioscience Database [Internet]*; Landes Bioscience: Austin, TX, USA, 2013.
67. Rakus, K.; Ronsmans, M.; Forlenza, M.; Boutier, M.; Piazzon, M.C.; Jazowiecka-Rakus, J.; Gatherer, D.; Athanasiadis, A.; Farnir, F.; Davison, A.J.; et al. Conserved Fever Pathways across Vertebrates: A Herpesvirus Expressed Decoy TNF-α Receptor Delays Behavioral Fever in Fish. *Cell Host Microbe* **2017**, *21*, 244–253. [CrossRef] [PubMed]
68. Chai, W.; Qi, L.; Zhang, Y.; Hong, M.; Jin, L.; Li, L.; Yuan, J. Evaluation of Cyprinid Herpesvirus 2 Latency and Reactivation in *Carassius gibel*. *Microorganisms* **2020**, *8*, 445. [CrossRef] [PubMed]
69. Liang, L.-G.; Xie, J.; Chen, K.; Bing, X. Pathogenicity and Biological Characteristics of CyHV-2. *Fish Pathol.* **2015**, *35*, 85–93.
70. Ouyang, P.; Zhou, Y.; Wang, K.; Geng, Y.; Lai, W.; Huang, X.; Chen, D.; Guo, H.; Fang, J.; Chen, Z.; et al. First Report of Cyprinid Herpesvirus 2 Outbreak in Cultured Gibel Carp, *Carassius auratus* Gibelio at Low Temperature. *J. World Aquac. Soc.* **2020**, *51*, 1208–1220. [CrossRef]
71. Wen, J.; Xu, Y.; Su, M.; Lu, L.; Wang, H. Susceptibility of Goldfish to Cyprinid Herpesvirus 2 (CyHV-2) SH01 Isolated from Cultured Crucian Carp. *Viruses* **2021**, *13*, 1761. [CrossRef]
72. Løkka, G.; Gamil, A.A.A.; Evensen, Ø.; Kortner, T.M. Establishment of an In Vitro Model to Study Viral Infections of the Fish Intestinal Epithelium. *Cells* **2023**, *12*, 1531. [CrossRef]

viruses MDPI

Article

Development of IgY-Based Passive Immunization Against Tilapia Lake Virus: Development and In Vitro Neutralization Assays

Piyathip Setthawong [1], Jidapa Yamkasem [2], Matepiya Khemthong [2], Puntanat Tattiyapong [2], Pornphimon Metheenukul [3], Noppadol Prasertsincharoen [3], Tuchakorn Lertwanakarn [1], Naris Thengchaisri [4] and Win Surachetpong [2,5,*]

1 Department of Physiology, Faculty of Veterinary Medicine, Kasetsart University, Bangkok 10900, Thailand; piyathip.s@ku.th (P.S.); tuchakorn.l@ku.th (T.L.)
2 Department of Veterinary Microbiology and Immunology, Faculty of Veterinary Medicine, Kasetsart University, Bangkok 10900, Thailand; jidapa.yam@ku.th (J.Y.); matepiya.kh@ku.th (M.K.); been_best@yahoo.com (P.T.)
3 Department of Veterinary Technology, Faculty of Veterinary Technology, Kasetsart University, Bangkok 10900, Thailand; pornphimon.m@ku.th (P.M.); noppadol.p@ku.th (N.P.)
4 Department of Companion Animal, Faculty of Veterinary Medicine, Kasetsart University, Bangkok 10900, Thailand; ajnaris@yahoo.com
5 Laboratory of Biotechnology, Chulabhorn Research Institute, Bangkok 10210, Thailand
* Correspondence: win.s@ku.th

Abstract: Tilapia lake virus (TiLV) poses a major threat to global tilapia aquaculture and contributes to significant economic losses due to the absence of effective vaccines and treatments. Given the high mortality rates and severe pathological effects of TiLV on tilapia, alternative strategies, such as immunoglobulin-based therapies, are being considered for disease control. In this study, we developed specific immunoglobulin Y (IgY) antibodies against TiLV and evaluated their neutralization activity. Laying hens were immunized via intramuscular injections of recombinant TiLV segment 4 protein, and IgY antibodies were extracted and purified from their egg yolks using polyethylene glycol precipitation. Western blot analysis confirmed the specificity of the IgY, which demonstrated no cross-reactivity with nontarget proteins. Neutralization assays revealed a dose-dependent reduction in TiLV infectivity, which declined from 5.01×10^6 $TCID_{50}$/mL to 5.01×10^4–1.26×10^5 $TCID_{50}$/mL, with the highest efficacy observed at a 1:2 dilution. Despite the variability in neutralization infectivity among the different hens, IgY effectively inhibited TiLV-induced cytopathic effects. Immunofluorescence assays further confirmed a significant reduction in the TiLV antigen levels in IgY-treated RHTiB cells. Our findings highlight IgY as a promising strategy for TiLV control and suggest its potential application in the prevention of emerging viruses.

Keywords: tilapia lake virus; passive immunization; immunoglobulin; IgY; chicken eggs; disease management

1. Introduction

Since its emergence in 2014, tilapia lake virus (TiLV) has posed a significant threat to global tilapia aquaculture. TiLV causes disease across various tilapia species, including wild *Sarotherodon galilaeus*, farmed *Oreochromis niloticus*, and commercially hybrid tilapia (*O. niloticus* × *O. aureus*), with markedly adverse economic and ecological impacts [1,2]. TiLV, a segmented RNA virus with 10 genome segments [3], was recently classified under the family *Amnoonviridae* [4] and recognized as a notifiable disease by the World Organization for Animal Health [5]. Infected tilapia typically show clinical signs, such as abnormal swimming, exophthalmia, skin congestion, scale protrusion, and abdominal swelling, with mortality rates exceeding 30% within a week of the initial infection [1]. TiLV spreads through direct fish-to-fish contact [6,7], vertical transmission from infected broodstock to offspring [8,9], and environmental pathways. The virus has been detected in water where infected fish reside, but the infectivity of the virus outside the host remains to be determined [10,11]. Notably, TiLV genomic material can persist in frozen tilapia fillets for up to 28 days at −20 °C, although the likelihood of this material causing wider infection in such fillets is extremely low [12]. With its presence now reported in 18 countries, TiLV continues to be of significant concern to the global tilapia aquaculture industry [2,13].

To mitigate the impact of TiLV, researchers have explored different strategies, among them, rapid diagnostics, biosecurity measures, disinfectants, antiviral agents, the selection of fish that are genetically resistant to the virus, and vaccine development [14–17]. However, despite these efforts, no commercial vaccine is yet available, and existing interventions have failed to fully contain the spread of the virus [5,18,19]. Novel approaches, such as passive immunization using antibodies, are being explored as potential strategies to manage disease in fish farms [20] and offer promising potential for reducing the impact of TiLV. Passive immunization, which involves the administration of preformed antibodies to confer immediate protection, holds promise for preventing infections and mitigating disease severity [21,22]. However, conventional antibody production often relies on animals such as rabbits and horses and requires invasive blood collection. This practice raises ethical concerns related to animal welfare, induces stress in the animals, and presents challenges in terms of the cost effectiveness for large-scale production [23,24]. These limitations highlight the need for alternative antibody production methods that are more sustainable and ethically responsible [25,26].

Immunoglobulin Y (IgY), an immunoglobulin isotype found in birds, reptiles, and amphibians, is considered the functional equivalent of mammalian IgG [27]. IgY antibodies can be produced in large quantities by immunizing laying hens with specific antigens, followed by the extraction of the antibody from their egg yolks [28]. This approach offers several advantages, including its noninvasive nature, which minimizes animal stress, and its cost effectiveness for large-scale antibody production [22,23]. Moreover, the accumulation of IgY in egg yolks enables its easy transfer to target organisms through ingestion, thereby facilitating its use in various applications. The efficacy of IgY antibodies in preventing the replication and spread of bacterial and viral pathogens is well documented for both terrestrial and aquatic animals [29,30]. In aquaculture, passive immunization with IgY has been successfully applied to combat infectious diseases caused by pathogens, such as *Edwardsiella tarda* in Japanese eels (*Anguilla japonica*) [31], *Yersinia ruckeri* and *Vibrio anguillarum* in rainbow trout (*Oncorhynchus mykiss*) [32,33], *Pseudomonas fluorescens* and *Vibrio fluvialis* in goldfish (*Carassius auratus*) [34,35], and *Aeromonas salmonicida* in koi carp (*Cyprinus carpio koi*) [36]. Despite its proven effectiveness against bacterial pathogens, the application of IgY to address viral infections in aquaculture has been relatively limited and has not been applied to emerging viral pathogens. Previous examples include research on the use of oral IgY against cyprinid herpesvirus 3 in common carp (*Cyprinus carpio*), which reduced mortality from 50% to 85% when fish were challenged with a lethal viral dose of 40 $TCID_{50}$/fish [37]. Furthermore, IgY has been shown to inhibit the replication of red-spotted grouper nervous necrosis virus in cell cultures and demonstrated a protective effect in vivo [38,39]. In this study, we aimed to develop and produce TiLV-specific IgY antibodies by immunizing laying hens with TiLV antigens and to evaluate their efficacy in inhibiting viral activity. The findings of this study could offer a novel and sustainable strategy for managing TiLV infections.

2. Materials and Methods

2.1. Preparation of a Recombinant TiLV-S4 Antigen

The TiLV antigen was prepared from the tissue of moribund red hybrid tilapia infected with TiLV in Ayutthaya Province, Thailand, in 2021. To confirm the infection, the total RNA was extracted from the livers of the red hybrid tilapia using GENEzol™ reagent (Geneaid Biotech, New Taipei City, Taiwan), and in line with a previous protocol [40], the TiLV RNA was detected using reverse transcription quantitative polymerase chain reaction (RT-qPCR) primers targeting TiLV segment 3. The extracted RNA subsequently served as the template for cDNA synthesis and the amplification of TiLV segment 4 (TiLV-S4) via PCR using specific primer pairs (forward primer 5′-GGATCCATATGGTGAGAACTACAAAGAC-3′ and reverse primer 5′-GTCGACTCGAGCTATCTTCCAACAGCCCC-3′). The primers were designed based on the sequence with GenBank accession number MK425013.1. The PCR product was cloned into the pET28a expression vector (Novagen, Tokyo, Japan). The pET28a–TiLV-S4 construct was subjected to DNA sequencing (Macrogen, Seoul, Republic of Korea). The recombinant TiLV segment 4 protein (rTiLV-S4) was expressed in the *Escherichia coli* strain BL21 (DE3) by induction with 1 mM isopropyl β-D-thiogalactopyranoside (Fermentas, Waltham, MA, USA) at 18 °C for 6 h while shaking at 225 rpm. The cell pellets were collected by centrifugation at 3000× *g* at 4 °C for 10 min and resuspended in a phosphate buffer (NaH_2PO_4 and Na_2HPO_4, pH 7.4) that contained 1 mM phenylmethylsulfonyl fluoride. The cell suspension was then sonicated using an XL2020 Sonicator Ultrasonic Processor XL (Misonix, Farmingdale, NY, USA). The crude protein containing the rTiLV-S4, as previously described [41], was collected from the supernatant after centrifugation at 12,000× *g* at 4 °C for 20 min. The concentration of the crude protein was determined using a bicinchoninic acid assay (Fermentas, Waltham, MA, USA).

2.2. Chicken Immunization and Egg Collection

To promote IgY production against TiLV, two hens were immunized with rTiLV-S4. The first antigen immunization contained 2 mg/mL of rTiLV-S4 in Freund's complete adjuvant (Sigma-Aldrich, St. Louis, MO, USA) and was followed by the second and third boosters, which contained rTiLV mixed with Freund's incomplete adjuvant (Sigma-Aldrich, St. Louis, MO, USA). The injections were administered at one-week intervals, with 1 mL of the mixture injected intramuscularly into three different sites in the pectoral musculature. Eggs were collected from the hens prior to immunization (T0) and one (T1) or two weeks (T2) after the last immunization and stored at 4 °C until use (Supplementary Figure S1). The procedures were approved by the Institutional Animal Care and Use Committee at Kasetsart University under protocol number ACKU65-VET-088. The principles of replacement, reduction, and refinement were followed to ensure the ethical and humane treatment of the chickens involved in this study.

2.3. Total IgY Extraction

The total IgY was extracted from the egg yolks using gradients of the polyethylene glycol (PEG) precipitation techniques described elsewhere [42–44]. The eggshell was carefully cracked, and the yolk was separated from the egg white. After removing the remaining egg white with filter paper, the yolk membrane was punctured with a pipette tip, and the yolk was transferred to a 50 mL tube. The yolk was mixed with phosphate buffered saline (PBS) at twice the volume of the yolk. Subsequently, 3.5% PEG 6000 (*w*/*v*) (Sigma-Aldrich, St. Louis, MO, USA) was added, and the mixture was vortexed for 10 min.

This step separated the suspension into two fractions: one containing the yolk and fatty substances and the other a liquid phase with IgY and other proteins. The tubes were centrifuged at 13,000× g for 20 min at 4 °C. The supernatant was filtered through paper and transferred to a new 50 mL tube. Next, 8.5% PEG 6000 (w/v) was added to the tube, mixed by a brief vortexing, and centrifuged at 13,000× g for 20 min at 4 °C. The supernatant was discarded, and the pellet was dissolved in 10 mL PBS and 12% PEG 6000 (w/v). The IgY extract was dialyzed overnight in 0.1% NaCl, followed by an additional 3 h in PBS at 4 °C, before being stored at −20 °C for further analysis.

2.4. IgY Characterization by SDS-PAGE and Western Blot Analysis

We initially confirmed the presence of IgY in the egg yolks using sodium dodecyl sulfate-polyacrylamide gel electrophoresis (SDS-PAGE) with a Protean II electrophoresis system (Bio-Rad, Hercules, CA, USA) by following the discontinuous buffer system method [45]. The collected egg yolks were electrophoresed in a 15% resolving SDS-PAGE gel under reducing conditions at 120 V for 70 min, and the protein bands were visualized using a Coomassie blue staining solution (Bio-Rad, Hercules, CA, USA). The protein band sizes were determined using protein molecular marker standards (AccuProtein Chroma range 16–250 kDa, Enzmart Biotech, Bangkok, Thailand).

Western blot analysis was subsequently conducted to confirm the immunogenic property of anti-TiLV-S4 IgY against the purified virus. One microgram of TiLV, which had been purified using a sucrose gradient [46], was electrophoresed in 12% SDS-PAGE at 120 V for 80 min and transferred to a polyvinylidene fluoride membrane (Bio-Rad, Hercules, CA, USA) using a mini blot transfer system. The membrane was washed three times with 0.1% Tween-20 in PBS (PBS-T) and then incubated overnight at 4 °C with a blocking buffer comprising 3% bovine serum albumin (BSA) in PBS-T. Following three washes with PBS-T, the membrane was incubated with anti-TiLV-S4 IgY at a dilution of 1:100 for 1 h at room temperature. The membrane was washed again with PBS-T and incubated with a horseradish peroxidase-labelled goat anti-chicken IgY antibody (Abcam, Carlsbad, CA, USA) at a dilution of 1:2000 for 1 h at room temperature. After three final washes with PBS-T, the enhanced chemiluminescence substrate (Lumiflash^{TM}, Visual Protein, Taipei, Taiwan) was applied and visualized using the ChemiDoc MP^{TM} Imaging System (Bio-Rad, Hercules, CA, USA).

2.5. In Vitro TiLV Neutralization Using Anti-TiLV-S4 IgY Antibodies

The anti-TiLV-S4 IgY antibody was serially diluted two-fold with Leibovitz's L-15 medium (Sigma-Aldrich, St. Louis, MO, USA) to obtain dilutions of 1:2, 1:5, and 1:10. The prepared antibody was mixed with 10^3 $TCID_{50}/mL$ of TiLV strain VET-KUTV08 at a ratio of 1:1. The mixture was incubated in an Eppendorf® ThermoMixer C (Eppendorf, Hamburg, Germany) at 25 °C for 2 h with continuous shaking at 400 rpm. The positive control consisted of 10^3 $TCID_{50}/mL$ TiLV mixed with an L-15 medium at a ratio of 1:1, and the negative control was the L-15 medium only. One hundred microliters of these mixtures and controls were added to a 96-well plate containing confluent E-11 cells, with four replicates per dilution. The cells were incubated at 25 °C for 1 h, then the mixtures were discarded, and a new L-15 medium supplemented with 2% FBS was added to each well. The cells were grown continuously for 7–12 days at 25 °C and monitored daily for the cytopathic effect (CPE) under an inverted microscope (CKX53, Olympus, Tokyo, Japan).

2.6. Virus Titration

To quantify the viral concentration following inhibition by the anti-TiLV IgY antibodies, the cells were initially lysed using a freeze–thaw process. The media from each replicate well of each dilution were pooled and transferred to 1.5 mL microcentrifuge tubes. The samples were centrifuged at 3000× g for 10 min at 4 °C. The supernatant was collected and serially diluted 10-fold to obtain dilutions ranging from 10^{-1} to 10^{-8}. The diluted samples were added to 96-well plates that contained confluent E-11 cells with five replicate wells per sample. The plates were incubated at 25 °C for 1 h. Following incubation, the media in each well were replaced with an L-15 media supplemented with 2% FBS, and the plates were incubated at 25 °C for 7–12 days. The plates were observed daily for CPEs. The 50% tissue culture infectious dose ($TCID_{50}$) was calculated using the Reed and Muench method [47].

2.7. Immunofluorescence Assay to Detect TiLV in the RHTiB Cell Line

An immunofluorescence assay (IFA) targeting TiLV was used to confirm the inhibition of viral entry into RHTiB cells (a cell line from the brain tissue of red hybrid tilapia) [48]. Briefly, the RHTiB cells were cultured in Leibovitz's L-15 medium supplemented with 10% FBS, 100 U/mL penicillin, 0.1 mg/mL streptomycin, and 0.25 µg/mL amphotericin B at a pH of 7.4 at 25 °C without CO_2. When the cells achieved 70% confluence, they were trypsinized, counted, and diluted to a concentration of 2×10^5 cells/mL. Subsequently, 500 µL of the cell suspension was seeded onto cell culture slides (SPL Life Science, Pocheon-si, Gyeonggi-do, Republic of Korea) in an L-15 medium containing 10% FBS and incubated until 80–90% confluence was achieved. The cells were infected using different methods: 100 µL TiLV mixed with 100 µL L-15 medium (positive control), TiLV mixed with a 1:2 or 1:10 IgY solution, TiLV mixed with IgY at T0, control egg, and control L-15 medium (negative control). These mixtures were incubated at 25 °C with continuous shaking at 400 rpm for 2 h (Eppendorf® ThermoMixer® C, AG, Darmstadt, Germany). The cells were washed twice with an L-15 medium without FBS, incubated with the virus mixtures for 1 h, and subsequently replaced with an L-15 medium containing 2% FBS, followed by incubation at 25 °C for 24 h.

The cells were fixed with ice-cold 100% methanol for 10 min and washed twice with PBS. They were then treated with 0.3% Triton X-100 in PBS for 10 min and washed again with PBS. The membrane was blocked with 2% BSA in PBS for 30 min to prevent nonspecific binding, followed by overnight incubation at 4 °C with the primary antibody (IgG of TiLV) in a blocking solution at a 1:100 dilution. The cells were then incubated with the secondary antibody (goat anti-rabbit IgG H&L Alexa Fluor™ 488; Abcam, Carlsbad, CA, USA) in PBS at a dilution of 1:500 for 1 h at room temperature and washed with PBS. The cell nuclei were stained with 4′,6-diamidino-2-phenylindole (DAPI; Sigma-Aldrich, St. Louis, MO, USA) at a 1:1000 dilution for 15 min, washed with PBS, mounted with ProLong™ Gold Antifade reagent (Invitrogen, Thermo Fisher Scientific, Waltham, MA, USA) on glass slides, and then a cover glass was placed over them. All images were captured using a confocal microscope (Fluoview 3000, Olympus, Tokyo, Japan), which confirmed the specific binding of IgY to the TiLV-infected cells through the colocalization of the DAPI and Alexa Fluor signals. The fluorescence intensity values were quantified using cellSens Dimension software version 2.3 (Olympus, Tokyo, Japan). Five areas were randomly selected, and the fluorescence intensity of the green signal, which indicated positive TiLV-infected cells, was measured. The green signal intensity was analyzed as mean ± standard deviation (SD) and compared with the positive control, egg control, and TiLV mixed with the 1:2 or 1:10 IgY solutions.

2.8. Statistical Analysis

The viral titer and fluorescence intensity were presented as mean ± SD. The mean fluorescence intensity data between different IgY concentrations were assessed using one-way analysis of variance (ANOVA), and the $TCID_{50}$ of the virus between different treatment and time points of infection was analyzed using two-way ANOVA. GraphPad

Prism software version 8.0 was used in the analysis, and a p-value less than 0.05 was considered statistically significant.

3. Results

3.1. Immunization and Preparation of Chicken IgY

All the hens immunized with crude protein containing rTiLV-S4 via intramuscular injections remained healthy with no observed abnormalities or mortality throughout the study period. Furthermore, no signs of inflammation, adverse reactions, or pathological changes were detected at the injection sites. The IgY antibody purification process revealed protein bands at approximately 20 kDa and 60 kDa, which corresponded to light (LC) and heavy chains (HC) of immunoglobulin, respectively (Supplementary Figure S2). Additionally, an impurity band at approximately 35 kDa, identified as the C-terminal fragment of the vitellogenin II precursor, was also observed [49].

The specific binding of the purified IgY antibodies against rTiLV-S4 extracted from the eggs of two hens was evaluated using Western blot analysis at T0 and T1 (Figure 1). The analysis confirmed the presence of specific binding in the post-immunization samples, which was indicated by a distinct 38 kDa band corresponding to TiLV. Notably, the binding signal was stronger in the samples from the first hen (Hen 1) compared to the second hen (Hen 2). Importantly, no nonspecific binding was observed in any of the samples, which suggested the specificity of the purified IgY antibodies.

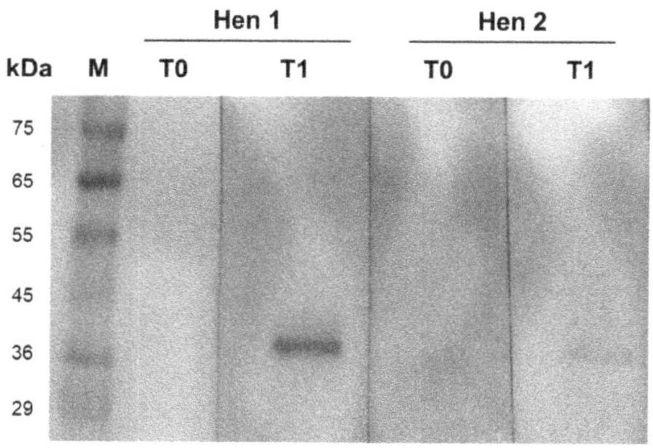

Figure 1. Western blot analysis showing specific binding of purified IgY from two individual chickens (Hen 1 and Hen 2) to the recombinant TiLV segment 4 protein (rTiLV-S4). No binding was observed at the pre-immunization stage (T0). However, distinct bands corresponding to rTiLV-S4 at 38 kDa were detected one-week post-immunization (T1) in the samples from both hens. Lane M represents the protein marker.

3.2. Neutralization of IgY Against TiLV

The neutralization activity of the purified IgY dilutions obtained from the eggs of the two hens at T1 and T2 post-immunization was evaluated at various dilutions (1:2, 1:5, and 1:10) against TiLV infectivity in E-11 cells (Figure 2). All the E-11 cells pretreated with anti-TiLV-S4 IgY, irrespective of the dilution, exhibited significantly reduced TiLV infection levels compared to the positive control, which had an infectivity of 5.01×10^6 $TCID_{50}$/mL. Of note, no statistical differences in TiLV infectivity were apparent between the cells neutralized with IgY from the T1 and T2 groups of either hen. For the IgY purified from Hen 1 at T2, the 1:2 dilution resulted in a significantly lower infectivity level of 7.94×10^4 $TCID_{50}$/mL compared to the 1:10 dilution, which yielded 1.26×10^5 $TCID_{50}$/mL. Similarly, for the IgY purified from Hen 2 at T1, the 1:2 dilution achieved a TiLV infectivity level of 5.01 × 10^4 $TCID_{50}$/mL, which was significantly lower than the 7.94 × 10^4 $TCID_{50}$/mL observed with the 1:5 dilution.

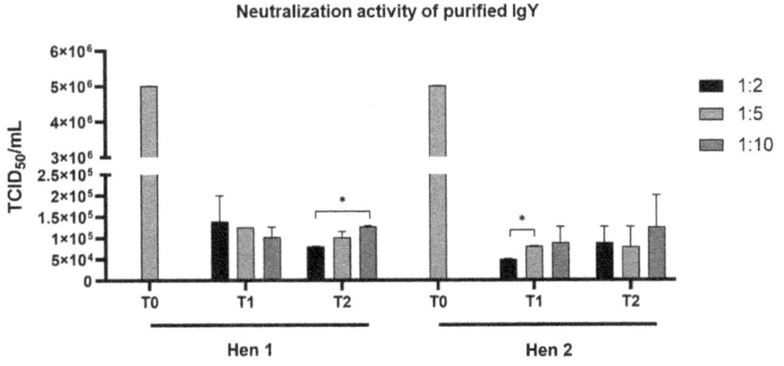

Figure 2. Neutralization activity of purified IgY against rTiLV-S4, assessed by the 50% tissue culture infective dose ($TCID_{50}$)/mL. The E-11 cells treated with purified IgY from the eggs of two hens (Hen 1 and Hen 2) collected after one week (T1) and two weeks (T2) post-immunization and infected with TiLV showed significantly lower infectivity compared to the control group (pre-immunization). Bars represent the mean ± standard deviation (SD) of the measured neutralization activity. * $p < 0.05$.

The neutralization activity of the purified IgY against TiLV was further assessed based on the presence of the CPE in the E-11 cells (Figure 3). In comparison to the uninfected E-11 cells, which exhibited no signs of the CPE (Figure 3A,B), the cells inoculated with TiLV as a positive control displayed a noticeable CPE at 7 days postinfection (dpi) (Figure 3C,D). When TiLV was incubated with IgY at different dilutions, a dose-dependent neutralization effect was observed. Specifically, the lowest level of CPE was detected in the group treated with IgY diluted at a ratio of 1:2, which demonstrated the highest neutralization efficiency (Figure 3E,F). In contrast, the cells treated with IgY diluted at a ratio of 1:5 showed an increased level of the CPE (Figure 3G,H), while those treated with IgY diluted at a ratio of 1:10 had CPE levels comparable to those of the positive control (Figure 3I,J). Based on these findings, the IgY purified from Hen 2 at T1 with a 1:2 dilution demonstrated the lowest TiLV infectivity in the E-11 cells and was selected for further study.

3.3. Immunofluorescence Assay

The neutralizing activities of the purified IgY in inhibiting TiLV entry into the RHTiB cells were further evaluated using an immunofluorescence assay (Figure 4). In comparison to the uninfected control cells, which showed no green fluorescence signals (Figure 4A–C), the TiLV-infected cells without IgY treatment (positive control) showed strong green fluorescence within the RHTiB cells at 1 dpi (Figure 4D–F). The infected cells incubated with the IgY samples from the pre-immunization period (egg control) exhibited similar levels of the green fluorescence signal (Figure 4G–I). Interestingly, treatment with the purified IgY remarkably reduced the presence of TiLV in the RHTiB cells. The cells treated with IgY at a dilution of 1:2 demonstrated a substantial reduction in green fluorescence intensity compared to the positive control (Figure 4J–L). However, although the cells treated with IgY at a 1:10 dilution showed higher fluorescence intensity relative to the 1:2 dilution, the intensity remained considerably lower than that of the positive control (Figure 4M–O).

When comparing the mean fluorescence intensity (MFI) values among the positive control, egg control, and IgY-treated groups (dilutions of 1:2 and 1:10), the positive control showed the highest MFI (1681 ± 171.5), followed by the egg control (1586 ± 19.78). The group treated with IgY at a 1:2 dilution showed an MFI of 1497 ± 17.18, while the group treated with IgY at a 1:10 dilution had an MFI of 1576 ± 13.59. Further statistical analysis revealed that the cells treated with IgY at a 1:2 dilution had significantly lower MFI values compared to the egg control and cells treated with IgY at 1:10 dilution groups (* $p < 0.05$ and *** $p < 0.001$, respectively) (Figure 4P).

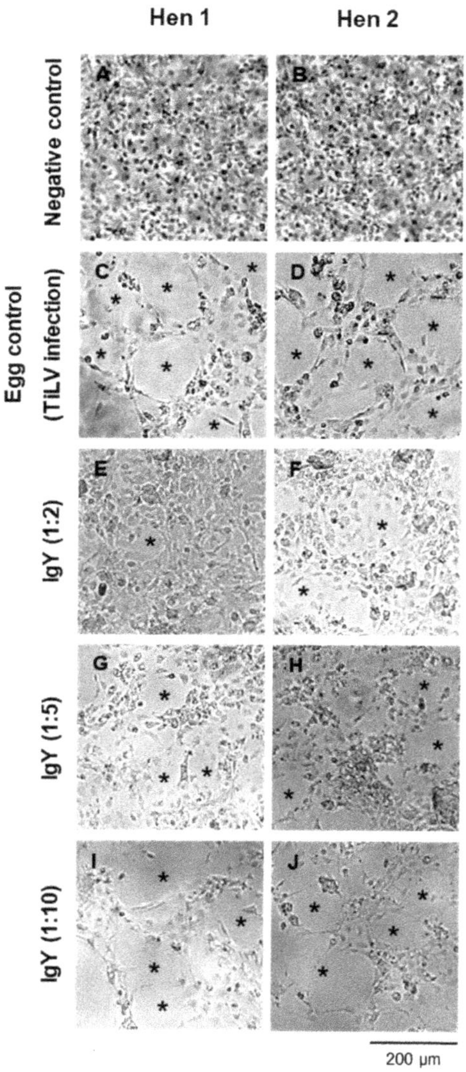

Figure 3. Evaluation of the neutralization activity of purified IgY from Hen 1 and Hen 2 against TiLV infectivity in E-11 cells at 7 days postinfection. The cellular morphologies are shown. (**A**,**B**) Negative control cells; (**C**,**D**) TiLV-infected cells treated with pre-immunized IgY (egg control); (**E**,**F**) infected cells treated with purified IgY at a 1:2 dilution; (**G**,**H**) 1:5 dilution; (**I**,**J**) 1:10 dilution. Asterisks (*) indicate the presence of cytopathic effects (CPE). Scale bar: 200 μm (10× magnification).

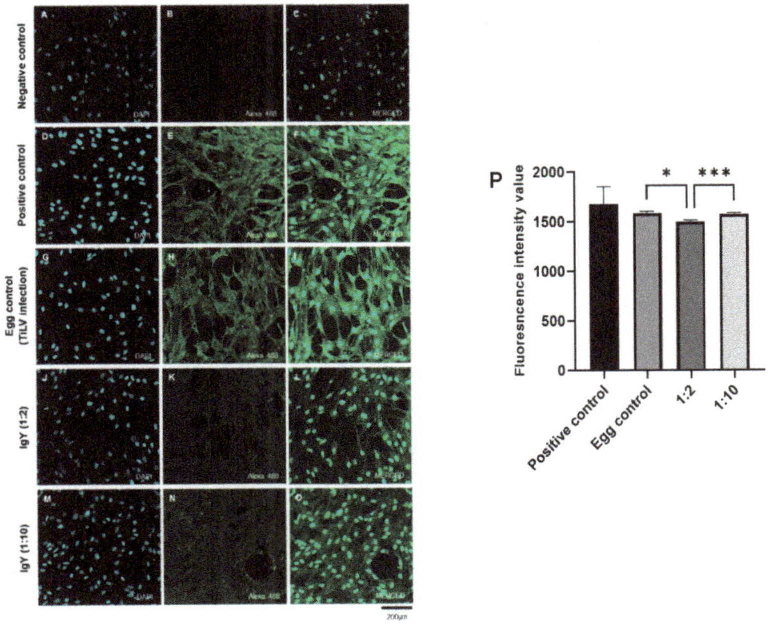

Figure 4. Neutralization activity of purified IgY in reducing TiLV antigen levels in RHTiB cells assessed using an immunofluorescence assay (IFA). (**A–C**) Negative control cells demonstrate only blue 4′,6-diamidino-2-phenylindole (DAPI) staining. (**D–F**) TiLV-infected cells (positive control) show colocalization of the TiLV antigen (green fluorescence) within the cytoplasms of the RHTiB cells. (**G–I**) Egg control group, namely RHTiB cells treated with purified IgY extracted from egg yolk during the pre-immunization period. (**J–L**) RHTiB cells treated with purified IgY at a 1:2 dilution and infected with TiLV. (**M–O**) RHTiB cells treated with purified IgY at a 1:10 dilution and infected with TiLV. (**P**) Graph showing the mean fluorescence intensity (MFI) values as mean ± SD for the positive control, egg control, and IgY-treated groups (1:2 and 1:10 dilutions). Statistically significant differences are indicated as * $p < 0.05$ and *** $p < 0.001$.

4. Discussion

Infection with TiLV causes serious disease in tilapia and poses a major challenge to global tilapia aquaculture. Various strategies are being explored to prevent the spread of infection and reduce the impacts of TiLV disease. Current approaches to TiLV mitigation include vaccine development [41], the application of probiotics [50], and the use of feed additives [51]. However, these strategies either require prolonged development and regulatory approval or may not provide immediate protection during outbreaks. Passive immunization using IgY offers a complementary approach characterized by noninvasive production, high specificity, cost effectiveness, and minimal ethical concerns. These attributes position IgY as a valuable tool for enhancing biosecurity and sustainability in aquaculture [28]. In this study, we successfully generated TiLV-specific IgY antibodies by immunizing laying hens with the rTiLV-S4 protein expressed in *E. coli* [41]. The immunization protocol involved the intramuscular administration of rTiLV-S4 formulated with Freund's complete adjuvant for the initial dose, followed by booster doses with Freund's incomplete adjuvant [44]. A significant increase in IgY levels was detected in the eggs of the hens immunized with the antigen, thereby confirming that this immunization strategy effectively stimulated robust IgY production, which is consistent with the findings of previous studies [52]. The specificity of the purified IgY was confirmed by Western blot analysis, which demonstrated strong binding to the TiLV-S4 without nonspecific interactions. Upon antigen exposure, the immune systems of the hens produced specific antibodies, which were then transferred to the egg yolks [29]. To purify these IgY antibodies, the egg yolks were subjected to PEG precipitation, a widely used and cost-effective method for large-scale antibody extraction [53]. This process effectively removed lipids and enriched the IgY concentration, which ensured antibody stability and solubility [54]. Optimized PEG precipitation techniques have been shown to improve IgY purity and yield, which supports the feasibility of IgY for broader aquaculture applications [55]. In our study, both HCs and LCs were detected, and a specific band demonstrating the interaction between the antibody and rTiLV-S4 was confirmed. After PEG precipitation, the neutralizing activity of the purified IgY against TiLV was proved in a continuous cell line. Interestingly, neutralization assays using different concentrations of IgY revealed the dose-dependent inhibition of TiLV, with the highest antiviral efficacy at a 1:2 dilution. These findings are consistent with previous studies demonstrating the dose-dependent effects of IgY-mediated inhibition on viral and bacterial pathogens in fish [21,39]. In particular, our study showed that IgY antibodies effectively reduced the viral load and mitigated CPE, which highlights their potential as a therapeutic approach against this emerging virus. Although variations in infectivity were detected when tissue was treated with IgY from different hens, its ability to reduce the CPE in TiLV-infected E-11 cells remained consistent, which highlights the antiviral potential of IgY. It should be considered that the variation in neutralization efficiency among the IgY samples may have been influenced by individual differences in immune responses to the same antigenic stimulation. Factors such as genetic background, age, breed, and egg-laying capacity have been reported to influence antibody production in laying hens [23,29]. These findings emphasize the need for further optimization of immunization protocols to ensure the consistency and potency of IgY antibodies against TiLV.

The neutralization activity of IgY was further confirmed using an immunofluorescence assay, which demonstrated a significant reduction in TiLV antigens in RHTiB cells, with the highest inhibition observed at a 1:2 dilution of purified IgY. Mechanistically, IgY may inhibit viral replication by blocking viral entry into host cells through specific binding to viral proteins, thereby preventing their attachment to the host cell receptors [38,56]. This evidence supports the potential of IgY as an effective antiviral strategy for controlling TiLV infection. Despite these promising findings, however, challenges remain in optimizing IgY deployment for field applications. Effective implementation in aquaculture settings requires key factors, such as optimal fish size, dosage, and delivery methods, to be addressed [23]. Such factors must be thoroughly researched and validated before IgY can be adopted for use in fish farms. For example, while the administration methods of intraperitoneal and intramuscular injections ensure systemic immunity, they are labor-intensive and impractical for large-scale aquaculture operations [57]. As a noninvasive alternative, oral administration may be suitable for targeting gastrointestinal and systemic pathogens; however, its efficacy may be limited by enzymatic degradation in the digestive tract [58]. Recent advancements in encapsulation technologies, such as chitosan–alginate microcapsules, have improved IgY stability and bioavailability, which suggests the future feasibility of oral delivery for large-scale aquaculture applications [59]. Therefore, encapsulating IgY in coated feed pellets or specialized formulations to enhance its stability and protect against enzymatic degradation presents a promising approach for effective oral delivery in fish farming. Nevertheless, further research is needed to establish standardized protocols for TiLV-specific IgY production and delivery to ensure consistency, efficacy, and practical implementation in the management of TiLV infection using IgY.

5. Conclusions

Passive immunization with IgY presents a sustainable and ethical approach to disease control in aquaculture. In our study, we demonstrated the potential application of IgY antibodies to mitigate TiLV infections in tilapia. The rTiLV-S4-specific IgY antibodies derived from laying hens had a dose-dependent neutralization effect and significantly reduced viral infectivity and the CPE in E-11 cells. Immunofluorescence assays further confirmed the inhibitory properties of IgY and revealed its ability to prevent TiLV infection in RHTiB cells. Despite these advantages, challenges remain in optimizing IgY deployment, particularly in determining the most effective administration routes and ensuring long-term stability. Further studies should focus on developing scalable application strategies to maximize the impact of IgY immunization on disease prevention in commercial fish farming.

Supplementary Materials: The following supporting information can be downloaded at https://www.mdpi.com/article/10.3390/v17030448/s1, Figure S1: The schematic illustrates the immunization protocol and egg collection schedule for the production of chicken egg polyclonal IgY against tilapia lake virus (TiLV); Figure S2: Gel electrophoresis of the IgY samples purified using polyethylene glycol (PEG) precipitation at different stages.

Author Contributions: Conceptualization, P.M., N.P., P.S., T.L., N.T. and W.S.; methodology, P.S., J.Y., P.T., M.K., P.M., N.P. and T.L.; validation, P.S., P.T., J.Y., M.K. and T.L.; formal analysis, P.S. and T.L.; investigation, P.S., P.T., J.Y., M.K. and T.L.; resources, W.S.; data curation, P.S., T.L. and W.S.; writing—original draft preparation, P.S., J.Y., T.L., N.T. and W.S.; writing—review and editing, P.S., J.Y., P.M., N.P., T.L., N.T. and W.S.; visualization, P.S., T.L. and W.S.; supervision, N.T. and W.S.; project administration, N.T. and W.S.; funding acquisition, W.S. All authors have read and agreed to the published version of the manuscript.

Funding: This project received financial support from the Faculty of Veterinary Medicine, Kasetsart University. The research was partly supported by the Kasetsart University Research and Development Institute under project number FF(KU)51.67.

Institutional Review Board Statement: The animal study protocol was approved by the Institutional Animal Care and Use Committee of Kasetsart University (protocol number ACKU65-VET-088) for animal handling and care.

Informed Consent Statement: Not applicable.

Data Availability Statement: The data presented in this study are available on request from the corresponding author. The data are not publicly available due to the anonymity granted to all participating parties.

Acknowledgments: We would like to thank the Central Laboratory (CTL), Center for Veterinary Research and Innovation, Faculty of Veterinary Medicine, Kasetsart University.

Conflicts of Interest: The authors declare no conflicts of interest.

References

1. Surachetpong, W.; Roy, S.R.K.; Nicholson, P. Tilapia lake virus: The story so far. *J. Fish Dis.* **2020**, *43*, 1115–1132. [CrossRef]
2. Kembou-Ringert, J.E.; Steinhagen, D.; Readman, J.; Daly, J.M.; Adamek, M. Tilapia Lake Virus Vaccine Development: A Review on the Recent Advances. *Vaccines* **2023**, *11*, 251. [CrossRef] [PubMed]
3. Bacharach, E.; Mishra, N.; Briese, T.; Zody, M.C.; Tsofack, J.E.K.; Zamostiano, R.; Berkowitz, A.; Ng, J.; Nitido, A.; Corvelo, A.; et al. Characterization of a Novel Orthomyxo-like Virus Causing Mass Die-Offs of Tilapia. *mBio* **2016**, *7*, e00431-16. [CrossRef] [PubMed]
4. Koonin, E.V.; Krupovic, M.; Surachetpong, W.; Wolf, Y.I.; Kuhn, J.H. ICTV Virus Taxonomy Profile: *Amnoonviridae* 2023. *J. Gen. Virol.* **2023**, *104*, 001903. [CrossRef] [PubMed]
5. Clyde, C.W.; Tan, J.P.; Yeap, S.K.; Yong, C.Y. Current updates on viral infections affecting tilapia. In *Aquaculture and Fisheries*; Elsevier: Amsterdam, The Netherlands, 2024. [CrossRef]
6. Liamnimitr, P.; Thammatorn, W.; U-thoomporn, S.; Tattiyapong, P.; Surachetpong, W. Non-lethal sampling for Tilapia Lake Virus detection by RT-qPCR and cell culture. *Aquaculture* **2018**, *486*, 75–80. [CrossRef]

7. Kembou-Ringert, J.E.; Hotio, F.N.; Steinhagen, D.; Thompson, K.D.; Surachetpong, W.; Rakus, K.; Daly, J.M.; Goonawardane, N.; Adamek, M. Knowns and unknowns of TiLV-associated neuronal disease. *Virulence* **2024**, *15*, 2329568. [CrossRef]
8. Yamkasem, J.; Tattiyapong, P.; Kamlangdee, A.; Surachetpong, W. Evidence of potential vertical transmission of tilapia lake virus. *J. Fish Dis.* **2019**, *42*, 1293–1300. [CrossRef]
9. Dong, H.T.; Senapin, S.; Gangnonngiw, W.; Nguyen, V.V.; Rodkhum, C.; Debnath, P.P.; Delamare-Deboutteville, J.; Mohan, C.V. Experimental infection reveals transmission of tilapia lake virus (TiLV) from tilapia broodstock to their reproductive organs and fertilized eggs. *Aquaculture* **2020**, *515*, 734541. [CrossRef]
10. Taengphu, S.; Kayansamruaj, P.; Kawato, Y.; Delamare-Deboutteville, J.; Mohan, C.V.; Dong, H.T.; Senapin, S. Concentration and quantification of *Tilapia tilapinevirus* from water using a simple iron flocculation coupled with probe-based RT-qPCR. *PeerJ* **2022**, *10*, e13157. [CrossRef]
11. Prasartset, T.; Surachetpong, W. Simultaneous detection of three important viruses affecting tilapia using a multiplex PCR assay. *J. Fish Dis.* **2023**, *46*, 459–464. [CrossRef]
12. Thammatorn, W.; Rawiwan, P.; Surachetpong, W. Minimal risk of tilapia lake virus transmission via frozen tilapia fillets. *J. Fish Dis.* **2019**, *42*, 3–9. [CrossRef] [PubMed]
13. Aich, N.; Paul, A.; Choudhury, T.G.; Saha, H. Tilapia Lake Virus (TiLV) disease: Current status of understanding. *Aquac. Fish.* **2022**, *7*, 7–17. [CrossRef]
14. Jaemwimol, P.; Sirikanchana, K.; Tattiyapong, P.; Mongkolsuk, S.; Surachetpong, W. Virucidal effects of common disinfectants against tilapia lake virus. *J. Fish Dis.* **2019**, *42*, 1383–1389. [CrossRef] [PubMed]
15. Barría, A.; Trinh, T.Q.; Mahmuddin, M.; Benzie, J.A.H.; Chadag, V.M.; Houston, R.D. Genetic parameters for resistance to Tilapia Lake Virus (TiLV) in Nile tilapia (*Oreochromis niloticus*). *Aquaculture* **2020**, *522*, 735126. [CrossRef]
16. Lertwanakarn, T.; Trongwongsa, P.; Yingsakmongkol, S.; Khemthong, M.; Tattiyapong, P.; Surachetpong, W. Antiviral Activity of Ribavirin against *Tilapia tilapinevirus* in Fish Cells. *Pathogens* **2021**, *10*, 1616. [CrossRef] [PubMed]
17. Tattiyapong, P.; Kitiyodom, S.; Yata, T.; Jantharadej, K.; Adamek, M.; Surachetpong, W. Chitosan nanoparticle immersion vaccine offers protection against tilapia lake virus in laboratory and field studies. *Fish Shellfish Immunol.* **2022**, *131*, 972–979. [CrossRef]
18. He, T.; Zhang, Y.-Z.; Gao, L.-H.; Miao, B.; Zheng, J.-S.; Pu, D.-C.; Zhang, Q.-Q.; Zeng, W.-W.; Wang, D.-S.; Su, S.-Q.; et al. Identification and pathogenetic study of tilapia lake virus (TiLV) isolated from naturally diseased tilapia. *Aquaculture* **2023**, *565*, 739166. [CrossRef]
19. Wang, B.; Thompson, K.D.; Wangkahart, E.; Yamkasem, J.; Bondad-Reantaso, M.G.; Tattiyapong, P.; Jian, J.; Surachetpong, W. Strategies to enhance tilapia immunity to improve their health in aquaculture. *Rev. Aquac.* **2023**, *15*, 41–56. [CrossRef]
20. Sparrow, E.; Friede, M.; Sheikh, M.; Torvaldsen, S. Therapeutic antibodies for infectious diseases. *Bull. World Health Organ.* **2017**, *95*, 235–237. [CrossRef]
21. Zhang, M.; Geng, H.; Tariq Javed, M.; Xu, L.; Li, X.; Wang, L.; Li, S.; Xu, Y. Passive protection of Japanese pufferfish (*Takifugu rubripes*) against *Vibrio harveyi* infection using chicken egg yolk immunoglobulins (IgY). *Aquaculture* **2021**, *532*, 736009. [CrossRef]
22. Wang, H.; Zhong, Q.; Lin, J. Egg Yolk Antibody for Passive Immunization: Status, Challenges, and Prospects. *J. Agric. Food Chem.* **2023**, *71*, 5053–5061. [CrossRef] [PubMed]
23. Kovacs-Nolan, J.; Mine, Y. Egg Yolk Antibodies for Passive Immunity. *Annu. Rev. Food Sci. Technol.* **2012**, *3*, 163–182. [CrossRef] [PubMed]
24. Pantaleo, G.; Correia, B.; Fenwick, C.; Joo, V.S.; Perez, L. Antibodies to combat viral infections: Development strategies and progress. *Nat. Rev. Drug Discov.* **2022**, *21*, 676–696. [CrossRef] [PubMed]
25. Hanly, W.C.; Artwohl, J.E.; Bennett, B.T. Review of Polyclonal Antibody Production Procedures in Mammals and Poultry. *ILAR J.* **1995**, *37*, 93–118. [CrossRef]
26. El-Kafrawy, S.A.; Abbas, A.T.; Oelkrug, C.; Tahoon, M.; Ezzat, S.; Zumla, A.; Azhar, E.I. IgY antibodies: The promising potential to overcome antibiotic resistance. *Front. Immunol.* **2023**, *14*, 1065353. [CrossRef]
27. Härtle, S.; Magor, K.E.; Göbel, T.W.; Davison, F.; Kaspers, B. Chapter 6—Structure and Evolution of Avian Immunoglobulins. In *Avian Immunology*, 2nd ed.; Schat, K.A., Kaspers, B., Kaiser, P., Eds.; Academic Press: Boston, MA, USA, 2014; pp. 103–120.
28. Pereira, E.P.V.; van Tilburg, M.F.; Florean, E.O.P.T.; Guedes, M.I.F. Egg yolk antibodies (IgY) and their applications in human and veterinary health: A review. *Int. Immunopharmacol.* **2019**, *73*, 293–303. [CrossRef]
29. Schade, R.; Calzado, E.G.; Sarmiento, R.; Chacana, P.A.; Porankiewicz-Asplund, J.; Terzolo, H.R. Chicken Egg Yolk Antibodies (IgY-technology): A Review of Progress in Production and Use in Research and Human and Veterinary Medicine. *Altern. Lab. Anim.* **2005**, *33*, 129–154. [CrossRef]
30. Xu, Y.; Li, X.; Jin, L.; Zhen, Y.; Lu, Y.; Li, S.; You, J.; Wang, L. Application of chicken egg yolk immunoglobulins in the control of terrestrial and aquatic animal diseases: A review. *Biotechnol. Adv.* **2011**, *29*, 860–868. [CrossRef]
31. Gutierrez, M.A.; Miyazaki, T.; Hatta, H.; Kim, M. Protective properties of egg yolk IgY containing anti-*Edwardsiella tarda* antibody against paracolo disease in the Japanese eel, *Anguilla japonica* Temminck & Schlegel. *J. Fish Dis.* **1993**, *16*, 113–122. [CrossRef]

32. Norqvist, A.; Hagström, A.; Wolf-Watz, H. Protection of rainbow trout against vibriosis and furunculosis by the use of attenuated strains of *Vibrio anguillarum*. *Appl. Environ. Microbiol.* **1989**, *55*, 1400–1405. [CrossRef]
33. Lee, S.B.; Mine, Y.; Stevenson, R.M.W. Effects of Hen Egg Yolk Immunoglobulin in Passive Protection of Rainbow Trout against *Yersinia ruckeri*. *J. Agric. Food Chem.* **2000**, *48*, 110–115. [CrossRef] [PubMed]
34. Liu, X.; Xiao, H.; Chao, J.; Jian, S.; Wu, X.; Lu, J.; Wang, J.; Chen, C.; Liu, Y. Polyvalent passive vaccine candidates from egg yolk antibodies (IgY) of important outer membrane proteins (PF1380 and ExbB) of *Pseudomonas fluorescens* in fish. *Fish Shellfish Immunol.* **2023**, *143*, 109211. [CrossRef] [PubMed]
35. Liu, X.; Xiao, H.; Cui, P.; Chen, J.; Chao, J.; Wu, X.; Lu, J.; Zhang, X.; Xu, G.; Liu, Y. Differential polyvalent passive immune protection of egg yolk antibodies (IgY) against live and inactivated *Vibrio fluvialis* in fish. *Fish Shellfish Immunol.* **2024**, *151*, 109751. [CrossRef] [PubMed]
36. Gan, H.; He, H.; Sato, A.; Hatta, H.; Nakao, M.; Somamoto, T. Ulcer disease prophylaxis in koi carp by bath immersion with chicken egg yolk containing anti-*Aeromonas salmonicida* IgY. *Res. Vet. Sci.* **2015**, *99*, 82–86. [CrossRef]
37. Zhenxing, L.; Ke, H.; Yanping, M.; Le, H.T.H.; Feng, G.-q.; Jiangyao, M.; Zhiling, L.; Li, Y. Oral Passive Immunization of Carp *Cyprinus carpio* with Anti-CyHV-3 Chicken Egg Yolk Immunoglobulin (IgY). *Fish Pathol.* **2014**, *49*, 113–120.
38. Yi, L.; Qin, Z.; Lin, H.; Zhou, Y.; Li, J.; Xu, Z.; Babu, V.S.; Lin, L. Features of chicken egg yolk immunoglobulin (IgY) against the infection of red-spotted grouper nervous necrosis virus. *Fish Shellfish Immunol.* **2018**, *80*, 534–539. [CrossRef]
39. Liu, J.; Qin, Y.; Yan, L.; Liu, W.; Shi, H.; Lu, Y.; Liu, X. Protective effects of egg yolk immunoglobulins (IgY) on juvenile groupers (*Epinephelus fuscoguttatus* × *Epinephelus lanceolatus*) with red-spotted grouper nervous necrosis virus infection. *Aquaculture* **2021**, *545*, 737218. [CrossRef]
40. Tattiyapong, P.; Sirikanchana, K.; Surachetpong, W. Development and validation of a reverse transcription quantitative polymerase chain reaction for tilapia lake virus detection in clinical samples and experimentally challenged fish. *J. Fish Dis.* **2018**, *41*, 255–261. [CrossRef]
41. Sanyalukruechai, C.; Watthanasakphuban, N.; Khemthong, M.; Surachetpong, W.; Rattanaporn, K. Expression and purification of recombinant tilapia lake virus segment 4 protein and its in-vitro biological activity for potential use in vaccine development. *Sci. Rep.* **2024**, *14*, 31529. [CrossRef]
42. Polson, A.; von Wechmar, M.B.; van Regenmortel, M.H.V. Isolation of Viral IgY Antibodies from Yolks of Immunized Hens. *Immunol. Commun.* **1980**, *9*, 475–493. [CrossRef]
43. Pauly, D.; Chacana, P.; Gutierrez Calzado, E.J.; Brembs, B.; Schade, R. IgY Technology: Extraction of Chicken Antibodies from Egg Yolk by Polyethylene Glycol (PEG) Precipitation. *J. Vis. Exp.* **2011**, *51*, e3084. [CrossRef]
44. Metheenukul, P.; Surachetpong, W.; Prasertsincharoen, N.; Arreesrisom, P.; Thengchaisri, N. Comparison of immunoglobulin Y antibody production in new and spent laying hens. *Vet. World* **2024**, *17*, 2177–2184. [CrossRef] [PubMed]
45. Laemmli, U.K. Cleavage of Structural Proteins during the Assembly of the Head of Bacteriophage T4. *Nature* **1970**, *227*, 680–685. [CrossRef]
46. Tattiyapong, P.; Dechavichitlead, W.; Waltzek, T.B.; Surachetpong, W. Tilapia develop protective immunity including a humoral response following exposure to tilapia lake virus. *Fish Shellfish Immunol.* **2020**, *106*, 666–674. [CrossRef] [PubMed]
47. Reed, L.J.; Muench, H. A simple method of estimating fifty per cent endpoints. *Am. J. Epidemiol.* **1938**, *27*, 493–497. [CrossRef]
48. Mohamad, A.; Khemthong, M.; Trongwongsa, P.; Lertwanakarn, T.; Setthawong, P.; Surachetpong, W. A New Cell Line from the Brain of Red Hybrid Tilapia (*Oreochromis* spp.) for Tilapia Lake Virus Propagation. *Animals* **2024**, *14*, 1522. [CrossRef]
49. Klimentzou, P.; Paravatou-Petsotas, M.; Zikos, C.; Beck, A.; Skopeliti, M.; Czarnecki, J.; Tsitsilonis, O.; Voelter, W.; Livaniou, E.; Evangelatos, G.P. Development and immunochemical evaluation of antibodies Y for the poorly immunogenic polypeptide prothymosin alpha. *Peptides* **2006**, *27*, 183–193. [CrossRef]
50. Yang, Y.-F.; Yamkasem, J.; Surachetpong, W.; Lin, Y.-J.; You, S.-H.; Lu, T.-H.; Chen, C.-Y.; Wang, W.-M.; Liao, C.-M. Assessing the effect of probiotics on tilapia lake virus-infected tilapia: Transmission and immune response. *J. Fish Dis.* **2022**, *45*, 1117–1132. [CrossRef]
51. Mohamad, A.; Yamkasem, J.; Paimeeka, S.; Khemthong, M.; Lertwanakarn, T.; Setthawong, P.; Nuez-Ortin, W.G.; Isern Subich, M.M.; Surachetpong, W. Efficacy of Feed Additives on Immune Modulation and Disease Resistance in Tilapia in Coinfection Model with Tilapia Lake Virus and *Aeromonas hydrophila*. *Biology* **2024**, *13*, 938. [CrossRef]
52. Tabll, A.A.; Shahein, Y.E.; Omran, M.M.; Hussein, N.A.; El-Shershaby, A.; Petrovic, A.; Glasnovic, M.; Smolic, R.; Smolic, M. Monoclonal IgY antibodies: Advancements and limitations for immunodiagnosis and immunotherapy applications. *Ther. Adv. Vaccines Immunother.* **2024**, *12*, 25151355241264520. [CrossRef]
53. Sugino, H.; Nitoda, T.; Juneja, L.R. General Chemical Composition of Hen Eggs. In *Hen Eggs*; CRC: Boca Raton, FL, USA, 2018; pp. 13–24.
54. Zhang, L.; Lin, L.; Qin, Z. A review on the application of chicken immunoglobulin Y in aquaculture. *Rev. Aquac.* **2024**, *16*, 536–551. [CrossRef]

55. Madera-Contreras, A.M.; Solano-Texta, R.; Cisneros-Sarabia, A.; Bautista-Santos, I.; Vences-Velázquez, G.; Vences-Velázquez, A.; Cortés-Sarabia, K. Optimized method for the extraction of contaminant-free IgY antibodies from egg yolk using PEG 6000. *MethodsX* **2022**, *9*, 101874. [CrossRef] [PubMed]
56. Yang, Y.-e.; Wen, J.; Zhao, S.; Zhang, K.; Zhou, Y. Prophylaxis and therapy of pandemic H1N1 virus infection using egg yolk antibody. *J. Virol. Methods* **2014**, *206*, 19–26. [CrossRef] [PubMed]
57. Arasteh, N.; Aminirissehei, A.H.; Yousif, A.N.; Albright, L.J.; Durance, T.D. Passive immunization of rainbow trout (*Oncorhynchus mykiss*) with chicken egg yolk immunoglobulins (IgY). *Aquaculture* **2004**, *231*, 23–36. [CrossRef]
58. Wang, H.; Zeng, X.; Lin, J. Ex Vivo Evaluation of Egg Yolk IgY Degradation in Chicken Gastrointestinal Tract. *Front. Immunol.* **2021**, *12*, 746831. [CrossRef]
59. Li, X.-Y.; Jin, L.-J.; McAllister, T.A.; Stanford, K.; Xu, J.-Y.; Lu, Y.-N.; Zhen, Y.-H.; Sun, Y.-X.; Xu, Y.-P. Chitosan–Alginate Microcapsules for Oral Delivery of Egg Yolk Immunoglobulin (IgY). *J. Agric. Food Chem.* **2007**, *55*, 2911–2917. [CrossRef]

viruses

MDPI

Article

PRV-1 Virulence in Atlantic Salmon Is Affected by Host Genotype

Mark Polinski [1,*], Lynden Gross [2], David Groman [3], Marta Alarcón [4], Mark Braceland [5], Marije Booman [5], Delphine Ditlecadet [6], Samuel May [1], Nellie Gagné [6] and Kyle Garver [2,*]

1. U.S. Department of Agriculture National Coldwater Marine Aquaculture Center, Orono, ME 04473, USA; samuel.may@usda.gov
2. Fisheries and Oceans Canada Pacific Biological Station, Nanaimo, BC V9T6N7, Canada; lynden.gross@dfo-mpo.gc.ca
3. University of Prince Edward Island Atlantic Veterinary College, Charlottetown, PEI C1A4P3, Canada; groman@upei.ca
4. Pharmaq Analytiq, 5008 Bergen, Norway; marta.alarcon@zoetis.com
5. Onda, Souris, PEI C0A2B0, Canada; mark.braceland@wellfishtech.com (M.B.); mbooman@onda.ca (M.B.)
6. Fisheries and Oceans Canada Gulf Fisheries Research Centre, Moncton, NB E1C5K4, Canada; delphine.ditlecadet@dfo-mpo.gc.ca (D.D.); nellie.gagne@dfo-mpo.gc.ca (N.G.)

* Correspondence: mark.polinski1@usda.gov (M.P.); kyle.garver@dfo-mpo.gc.ca (K.G.)

Abstract: Heart and skeletal muscle inflammation (HSMI) is a significant disease affecting Atlantic salmon (*Salmo salar*) production in Norway but has had limited impact to production in North America. The causative agent of HSMI is piscine orthoreovirus genotype 1 (PRV-1), and disease variation between regions is suggested to be at least partially driven by genetic variation of the virus. Using controlled laboratory injection challenges, we corroborate variations in disease outcomes for three PRV-1 isolates (PRV-1a from the eastern Pacific, PRV-1a from the western Atlantic, and PRV-1b from the Norwegian sea); however, virus replication dynamics, host recognition, and PRV-1-associated heart inflammation were also discrete relative to the Atlantic salmon stock challenged, irrespective of the viral isolate used. Specifically, New Brunswick Tobique River Atlantic salmon had less ($p < 0.01$) heart inflammation relative to Mowi-McConnell Atlantic salmon of Western Canada which, in turn, had less ($p < 0.01$) heart inflammation than Mowi Atlantic salmon of Scotland when cumulatively considering challenges using all three PRV-1 isolates. These data indicate that the presence of PRV-1a or PRV-1b alone is not sufficient to reliably predict disease and highlights at least one potential mechanism (host genotype) for reducing HSMI disease severity.

Keywords: piscine orthoreovirus (PRV); atlantic salmon; virulence; laboratory disease challenge; genotype

1. Introduction

The primary drivers of piscine orthoreovirus genogroup 1 (PRV-1) virulence are largely unknown. Despite the establishment of a causative link between PRV-1 and the disease of farmed Atlantic salmon known as heart and skeletal muscle inflammation (HSMI) [1], infection with PRV-1 does not always result in disease [2]. In fact, the virus is relatively commonplace across countries with Atlantic salmon farming, but the occurrence of HSMI is not [3]. For example, in Norway—where PRV-1 and HSMI were first identified [4–6]—PRV-1 is ubiquitous among farmed Atlantic salmon and HSMI poses a significant and widespread production concern for the industry (e.g., 150–180 farm-level HSMI cases per year across >600 active farms) [7]. In contrast, in Pacific Canada, PRV-1 is also ubiquitous across the Atlantic salmon industry, reaching high viral loads, while the occurrence of HSMI-like disease is rare (e.g., typically 0 or occasionally 1 farm-level HSMI-like case per year across >70 farms) [2,8–10]. Differential disease outcomes of PRV-1 infection have also been observed across controlled laboratory exposure studies with similar environmental parameters where the universal establishment of high load systemic infections have produced varying amounts of heart inflammation [1,2,11,12].

Current data indicate that PRV-1 virulence is at least partially driven by genetic variation within the virus. This is supported by a side-by-side controlled laboratory comparison of six PRV-1 genetically unique isolates which differed in their ability to induce HSMI-like heart inflammation in a European stock of Atlantic salmon [12]. Specifically, isolates categorized in the PRV-1a genotype based on S1 genome segment phylogeny were demonstrated to have lower virulence than two PRV-1b isolates, and this has been supported by HSMI field observations [13]. Additional consideration of other genome segments has further led to the hypothesis that viral reassortment of two ancestral genotypes in Norwegian aquaculture gave rise (and may continue to give rise) to virulent phenotype(s) of PRV-1, and that genotypic variation in segments beyond S1 are required for increased virulence [14,15]. One current unexplained caveat is that the PRV-1 genotype is not always explanatory of disease—as evidenced by occasional farm-level diagnoses of HSMI or HSMI-like disease with PRV-1 infections that are considered to be "low-virulence" genotypes [2,9,16].

Host and environmental factors also implicitly influence disease outcome; yet fewer data are available for implicating either host or environmental conditions in the establishment of HSMI. This disease appears to be associated with an adaptive cytotoxic T cell response by Atlantic salmon to PRV-1 antigen (reviewed by [3]) and, thus, modulation of salmon T cell responses would be expected to impact disease outcomes. This is supported by investigations involving the dietary modulation of Atlantic salmon immune responses that may at least partially mitigate HSMI [17–19]. Furthermore, some Atlantic salmon stocks may inherently be more (or less) predisposed to PRV associated disease, as suggested by the development of a strain of Atlantic Salmon in Norway that has been claimed to have less heart damage and higher survival in association with HSMI yet is equally susceptible to PRV-1 infection [20,21].

Laboratory vs. field observations for HSMI severity also clearly indicate an environmental component associated with this disease [3], yet the particular mechanisms in field environments that enhance HSMI outcomes are unknown. Although "stress" associated with production environments has been commonly hypothesized to contribute to HSMI, application of maximal glucocorticoid, forced transition to saltwater, or viral coinfection stressors in laboratory environments have been insufficient to recreate equivalent disease [11,22,23].

In this study, we aimed to investigate the degree to which host genetic differences affect PRV-1-associated disease and compare the influence of host Atlantic salmon strain directly to PRV-1 genotype to see which factor had a greater impact on PRV-1-associated disease. We hypothesized that both the host strain of Atlantic salmon and the viral genotype would influence disease severity following the disease triangle principle [24], but that viral genotype would have a greater influence based on currently available data that suggest at least some PRV-1b isolates appear to be more strongly associated with disease in both the laboratory and field.

2. Materials and Methods

2.1. PRV-1 Isolates

Three PRV-1 isolates were used in this study: BC-16-005ND, NOR-2018NL, and NB-2018-128 (Table 1). The first isolate, BC-16-005ND, originated from a commercial farmed Atlantic salmon from the Pacific coast of British Columbia, Canada [2]. Viral particles for this isolate were purified from infected whole blood as previously described [1]. Briefly, sonicated and clarified blood homogenate was combined with Vertrel XF (DuPont Chemicals, Wilmington, DE, USA) was layered onto a 1.22–1.45 g/mL cesium chloride gradient and centrifuged using a SW 40 TI rotor in an Optima LE 80k Ultracentrifuge (Beckman Coulter, Brea, CA, USA) at 40,000 rpm for 20 h. Viral particles were collected from the 1.22/1.45 interphase by needle puncture and dialyzed at 4 °C against fresh Dulbecco's PBS for 1, 3, and >12 h using a 7K Slide-A-Lyzer Dialysis Cassette (Thermo Fisher, Waltham, MA, USA) before 15% glycerol was added prior to freezing and storage at −80 °C. The next isolate, NOR-2018NL, was obtained as cesium chloride-purified particles from the Norwegian University of Life Sciences with demonstrated virulent potential in experimental challenge trials [12]. The third isolate, NB-2018-128, was obtained from a commercially farmed Atlantic salmon in New Brunswick, passed through Atlantic salmon (n = 6) by intraperitoneal injection of a 0.1 mL homogenized blood cell pellet, and harvested from whole blood via needle puncture of the caudal vein 6 weeks later. Although cesium chloride purification methods were attempted for NB-2018-128, the low viral load (as measured by qPCR) resulted in insufficient recovery of purified particles to conduct subsequent challenge trials. Therefore, sonicated and low-speed centrifuge-clarified blood homogenate diluted in PBS was used directly as the inoculate for challenge trials for this isolate, as previously described [2]. For each PRV-1 isolate, a 10 μL sample immediately prior to challenge was used to quantify genomic material by qPCR (see Section 2.4). All viral isolates were stored at −80 °C prior to use.

Table 1. Matrix of viral inoculates and fish stocks used in the two disease challenge trials presented in this study. PRV-1 isolates—BC-16-005ND (BC-PRV), NOR-2018NL (NOR-PRV), and NB-2018-128 (NB-PRV)—along with saline control inoculate (SC) are indicated, as well as the four fish stocks used: BC—Mowi-McConnell (BC salmon), New Brunswick St. John River (NB-SJR salmon), New Brunswick Tobique River (NB-TR salmon), and European Mowi from Scotland (EU salmon). Mean fish weights obtained from opportunistic subsampling at the start of the challenge are provided.

Challenge Trial	Viral Inoculate	Fish Stock	Mean Fish Weight (g)
1	BC-PRV (purified particles)	BC salmon	36
	SC	BC salmon	42
	BC-PRV (purified particles)	NB-SJR salmon	42
	SC	NB-SJR salmon	44
2	BC-PRV (purified particles)	BC salmon	59
	NB-PRV (blood homogenate)	BC salmon	61
	NOR-PRV (purified particles)	BC salmon	58
	SC	BC salmon	55
	BC-PRV (purified particles)	NB-TR salmon	43
	NB-PRV(blood homogenate)	NB-TR salmon	44
	NOR-PRV (purified particles)	NB-TR salmon	44
	SC	NB-TR salmon	48
	BC-PRV (purified particles)	EU salmon	67
	NB-PRV (blood homogenate)	EU salmon	74
	NOR-PRV (purified particles)	EU salmon	70
	SC	EU salmon	71

Challenge inoculates were prepared the day of challenge from either cesium chloride-purified (BC-16-005ND—Challenge 1 and 2; NOR-2018NL—Challenge 2) or clarified blood homogenate (NB-2018-128, Challenge 2) frozen stocks by diluting each to a final concentration of 1×10^8 PRV-1 genome copies per mL in refrigerated phosphate-buffered saline with a final concentration of 0.01% glycerol. Control inoculate was identically prepared without virus. For clarified blood homogenate inoculate (NB-2018-128), blood homogenate made up less than 10% (10 uL) of the total volume.

2.2. Atlantic Salmon—Sources and Husbandry

Atlantic salmon were sourced from four independent hatcheries representing four regionally distinct genetic strains that generally encompass the major stocks used in commercial production globally (Table 1). All fish were transported by 2-day courier directly to Onda, in Souris, Prince Edward Island. One strain was sourced as ~10 g parr from a commercial hatchery on Vancouver Island, British Columbia. These Atlantic salmon were of a Pacific-adapted Mowi-McConnell strain from British Columbia with at least 30 years isolation from the originating European stocks [25]. The second stock was sourced as ~10 g parr from a commercial hatchery in New Brunswick, Canada, of a St. John River strain that have been selectively bred for aquaculture production for more than 16 years [26]. The third stock of fish was obtained as ~10 g parr from Mactaquac Biodiversity Facility, New Brunswick, Canada, where hatchery-spawned wild-caught salmon from the Tobique River [27] were manually crossed to produce the F1 generation used in this study. The fourth stock of fish was obtained as eggs from a commercial aquaculture facility in Scotland, UK, rearing the domesticated Mowi Atlantic salmon strain heavily used in Norway and Scotland for commercial production. All stocks were of a mixed-sex, mixed-family composition.

Once at Onda, fish were maintained in UV-treated flow-through fresh well water (7–12 °C) for 3–6 months. Upon reaching approximately 40–50 g mean body weight, each stock was progressively acclimated over a 20-day period to UV-irradiated brackish well water (10–12 °C, 25 ppt NaCl). A 12 h light–12 h dark photoperiod was used except during saltwater acclimation, when a 24 h light cycle was temporarily employed. Fish were fed dry pellets (EWOS) at 2–5% body weight per day prior to the challenge. During both challenge trials, fish were maintained in UV-irradiated brackish well water (11 ± 1 °C, 25 ppt) and fed a ration of EWOS pellets at 1% body weight per day in biosafety level 3 (BSL-3) containment.

2.3. PRV-1 Challenges of Atlantic Salmon

Two challenge trials were conducted as part of this study. Challenge 1 utilized a single PRV-1 isolate (BC-16-005ND hereafter referred to as BC-PRV) administered to two strains of Atlantic salmon: BC Mowi-McConnell (BC salmon; mean weight 43.7 g) and New Brunswick St. John River (NB-SJR salmon; mean weight 43.3 g) (Table 1). An intraperitoneal injection of 100 µL was administered individually to 50 fish per tank at a targeted dose of 1×10^6 BC-PRV reverse-transcribed L1 genomic copies per 100 µL inoculate to duplicate tanks of either BC or NB-SJR salmon. Inoculate without virus was identically prepared and administered to duplicated tanks for each salmon strain to act as vehicle saline control (SC). All fish were anesthetized in 75 mg per L tricaine mesylate (MS-222) prior to intraperitoneal injection. Following injection, 6 fish per treatment from one replicate tank were immediately sampled following an overdose of MS-222 (200 mg per L). Subsequently, every two weeks post-challenge (wpc) 6 fish per treatment were sampled for a period totaling 14 weeks. Whole blood (100 µL) was collected via caudal vein puncture and immediately frozen at −80 °C for PRV screening and host gene expression analysis by qPCR. Hearts were bisected longitudinally, and one half preserved in 10% neutral buffered formalin for histopathology while the other half was immediately frozen at −80 °C for PRV screening and host gene expression (qPCR) analyses.

Challenge 2 utilized three PRV-1 isolates—BC-16-005ND (BC-PRV), NOR-2018NL (NOR-PRV), and NB-2018-128 (NB-PRV)—that were administered to three strains of Atlantic salmon: BC Mowi-McConnell (BC salmon; mean weight 59.5 g), New Brunswick Tobique River (NB-TR salmon; mean weight 44.5 g), and European Mowi (EU salmon; mean weight 70.7 g) from Scotland (Table 1). An intraperitoneal injection of 100 μL inoculate was administered individually to 50 fish per tank at a targeted dose of 1×10^6 PRV-1 reverse-transcribed L1 genomic copies per 100 μL to duplicate tanks of each salmon strain. Inoculate without virus was identically prepared and administered to duplicate tanks of each salmon strain to act as the vehicle saline control (SC). As per Challenge 1, fish were anesthetized in 75 mg per L MS-222 prior to intraperitoneal injection. Immediately following injection, four fish per treatment were lethally sampled (two from each duplicate tank) and every 2 wpc, eight fish per treatment were lethally sampled (four from each duplicate tank) for a period totaling 14 weeks. A portion of whole blood (100 μL) was collected via caudal vein puncture and immediately frozen at −80 °C for PRV screening by qPCR. A second portion of whole blood (up to 900 μL) was collected and spun at 2000× *g* for 10 min at 4 °C in a lithium–heparin-treated vacutainer (BD) from which 100 μL plasma was transferred to a fresh 1.5 mL tube and frozen at −80 °C for PRV screening. Approximately 200 mg section of red/white skeletal muscle was excised from the left lateral line at approximately the mid-body and preserved in 10% neutral buffered formalin (NBF). Hearts were bisected longitudinally, and one half was preserved in 10% neutral buffered formalin while the other half was immediately frozen at −80 °C for qPCR analyses. Inventories of the analyzed data for Challenge 1 and Challenge 2 are presented in Tables S1 and S2, respectively.

2.4. PRV-1 Detection and Quantification

PRV-1 L1 RNA was detected in whole blood, plasma, heart, and inoculate samples by real-time quantitative PCR (qPCR) following reverse transcription at the Fisheries and Oceans Canada Pacific Biological Station in Nanaimo, British Columbia, based on previously published methods [8]. Specifically, total RNA was extracted from 100 μL blood, approximately 50 mg tissues, or 10 μL purified viral inoculates in TRIzol Reagent (Life Technologies, Carlsbad, CA, USA) as per the manufacturer's instructions that implemented a 5 mm steel bead and TissueLyser II (Qiagen, Germantown, MD, USA) operating for 2 min (blood) or 4 min (heart) at 25 Hz. Nucleic acids were extracted from 10 μL of plasma diluted in 130 μL PBS using the QIAamp Viral RNA Mini Kit (Qiagen). A portion of eluted RNA (1.0 μg from whole blood or tissue; 10 μL for plasma and inoculate) was denatured for 5 min at 95 °C and immediately cooled to 4 °C. Denatured blood, tissue, and viral inoculate RNA were reverse transcribed using a high-capacity cDNA reverse transcription kit (Life Technologies) following the manufacturer's instructions. The resulting cDNA was used directly as a template for qPCR analysis in a StepOne-Plus real-time detection system (Applied Biosystems, Waltham, MA, USA) using previously described primers and TaqMan probe [4]. Each reaction contained 400 nM primers and 300 nM TaqMan probe, 1X TaqMan Universal Master Mix, and 1 μL cDNA template within each 15 μL reaction. Cycling conditions included an initial incubation of 95 °C for 10 min followed by 40 cycles of 95 °C for 10 s and 60 °C for 30 s. Denatured plasma RNA (5 μL) was reverse transcribed and assayed with the OneStep RT-PCR Kit (Qiagen) following the manufacturers protocol and using the same concentration of PRV L1 primers and TaqMan probe specified above, 1× OneStep RT-PCR Buffer, dNTP Mix (400 μM of each dNTP), 1× Q-Solution, 500 nM ROX reference dye, and 1 μL OneStep RT-PCR Enzyme Mix, within each 25 μL reaction. Reverse transcription and cycling occurred at 50 °C for 30 min and 95 °C for 15 min, with 40 cycles of 94 °C for 15 s, followed by 60 °C for 30 s and 72 °C for 30 s. All samples were assayed in duplicate and were considered positive if both technical replicates reported a Ct value < 40 cycles or negative if one or both technical replicates failed to fluoresce beyond the preset threshold (Δ Rn 0.01) during 40 cycles. PRV L1 RNA quantification was determined in each positive instance by serial dilution of a 482 bp double-stranded DNA gBLOCK fragment (Integrated DNA Technologies, Coralville, IA, USA) consisting of sequence targeted by the qPCR primer and probe [11] using a seven-step tenfold dilution series of the gBLOCK fragment spanning a dynamic range of 10–10^7 target copies per reaction.

2.5. Atlantic Salmon Gene Expression and Histopathology

Histopathology was used in this study to measure inflammatory disease to coincide with how HSMI is diagnosed in the field [28,29]. Tissues in 10% NBF were fixed for 24–48 h, transferred to 70% isopropanol, and paraffin-embedded following standard methods. Sections 3 μm thick were transferred to glass slides and stained routinely with hematoxylin and eosin for light microscopy. For Challenge 1, a single pathologist (Michael Pawlik, BC Ministry of Agriculture and Food) reviewed all histopathology samples following a previously published scoring metric ([2]; Table S1). For Challenge 2, a single pathologist (David Groman, University of Prince Edward Island) reviewed all histology samples following a slightly different published scoring metric ([29]; Table S2). To ensure inter-sample consistency as well as consistency relative to other published PRV studies, approximately 20% of samples collected as part of Challenge 2 (144 of 713) were also reviewed by a second pathologist (Marta Alarcón, Pharmaq Analytiq) with experience assessing HSMI in Norway (Table S2; Figure S1). All pathologists were blinded to the PRV exposure status, salmon stock, and scores provided by the other pathologist until after all scoring had been completed. Images of some heart tissues with associated scoring are provided in Figure S1.

For gene expression in Challenge 1, a portion (5 μg) of total RNA extracted from the blood and heart that was not reverse transcribed for the detection of the virus was purified using 2 U of DNase I (RNase free) (Life Technologies) at 37 °C for 45 min followed by RNeasy MinElute Cleanup (Qiagen) as per the manufacturer's instructions. For Challenge 2, a portion (10 μg) of total RNA from the heart samples was subjected to DNase treatment (Qiagen) at ambient temperature for 10 min and column-purified with the RNeasy Mini Kit (Qiagen) following the manufacturer's RNA Cleanup protocol. For both challenge trials, RNA (1.5 μg) was reverse transcribed using a high-capacity cDNA reverse transcription kit without RNase inhibitor in which the random primer mix was substituted with 5 μM Oligo d(T)16. All real-time qPCR analyses were conducted on a StepOne-Plus real-time detection system using SYBR green chemistry. Each reaction consisted of 1× Power SYBR master mix (Applied Biosystems), forward and reverse primers (800 nM each; [23]), and 1–2 μL cDNA template, to a final volume of 15 μL. Samples were assayed in duplicate with a five-step, fourfold dilution series of pooled cDNA included in each run to calculate amplification efficiency, linearity, and relative quantity, and to provide inter-run calibration. The cycling conditions consisted of an initial activation of DNA polymerase at 95 °C for 10 min, followed by 40 cycles of 95 °C for 5 s, 60 °C for 20 s, and 72 °C for 10 s. A melt-curve analysis was conducted for each run to ensure amplification specificity, and amplification efficiency for all assays was ensured to be greater than 90%. Five mRNA transcripts of immune-associated proteins were assessed for expression in this study. In Challenge 1, the *mx1* transcript which codes Myxovirus protein-1 was used to identify general viral recognition by host cells via activation of the type-1 interferon pathway. The *cd8a* and *gzma* transcripts code for the cluster of differentiation-8 protein and granzyme protein, respectively, and provided markers for CD8+ cytotoxic T cell occurrence and activation used in the host immune-directed killing of virus-infected cells. In Challenge 2, *cd8a* was again used to identify CD8+ cytotoxic T cell occurrence, while *ifna* and *ifng* transcripts which code for type-1 and type-2 interferon, respectively, were measured to differentiate general host cell recognition of the virus (type-1 interferon) relative to the professional immune cell recognition of the viral infection (type-2 interferon). All gene expression data were normalized to *actb* transcription that codes for the beta-actin protein which has previously shown stable expression following the PRV infection of salmon [22].

2.6. PRV-1 and Atlantic Salmon Genotyping

For PRV-1 genotyping, a portion (1 μg) of purified total RNA from 3 infected challenged fish (#176, #199, and #219) not used for qPCR analysis was pre-amplified using primer pairs designed in Primal Scheme (http://primalscheme.com; accessed on 18 February 2025) to target each of PRV-1′s 10 genomic segments (Table S3) and run in multiplex PCR, then cleaned and sequenced following previously published protocols [30] using a nanopore MinION sequencer. Reads were assembled against the published PRV-1 genome (GenBank GCA_002829625.1; [31]) using NanoPipe [32].

For Atlantic salmon genotyping, fin clips from 16 fish per stock during Challenge 2 were collected and preserved in ethanol—representing BC salmon, EU salmon, and NB-TR salmon. NB-SJR and BC salmon used in Challenge 1 were not fin clipped at the time of sampling; however, BC stocks were sourced from the same hatchery and breeding program as for Challenge 2 (only 1 cohort earlier), and 16 fin clips from SJR stocks held at the U.S. Department of Agriculture National Coldwater Marine Aquaculture Center in Franklin, ME, from the same cohort year and originating breeding program as those used in Challenge 1 were genotyped as representative proxies. All fin clips were sent to the Center of Aquaculture Technologies (San Diego, CA, USA) for DNA extraction and single nucleotide polymorphism (SNP) genotyping using a recently developed 50 k SNP array for North American Atlantic salmon ([33]; publicly available via Thermo Fisher Axiom USDA Sal, SKU# 551,627 and 551628). The resulting SNP microarray dataset was filtered using the adegenet package [34] in R (v.4.3.1; R Core Team). To ensure high-quality genotype data, loci were removed if genotyped in <90% of individuals. No individuals were genotyped at <90% of loci and, thus, they were not filtered. Monomorphic loci or SNPs with a minor allele frequency (MAF) ≤ 0.05 were removed to eliminate rare variants or possible genotyping errors. To remove non-neutral loci that were putatively under selection, exhibiting non-Mendelian inheritance, or possible null alleles, loci that significantly deviated from the Hardy–Weinberg equilibrium (HWE, estimated in pegas, v.0.13; [35]) or with an absolute difference between expected and observed heterozygosity ($|He-Ho|$) > 0.2 in any one of the four populations were removed. These filtering steps ensured that the retained loci were robust for neutral population genetic analyses.

2.7. Statistical Analyses

Phylogenetic comparison of concatenated segments of the three PRV-1 isolates used in this study was performed against 48 published whole genome sequences, as presented by Siah et al. [36] via a Tamura–Nei neighbor-joining method following Clustal Omega maximum-likelihood alignment in Geneious Prime 2024.0.5 using 1000 bootstrap iterations. Genetic differentiation among the four populations of Atlantic salmon was assessed using Weir and Cockerham's Fst, estimated in the hierfstat package [37]. Confidence intervals for Fst values were determined via non-parametric bootstrapping of loci with 1000 iterations. To visualize genetic variation among populations, a genetic principal component analysis (PCA) was implemented in adegenet. As this analysis does not tolerate missing data, missing genotypes were imputed using population medians.

For the Challenge 1 experiment, BC-PRV transcriptional blood load as well as fold change in Atlantic salmon *mx1*, *cd8a*, and *gzma* relative to mean SC timepoint- and salmon strain-matched expression in blood and heart was compared via two-way ANOVA, followed by Sidak's multiple comparison tests of log-transformed data. Categorical heart inflammatory scores for epicarditis and endocarditis were compared separately between treatments using a Kruskal–Wallis test followed by Dunn's multiple comparison tests.

For the Challenge 2 viral load comparisons, PRV-1 L1 RNA quantities were assessed at both the peak and persistent phases of infection by a one-way ANOVA and Tukey's multiple comparison tests of log-transformed quantities. As peak infections occurred at slightly different time points depending on the PRV-1 isolate × Atlantic salmon strain challenge (Figure S2), we considered the eight highest recorded values obtained between 6–10 wpc from each PRV-1 isolate × Atlantic salmon strain combination as being representative of the peak infection and categorized these data for comparison either by the PRV-1 isolate received (n = 24 per category) or by the Atlantic salmon strain challenged (n = 24 per category). Persistent infections were assessed at 14 wpc for all PRV-1 isolate × Atlantic salmon strain challenges (n = 8 per specific challenge) categorized in the same manner—either by PRV-1 isolate (n = 24) or Atlantic salmon strain (n = 24).

For the Challenge 2 heart *cd8a*, *ifna*, and *ifng* gene expression analyses, quantities of each target transcript were normalized to *actb*—a transcript previously demonstrated to have stable expression following PRV infection of salmon [11,22]—and scaled to the minimum observed value (i.e., the corrected normalized relative quantity; CNRQ). A Pearson r correlation was obtained for *cd8a* and *ifna* or *ifng* based on log-transformed values from the full dataset (n = 384 samples obtained between 6–12 wpc). The CNRQ of *cd8a* and *ifna* expression in relation to histopathological cardiac inflammation score was also assessed for the complete data set using the Kruskal–Wallis nonparametric test and Dunn's multiple comparison tests based on inflammation score. The CNRQs of *cd8a* transcripts from the entire data set sans SC samples (n = 288) were compared by two-way ANOVA and Šídák's multiple comparison tests of log-transformed CNRQ values categorized either by PRV-1 isolate (n = 24 per PRV-1 genotype per time point) or by Atlantic salmon strain (n = 24 per salmon genotype per time point).

For Challenge 2 histopathological observations, cumulative categorical heart inflammatory scores of PRV-1-challenged individuals between 2–14 wpc (n = 504) were categorized by either PRV-1 isolate (BC, NB, or NOR) or Atlantic salmon strain (BC, NB-TR or EU) and compared using the Kruskal–Wallis nonparametric test and Dunn's multiple comparison tests. All comparisons were performed using GraphPad Prism 10.3.0. As principal component analysis and direct variance comparisons suggested homogeneity between tank replicates in this study, samples collected from replicate treatment tanks were pooled for all the analyses described above.

3. Results

3.1. Phylogentic Variation in Atantic Salmon Strains and PRV-1 Isolates

Unique genetic structuring was observed for each Atlantic salmon strain used in this study. Filtering processes for the microarray panel data reduced the number of SNP loci from 55,044 to 32,381 SNPs. The retained loci were used to estimate pairwise Fst values among the four populations (Table S4), confirming varying degrees of genetic differentiation between populations and subspecies. Principal component analysis also demonstrated clear separation between the four populations (Figure 1A), supporting the Fst results. The first two principal components explained a substantial proportion of the genetic variance (35.71% and 3.77%, respectively), with clear clustering corresponding to each population, indicating distinct genetic structuring between the Atlantic salmon strains.

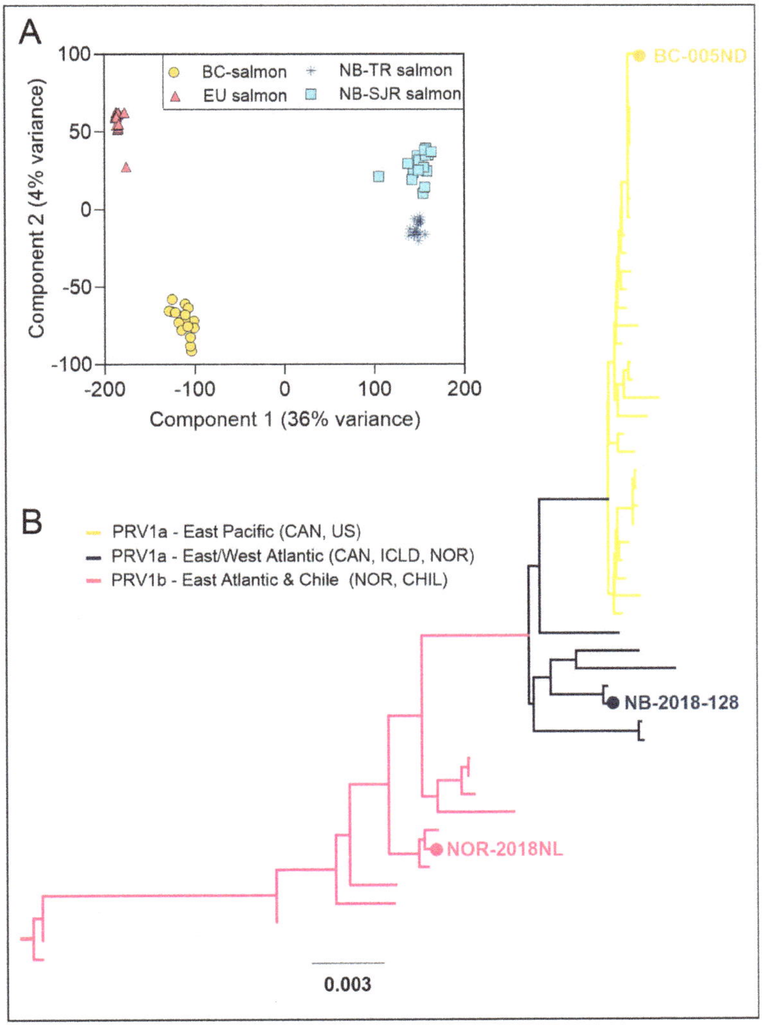

Figure 1. (**A**) Two-dimensional variance determined by a genetic principal component analysis of 50 k microarray genotypes among individual Atlantic salmon (n = 64) representing 4 geographically separate stains: British Columbia, Canada—Mowi-McConnell (BC salmon), European (Scotland) origin Mowi (EU salmon), New Brunswick, Canada—Tobique River (NB-TR salmon), and New Brunswick, Canada—Saint John River (NB-SJR salmon). (**B**) Tamura–Nei neighbor-joining phylogram indicating the genetic diversity of 48 concatenated PRV-1 genomes as presented by Siah et al. [36], as well as the 3 concatenated PRV-1 genomes sequenced in this study (highlighted with text) following Clustal Omega maximum-likelihood alignment. Phylogenetic groups (PRV-1b from Eastern Atlantic and Chile, PRV-1a from both Western and Eastern Atlantic, and PRV-1a from the Eastern Pacific) are indicated by branch lines color at 100% bootstrap consensus support. Scale bar indicates the genetic divergence as the average nucleotide substitutions per position.

Unique genetic structuring was also observed for each PRV-1 isolate used in this study. A whole concatenated genome comparison of the three PRV-1 isolates from this study to the 48 PRV-1 genomes considered by Siah et al., 2020 [36], identified similar phylogenetic ordination using Tamura–Nei neighbor-joining methods to the Bayesian phylogenetic analyses previously employed. The BC-PRV grouped together with all other Northeastern

Pacific PRV-1 isolates to form a monophyletic clade, NB-PRV grouped with other isolates from the East/West Atlantic, and NOR-PRV grouped with a subset of other Eastern Atlantic isolates from Norway and Chile (Figure 1B). Specific to the S1 component of these genomes, both BC- and NB-PRV were members of what has been previously designated as the PRV-1a group, while the NOR-PRV is within the PRV-1b group [31] shown to be associated with higher virulence in Norway [12]. Further virulence characterization of these isolates, as described by Vatne et al., 2021 [15] using 4 viral genomic segments (L1, L2, S1, and S4), classifies BC-PRV and NB-PRV within the "Low-2" and NOR-PRV within "High-1" genogroups. NCBI accession numbers for the PRV-1 sequences obtained in this study are shown in Table S5.

3.2. Challenge 1: PRV-1 Infection Dynamics and Host T Cell Responsiveness Can Vary Among Atlantic Salmon Strains

Identical aliquots of BC-PRV administered to BC and NB-SJR salmon in Challenge 1 produced different viral infection dynamics as well as host immune responsiveness in the two salmon strains. The BC-PRV took longer (by 4 weeks) to reach peak transcriptional loads in whole blood of NB-SJR fish (median 8.9×10^6 at 10 wpc) than in BC fish (median 4.7×10^7 at 6 weeks), with lower viral RNA blood loads at both 4 and 6 wpc for most NB-SJR individuals (Figure 2A). Not only did BC-PRV amplification appear to be delayed in NB-SJR salmon, but peak viral blood loads were also slightly lower than in BC salmon (mean 42% $\pm$ 19%, $p = 0.009$) when comparing the highest observed blood infections (n = 8) from each strain. Nevertheless, at peak replication, high loads were reached in nearly all individuals sampled in both salmon strains (mean 4.4 and 2.5×10^7 copies/ug blood RNA, respectively). Persistent blood loads at 12 and 14 wpc also remained substantial in both salmon strains (mean > 1×10^6 copies/ug blood RNA) (Figure 2A).

Systemic BC-PRV recognition as measured by whole blood *mx1* transcription—a marker for interferon pathway activation—became elevated (4–8-fold) in both BC and NB-SJR salmon, although this activation was considerably more prolonged in BC salmon (observed 6–14 wpc) relative to NB-SJR salmon (observed only at 10 wpc; Figure 2B). Whole blood transcription of *cd8*– a transcript marker for cytotoxic T cell presence, differentiation, and activation—did not appear to be affected by the presence of BC-PRV (Figure 2C), although *gzma*—a transcript associated with cytotoxic T cell-directed killing—was similarly elevated by 5–14-fold at 10 wpc in both salmon strains (Figure 2D). Together, these data suggest that BC-PRV was systemically recognized in blood by infected host cells (although for a longer period in BC salmon) with transient low-level increases in cytotoxic T cell killing activity, but without evidence of increased cytotoxic T cell presence.

In the heart, *mx1* transcriptional patterns mirrored those observed within whole blood, with 4–10-fold increases in transcription occurring in both salmon strains with more prolonged activation in BC salmon (Figure 2E). However, in contrast to *cd8* and *gzma* expression in the blood, heart expression of these transcripts was highly salmon strain-dependent. Transcription of both *cd8* and *gzma* were unaffected by BC-PRV in hearts of NB-SJR salmon, whereas mean *cd8* transcription became elevated 10–12-fold in hearts of BC salmon at 10–14 wpc (Figure 2F) and mean *gzma* expression was increased by approximately 100-fold at 14 wpc in BC salmon relative to controls (Figure 2G). Additionally, cumulative heart endocarditis (but not epicarditis) was elevated in BC-PRV-infected BC salmon, but not in BC-PRV-infected NB-SJR salmon relative to strain-matched SC fish (Figure 2H). These data indicate that although BC-PRV was recognized in infected hearts of both salmon genotypes, cytotoxic T cell presence and/or differentiation as well as activation of T cell-specific killing pathways were enhanced in BC salmon and not in NB-SJR salmon, and this T-cell activation in BC salmon was associated with elevated endocarditis.

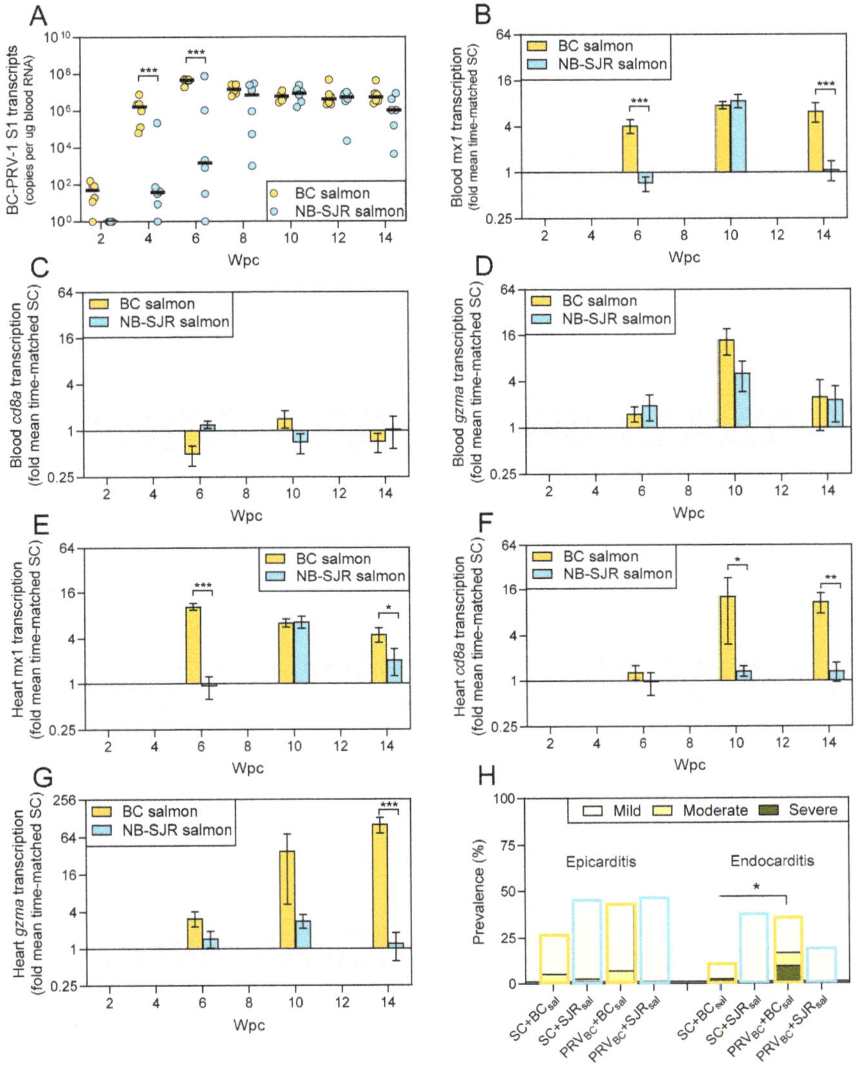

Figure 2. PRV-1 load, antiviral responsiveness, and inflammation in side-by-side challenge of BC Mowi-McConnell (BC) and New Brunswick St. John River (NB-SJR) Atlantic salmon. (**A**) Mean (line) and individual (dot) BC-PRV-1a L1 RNA loads measured 2–14 weeks post-challenge (wpc). Mean fold change (±SE) of (**B**) blood *mx1*, (**C**) blood *cd8a*, (**D**) blood *gzma*, (**E**) heart *mx1*, (**F**) heart *cd8a*, and (**G**) heart *gzma* transcripts relative to the mean of time-matched, strain-matched controls (SC); * $p < 0.05$, ** $p < 0.01$, *** $p < 0.001$ by two-way ANOVA and Šídák's multiple comparisons tests of log-transformed fold changes; minimum twofold change suggestive of biological relevance is shaded. (**H**) The cumulative prevalence of epicarditis and endocarditis in hearts of control (SC) and BC-PRV-1a challenged (PRV) fish within 14 wpc. * $p < 0.05$ by the Kruskal–Wallis and Dunn's multiple comparisons tests.

3.3. Challenge 2: PRV-1 Infection Dynamics Can Be Affected More by PRV-1 Isolate than Atlantic Salmon Strain

PRV-1 infection dynamics can be generally categorized into three temporal phases: early entry and dissemination, peak systemic replication, and long-term persistence [3]. We confirmed this general temporal pattern of systemic viral load in all challenges performed in this study, although the exact timing of each phase slightly varied based on host–virus combinations (Figure S2). Therefore, we assessed 72 individuals at both the peak (6–10 wpc depending on the PRV-1 isolate × salmon strain combination) and persistent (14 wpc for all PRV-1 × salmon combinations) phases of infection and compared these individuals based either on the PRV-1 isolate they were challenged with or by which salmon strain they belonged to. This revealed that the peak PRV-1 L1 RNA load in whole blood—a presumptive measure for intracellular systemic PRV-1 replication—was similar in fish administered with either BC- or NB-PRV genotypes (mean peak blood loads of 6.7 ± 0.6 SE and 5.3 ± 0.5 SE × 10^{11} copies/mL, respectively; $p = 0.59$), but was 4–5 times lower for fish administered NOR-PRV (1.1 ± 0.3 SE × 10^{11} copies/mL; $p < 0.0001$) when assessed cumulatively across the three salmon strains (Figure 3A). These quantitative relationships were mirrored in heart tissues, but at universally lower concentrations (Figure 3B).

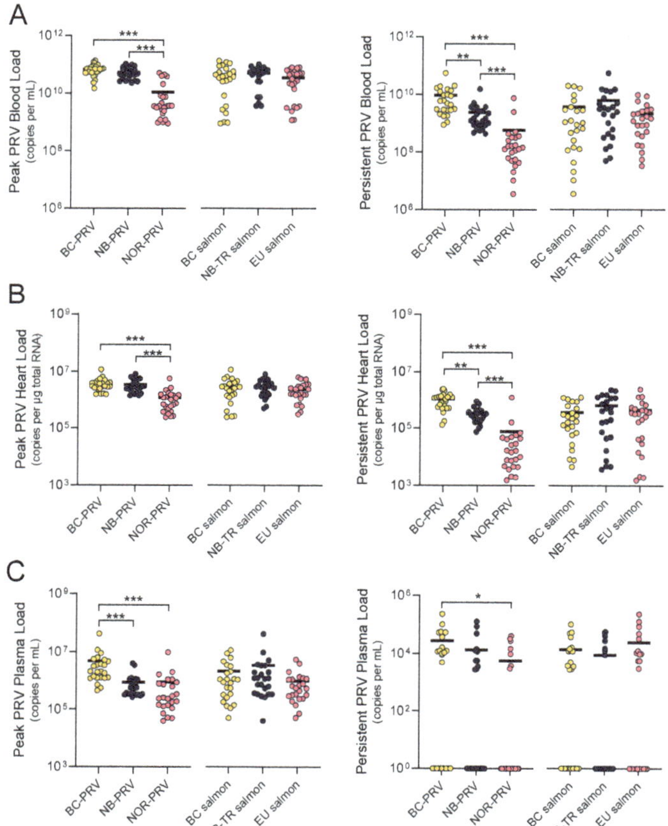

Figure 3. PRV-1 RNA loads of 72 individuals experiencing peak (6–10 wpc; left panels) or persistent (14 wpc; right panels) infections, categorized by either the strain of Atlantic salmon challenged (right component each panel) or the PRV-1 isolate administered (left component each panel). Mean (line) and individual (dot) PRV-1 L1 RNA loads identified in (**A**) whole blood, (**B**) the heart ventricle, or (**C**) plasma presented as the peak (eight highest recorded values 6–10 wpc from each salmon-virus challenge combination; n = 24 per category) or persistent (eight recorded values 14 wpc from each salmon-virus challenge combination; n = 24 per category) phases of infection. * $p < 0.05$, ** $p < 0.01$, *** $p < 0.001$ by one-way ANOVA and Tukey's multiple comparisons tests of log-transformed values.

By the persistent phase of infection, salmon injected with BC-PRV maintained the highest blood loads (9.3 ± 2.3 SE $\times 10^{10}$ copies/mL; $p < 0.002$), followed by fish injected with NB-PRV (2.4 ± 0.6 SE $\times 10^{10}$ copies/mL; $p < 0.002$) and then those injected with NOR-PRV (0.6 ± 0.3 SE $\times 10^{10}$ copies/mL; $p < 0.0001$) (Figure 3A). These quantitative relationships were also mirrored in heart tissue at universally lower concentrations (Figure 3A,B).

Specific to the plasma component of blood—a presumptive measure for systemically shed virus particles—peak PRV-1 RNA loads were highest for BC-PRV (4.8 ± 1.7 SE $\times 10^6$ copies/mL; $p < 0.001$) relative to either NB-PRV- or NOR-PRV-injected individuals, which were similar (8.5 ± 1.9 SE $\times 10^5$ copies/mL and 8.7 ± 4.0 SE $\times 10^5$ copies/mL, respectively; $p = 0.09$) (Figure 3C). The maintenance of higher BC-PRV plasma loads (at least relative to NOR PRV) was putatively suggested in the persistent phase of infection as well ($p < 0.03$); however, non-detections and relatively low quantities of RNA were observed for all three PRV-1 isolates, suggesting that these putative variations may be of limited biological relevance (Figure 3C).

In considering Atlantic salmon strain rather than PRV-1 isolate in these challenges, similar PRV-1 loads were recorded between the three salmon strains in the whole blood, heart, and plasma samples, at peak and persistent phases of infection (Figure 3). Cumulatively, these data suggest that PRV-1 isolate had somewhat unique infection dynamics in this study, whereby BC-PRV and to a more limited extent NB-PRV showed a higher capacity for viral replication, dissemination, and persistence than NOR-PRV. Conversely, these data demonstrated that variations in host Atlantic salmon strain had limited to no impact on PRV-1's capacity for replication, dissemination, and persistence.

3.4. Challenge 2: Atlantic Salmon T Cell Responsiveness Can Be Affected by Both PRV-1 Isolate and Atlantic Salmon Strain

A previous study has indicated that activation of CD8+ cytotoxic T cells is associated with heart inflammation during PRV-1 infections (reviewed by [3]). Here, we measured the transcription of *cd8*—a transcript marker for cytotoxic T cell presence, differentiation, and activation—along with *ifng*—a transcript marker for cytotoxic T cell and/or natural killer cell activation—and *infa*—a transcript marker for more general intracellular and/or local recognition of virus—to compare responses between Atlantic salmon genotypes and for different PRV-1 genotype exposures. We identified that during PRV-1 infection, *cd8a* transcription was highly correlated ($r = 0.86$; $p < 0.001$) with *ifng* cytokine transcription (Figure 4A), indirectly indicating that CD8+ T cells were likely becoming locally attracted/activated and generating gamma interferon in response to PRV-1 recognition. Expression of *cd8a* appeared to further be correlated, although to a lesser degree, with general cellular antiviral responsiveness within heart tissues, as measured by *ifna* transcription ($r = 0.41$; $p < 0.001$) (Figure 4A).

In considering CD8+ T cell involvement in PRV-1-associated heart inflammation, *cd8a* transcription—and by implication CD8+ T cell involvement—was highly associated with heart inflammation in this study. Mild, moderate, or severe heart inflammation was associated with a mean 6- to 17-fold higher *cd8a* transcription relative to fish with no heart inflammation (Figure 4B). In contrast, general intracellular virus recognition as measured by *ifna* transcription was similar in fish with or without heart inflammation (1.1- to 1.4-fold mean increase; Figure 4B). Furthermore, temporal assessments of *cd8a* transcription categorized by either PRV-1 isolate or Atlantic salmon strain indicated that both PRV-1 isolate and Atlantic salmon strain influenced the quantity of heart *cd8a* transcription in response to PRV-1 (Figure 4C,D). Cumulatively, these data support previous findings that recognition of PRV-1-infected cells by CD8+ T cells is likely the main driver of PRV-1-associated heart inflammation in Atlantic salmon and that HSMI is associated with an adaptive immune response. We further identify that PRV-1 activation of CD8+ T cells likely initiates interferon gamma signaling cascades, and is affected to some degree by both PRV-1 and host genotype, whereby BC-PRV was associated with lower CD8+ T cell responses relative to NB- or NOR-PRV, while EU salmon showed higher CD8+ T cell sensitivity than either BC or NB-TR Atlantic salmon during the peak to early persistent phases of infection (6–10 wpc).

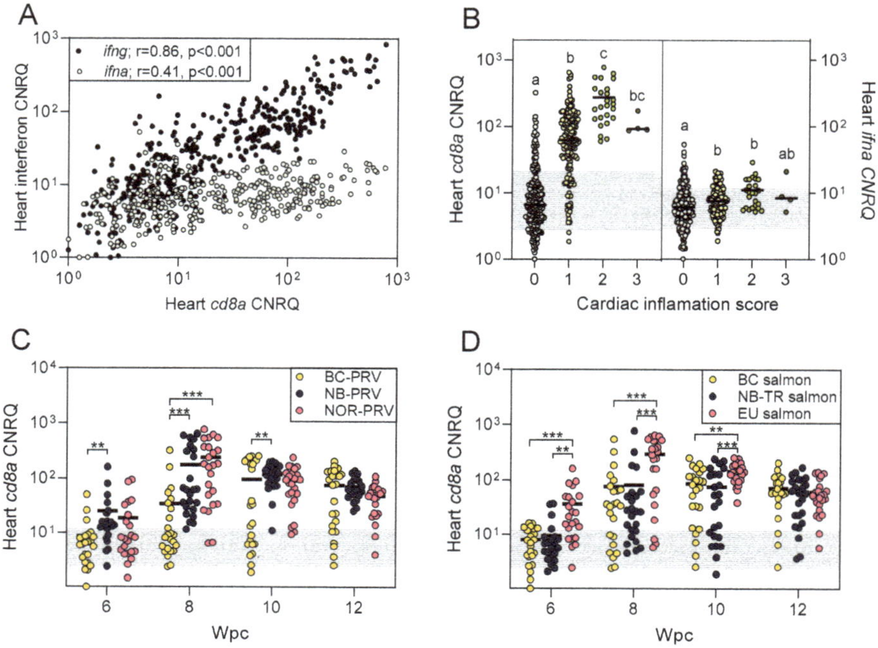

Figure 4. Heart *cd8a*, *ifna*, and *ifng* transcriptional expression following PRV-1 challenge. (**A**) Corrected normalized relative quantity (CNRQ) of *ifna* (light circles) and *ifng* (dark circles) are presented relative to *cd8a* transcriptional expression for all (n = 384) salmon sampled between 6–12 wpc in this study. Pearson r (r) and associated *p*-value is provided. (**B**) Mean (line) and individual (dots) heart *cd8a* (left side) and *ifna* (right side) CNRQ transcripts in relation to histopathological cardiac inflammation score, where 0 = no inflammation (n = 201), 1 = mild inflammation (n = 151), 2 = moderate inflammation (n = 25), and 3 = severe inflammation (n = 4) for the same 6–12 wpc dataset. Letters denote groupings for mean statistical similarity by the Kruskal–Wallis nonparametric test and Dunn's multiple comparisons tests at $p < 0.01$. Shaded areas present one standard deviation of target transcription in cardiac inflammation score 0 fish to provide a minimum threshold suggestive of biological relevance. (**C**) Mean (line) and individual (dots) of heart *cd8a* transcripts measured at 6–12 wpc from PRV-1-challenged salmon (n = 288) categorized relative to PRV-1 isolate administered (n = 24 per PRV-1 isolate per time point), or (**D**) relative to Atlantic salmon strain challenged (n = 24 per salmon strain per time point). ** $p < 0.01$, *** $p < 0.001$ by two-way ANOVA and Šídák's multiple comparisons tests of log-transformed CNRQ values; one standard deviation of *cd8a* expression recorded across all SC fish (n = 96) is shaded to suggest a minimum threshold for biological relevance.

3.5. Challenge 2: PRV-1-Associated Heart Inflammation Can Be Affected More by Atlantic Salmon Strain than PRV-1 Isolate, Whereas Skeletal Muscle Inflammation Is Influenced Similarly by Both Salmon Strain and PRV-1 Isolate

Regarding heart inflammation, the three PRV-1 isolates had similar cumulative impacts on heart inflammation across the three strains of Atlantic salmon used in this study, with approximately 40–50% of sampled fish showing some degree of heart inflammation within the 14-week study for all three PRV-1 isolates administered relative to approximately 3% in SC fish (Figure 5A). However, the impact of Atlantic salmon strain on heart inflammation indicated that NB-TR salmon had the lowest prevalence of heart inflammation (29%, $p < 0.002$), while BC salmon had an intermediate prevalence (45%, $p < 0.002$) and EU salmon had the most heart inflammation (62%, $p < 0.002$) following PRV-1 challenge (Figure 5B). NB-TR salmon also appeared to be the only salmon genotype to show some evidence of heart inflammation resolution within this 14-week study period (Figure 5C).

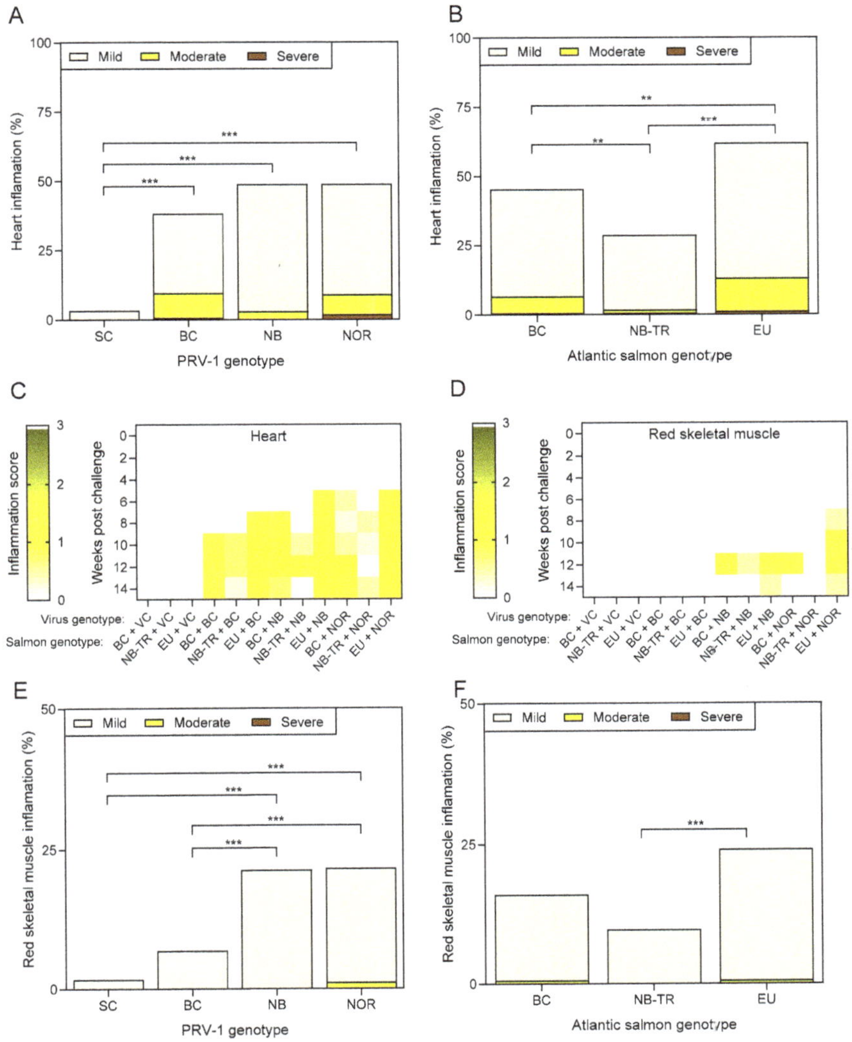

Figure 5. Heart and red skeletal muscle inflammation following PRV-1 challenge. The cumulative prevalence (%) of heart inflammation in PRV-1-challenged salmon over 14 weeks categorized by either (**A**) the PRV-1 isolate administered or (**B**) the strain of the recipient Atlantic salmon; ** $p < 0.01$, *** $p < 0.001$ by Kruskal–Wallis and Dunn's multiple comparisons tests. Heat maps of (**C**) heart and (**D**) red skeletal muscle presented as median inflammatory heart score in relation to each Atlantic salmon x PRV combination over a 14-week progression. Median inflammation score is defined by color, which ranges from 0 (no inflammation; white) to 3 (severe inflammation; brown). The cumulative prevalence (%) of red skeletal muscle inflammation in PRV-1-challenged salmon over the same 14 weeks categorized by either (**E**) the PRV-1 isolate administered or (**F**) the strain of recipient Atlantic salmon is also provided; *** $p < 0.001$ by Kruskal–Wallis and Dunn's multiple comparisons tests.

Skeletal muscle inflammation has been less directly attributable to PRV-1 infection relative to heart inflammation [29], particularly in laboratory challenge trials [1]. Nevertheless, increased red (but not white) skeletal muscle inflammation reached 21% across the three salmon strains following both NB- and NOR-PRV challenge, which was higher than for either the SC or BC-PRV challenge groups (2% and 7% prevalence, respectively; $p < 0.001$; Figure 5E). Median non-zero red skeletal muscle pathology scores at each time point were also only observed in NB- and NOR-PRV-challenged fish (n = 8 per treatment per time point) which occurred between 8–14 wpc (Figure 5D). Salmon strain also appeared to have some influence over red skeletal muscle inflammation in this study, where EU salmon showed a higher prevalence (24%) compared to NB-TR salmon (10%) (Figure 5F).

4. Discussion

In two laboratory challenge trials with strict environmental control, we identified that both Atlantic salmon host genetics and PRV-1 genotype influenced disease severity; however, our hypothesis for viral genotype being the main driver of heart inflammation was proved incorrect in this instance—variations in host Atlantic salmon genotype had a stronger influence on the occurrence and severity of heart inflammation than the PRV-1 genotype in these experimental investigations. Similar observations of a strong host phenotypic influence on disease outcomes have also been observed in other reovirus infections of teleost fish (e.g., during Grass Carp reovirus infections) [38].

The distinct phylogenetic ordinations for the three PRV-1 isolates used in this study with high bootstrap support are corroborated by previous Bayesian analyses conducted using virtually the same data set [36] as well as more broadly in PRV-1 phylogenetic investigations [13,39,40]. These data cumulatively lend strong support for a clear phylogenetic distinction between all Northern Pacific PRV-1a isolates sequenced to date including the BC-PRV isolate from PRV-1a isolates obtained or originating from the Atlantic Ocean or coastal Chile, including NB-PRV [40]. These data further support the well-established distinctions between PRV-1a and PRV-1b genotypes, as determined by S1 and whole genome sequencing [31], for which NOR-PRV is clearly classified as PRV-1b and BC-PRV and NB-PRV is clearly classified as PRV-1a.

Given the robust support for separate phylogenetic classifications of the three PRV-1 isolates in this study, it can be presumed that these genetic differences would be mirrored at least to some degree by phenotypic variations as well. We confirm this to be the case—both PRV-1 infection dynamics as well as virulence were at least partially dependent on the genotype of the PRV-1 isolate. Specific to Challenge 2 trials, BC-PRV had the highest capacity for replication, dissemination, and persistence across the three Atlantic salmon strains, while NOR-PRV had the lowest capacity. In a previous study comparing BC-PRV and NOR-PRV infection dynamics within a single strain of European Atlantic salmon [12], it was similarly observed that BC-PRV (i.e., CAN 16-005ND) reached higher systemic blood loads than NOR-PRV (i.e., NOR-2018NL). However, in that study, PRV-1 RNA reached equal if not slightly higher quantities in both the plasma and heart of NOR-PRV-infected fish relative to BC-PRV fish which was the opposite to what was observed in our current study. This suggests there are likely additional factors that influence PRV-1 infection dynamics beyond PRV-1 genetics, which requires further exploration.

In our present study, NOR-PRV demonstrated higher virulence for inducing heart and/or skeletal muscle inflammation relative to BC-PRV, which was consistent with previous challenges comparing these two genotypes [12]. In both previous challenges as well as the challenges conducted here, heart inflammation persisted longer, and median (or mean) heart inflammation scores reached the highest levels in NOR-PRV-infected salmon, although the present study identified this to be confounded to some degree by host genetic factors. Nevertheless, the fact that BC-PRV was able to reach 4–5x higher systemic loads relative to NOR-PRV and still be associated with less cytotoxic T cell responsiveness and heart inflammation in our study strongly supports the supposition that BC-PRV is less virulent than NOR-PRV, and that NB-PRV preliminarily appears to be somewhere in between. We also identified, for the first time, a differential association for generating skeletal muscle inflammation between PRV-1 isolates where both NOR-PRV and NB-PRV were associated with increased red skeletal muscle inflammation (20–25% prevalence), whereas BC-PRV did not notably induce skeletal muscle inflammation beyond control levels (e.g., <8% prevalence). This observation appears to be at least partially supported by field observations in British Columbia vs. Norway, where in BC, skeletal muscle inflammation has been rarely associated with PRV infection [2,8,10,41] whereas in Norway skeletal muscle inflammation has been considerably more common (i.e., common enough to be included in the disease name) [5,6,42]. This may be an important factor to consider in assigning PRV-1 disease pathology in different regions with distinct PRV-1 genotype occurrences.

Cytotoxic T cells have been shown to be highly associated with PRV-1b heart inflammation during HSMI [43], which has also been observed in PRV-1a with milder heart inflammation [23], implying that disease and inflammation associated with PRV-1 infections are generally a result of cytotoxic T cell signaling and directed killing rather than viral egress or pattern recognition receptor activation by infected cells [3]. We confirm that cytotoxic T cells appear to be highly associated with both PRV-1a and -1b heart inflammation in this study, as evidenced by increased *cd8a*, *gzma*, and *ifng* transcriptional signaling, and that this recognition is generally independent from systemic intracellular recognition of the virus by infected cells, as evidenced by a mild systemic and heart-specific *infa* transcriptional response. Thus, the main virulence factors for PRV-1 in Atlantic salmon appear to be how robustly host cytotoxic T cells recognize and respond to the virus in local tissues and not the degree to which tissue or blood cells intracellularly recognize infection. Further, variations in host salmon genotype appeared to be a primary driver of adaptive cytotoxic T cell recognition, where EU salmon mounted a more robust adaptive response to PRV-1 exposure than either BC or NB-TR salmon, indicating that something in the adaptive response of EU fish resulted in better virus recognition, or that viral dissemination into heart tissues and cells was more prevalent in this salmon strain.

Although we saw evidence in this study that PRV-1 genotype can influence cytotoxic T cell recognition and response, variations in viral genetics were clearly outweighed in this instance by genotypic variation in host Atlantic salmon strain. EU Atlantic salmon clearly generated more heart inflammation in response to PRV-1 than either BC or NB salmon, and NB salmon showed the lowest inflammatory responsiveness to PRV-1. These findings appeared to be at least partially consistent across both Challenge 1 and Challenge 2 experiments—BC salmon showed more heart inflammation relative to either of the closely related New Brunswick Atlantic salmon stocks used in this study (NB-SJR in Challenge 1 and NB-TR in Challenge 2). This finding is also consistent with multiple previous challenge trials using BC salmon and BC-PRV that generated less heart inflammation compared to similarly conducted trials that used NOR-PRV in domesticated EU salmon [1,2,11,12,23]. Assuming that this large host effect on determining PRV-1-associated heart inflammation and HSMI can be confirmed through further investigations, host genetics will be important to consider in designing future studies exploring PRV-1 virulence. Specifically, genetic assessments have the potential to identify markers in salmon for establishing resistance to HSMI, thereby reducing production-associated losses in areas where HSMI is problematic. Genome-wide association studies (GWAS) comparing genetic differences between salmon phenotypes for PRV-1 resistance may also be useful in evaluating risks associated with PRV-1 exposure. GWAS studies have shown promise for clarifying the host genetic role and defining risk factors for rheumatic heart disease occurrence in humans [44] with similar epidemiological patterns, which may prove advantageous in Atlantic salmon where GWAS are already relatively common within cultured populations and PRV-1 prevalence is high. Epigenetic factors, particularly relating to past exposure to pathogens, may also prove interesting avenues for investigation for explaining differential responses to PRV-1 by different salmon strains [45,46]. Interestingly, the body of disease prevention work relating to PRV-1—including this present study—indicates that GWAS and other strategies have promise in reducing HSMI-like disease but will likely have little effect on systemic PRV-1 infection dynamics (reviewed by [3]).

Environmental factors are also implicitly important for disease to occur [24]. Although the environment was tightly controlled in this study to allow us to focus on host and virus factors, we nevertheless observed that heart inflammation generated by BC-PRV in BC salmon in the two challenge trials of this study conducted at the Onda Prince Edward Island facility appeared anecdotally more severe when compared to previous challenge trials using the same or similar BC-PRV and BC salmon conducted at a Fisheries and Oceans Canada facility on Vancouver Island [2,11,23]. This would indicate that there are indeed potential differences in environment, even under controlled laboratory conditions, that can impact PRV-1 virulence and which warrant further exploration. Investigations to untangle how environmental impacts might or might not be synergistically tied to host or viral factors may also help to clarify the generally discrete field observations for minimal PRV-1-associated disease in Pacific and Atlantic Canada relative to coastal Norway.

In conclusion, this study identifies that host genetics can exert a majority influence on PRV-1 virulence and associated heart inflammation in controlled laboratory environments, demonstrating that the presence of either PRV-1a or PRV-1b alone is not sufficient to reliably predict disease. This finding also highlights host genotyping and putative genetic selection of Atlantic salmon may prove effective avenues for reducing HSMI disease severity in areas where it is problematic. This study also confirms variation in virulence potential and viral replication between PRV-1 genotypes, particularly regarding replication and an association with generating skeletal muscle inflammation, which may be important to consider in assessing PRV-associated disease or disease risk across geographic landscapes. Lastly, this study reiterates the importance of a host-generated cytotoxic T cell response as a major source of PRV-1-associated disease when it is observed, and, therefore, further research aimed at minimizing and better understanding this response will likely prove useful in mitigating disease in the areas or populations which appear to be at higher risk.

Supplementary Materials: The following supporting information can be downloaded at: https://www.mdpi.com/article/10.3390/v17020285/s1, Figure S1: Histopathology scoring report provided by M.A. and Fish Vet group including 10 images exemplifying inflammatory scoring. Figure S2: Challenge 2 14-week PRV-1 transcriptional blood load for each specific PRV-1 isolate x Atlantic salmon strain challenge; Table S1: Challenge 1 collected and generated data; Table S2: Challenge 2 collected and generated data; Table S3: Tiling primers used for PRV-1 sequencing; Table S4: Atlantic salmon functional sequence comparisons; Table S5: GenBank accession numbers for PRV-1 isolates used in this study.

Author Contributions: Conceptualization, M.P. and K.G.; methodology, M.P., M.B. (Mark Braceland) and K.G.; formal analysis, M.P., D.G., M.A., L.G., S.M. and M.B. (Marije Booman); investigation, M.P., L.G. and K.G.; resources, D.D., N.G. and K.G.; data curation, M.P., L.G., M.B. (Marije Booman) and M.B. (Mark Braceland); writing—original draft preparation, M.P. and K.G.; writing—review and editing, all authors; project administration, M.P., M.B. (Mark Braceland) and K.G.; funding acquisition, K.G. All authors have read and agreed to the published version of the manuscript.

Funding: This project was funded by the Department of Fisheries and Oceans Canada through the Program for Aquaculture Regulatory Research (PARR) to KG and congressionally funded within the U.S. Department of Agriculture—Agricultural Research Service (project 8030-31000-005-00D) to MP and SM.

Institutional Review Board Statement: Animal study protocols were approved by the Onda Animal Care Committee (AUP #'s: MB-0037 and MB-0102).

Informed Consent Statement: Not applicable.

Data Availability Statement: All data presented in this study are available in the manuscript, supplemental material, or freely accessible via NCBI.

Acknowledgments: The authors would like to thank Øystein Wessel and Espen Rimstad from the Norwegian University of Life Sciences for donating the purified NOR-PRV used in this study, and Mowi Global, Mowi Canada West, and the Fisheries and Oceans Canada Mactaquac Biodiversity Facility for donating eggs and/or fish. We also thank Haley Matkin and Julie Zhao for technical assistance and sample collection, Onda for conducting the challenge trials, and Michael Pawlick from the B.C Ministry of Agriculture and Food for his reading of the Challenge 1 histology slides. The opinions and conclusions in this publication are those of the authors and should not be construed to represent any official USDA or U.S. Government determination or policy. The USDA is an equal opportunity provider and employer.

Conflicts of Interest: Authors Marije Booman and Mark Braceland were employed by the company Onda. Author Marta Alarcón was employed by Pharmaq Analytiq. The remaining authors declare that the research was conducted in the absence of any commercial or financial relationships that could be construed as a potential conflict of interest. The funders had no role in the design of the study; in the collection, analyses, or interpretation of data; in the writing of the manuscript; or in the decision to publish the results.

References

1. Wessel, Ø.; Braaen, S.; Alarcon, M.; Haatveit, H.; Roos, N.; Markussen, T.; Tengs, T.; Dahle, M.K.; Rimstad, E. Infection with purified Piscine orthoreovirus demonstrates a causal relationship with heart and skeletal muscle inflammation in Atlantic salmon. *PLoS ONE* **2017**, *12*, e0183781. [CrossRef] [PubMed]
2. Polinski, M.P.; Marty, G.D.; Snyman, H.N.; Garver, K.A. Piscine othoreovirus demonstrates high infectivity but low virulence in Atlantic salmon of Pacific Canada. *Sci. Rep.* **2019**, *9*, 3297. [CrossRef]
3. Polinski, M.P.; Vendramin, N.; Cuenca, A.; Garver, K.A. Piscine orthoreovirus: Biology and distribution in farmed and wild fish. *J. Fish Dis.* **2020**, *43*, 1331–1352. [CrossRef]
4. Palacios, G.; Lovoll, M.; Tengs, T.; Hornig, M.; Hutchison, S.; Hui, J.; Kongtorp, R.-T.; Savji, N.; Bussetti, A.V.; Solovyov, A.; et al. Heart and Skeletal Muscle Inflammation of Farmed Salmon Is Associated with Infection with a Novel Reovirus. *PLoS ONE* **2010**, *5*, e11487. [CrossRef]
5. Kongtorp, R.T.; Kjerstad, A.; Taksdal, T.; Guttvik, A.; Falk, K. Heart and skeletal muscle inflammation in Atlantic salmon, *Salmo salar* L.: A new infectious disease. *J. Fish. Dis.* **2004**, *27*, 351–358. [CrossRef]
6. Kongtorp, R.T.; Taksdal, T.; Lyngøy, A. Pathology of heart and skeletal muscle inflammation (HSMI) in farmed Atlantic salmon *Salmo salar*. *Dis. Aquat. Org.* **2004**, *59*, 217–224. [CrossRef] [PubMed]
7. Sommerset, I.; Wiik-Nielsen, J.; Oliveira, V.H.S.; Moldal, T.; Bornø, G.; Haukaas, A.; Brun, E. *Norwegian Fish. Health Report 2022*; Norwegian Veterinary Institute: As, Norway, 2023; p. 218.
8. Polinski, M.P.; Gross, L.; Marty, G.D.; Garver, K.A. Heart inflammation and piscine orthoreovirus genotype-1 in Pacific Canada Atlantic salmon net-pen farms: 2016–2019. *BMC Vet. Res.* **2022**, *18*, 306. [CrossRef]
9. Di Cicco, E.; Ferguson, H.W.; Schulze, A.D.; Kaukinen, K.H.; Li, S.; Vanderstichel, R.; Wessel, Ø.; Rimstad, E.; Gardner, I.A.; Hammell, K.L. Heart and skeletal muscle inflammation (HSMI) disease diagnosed on a British Columbia salmon farm through a longitudinal farm study. *PLoS ONE* **2017**, *12*, e0171471. [CrossRef] [PubMed]
10. Marty, G.D.; Bidulka, J.; Joseph, T. Cross-sectional study of histopathology and piscine orthoreovirus during a marine production cycle of farmed Atlantic salmon (*Salmo salar* L.) in British Columbia, Canada. *J. Fish Dis.* **2020**, *43*, 1019–1028. [CrossRef] [PubMed]
11. Garver, K.A.; Johnson, S.C.; Polinski, M.P.; Bradshaw, J.C.; Marty, G.D.; Snyman, H.N.; Morrison, D.B.; Richard, J. Piscine orthoreovirus from western North America is transmissible to Atlantic salmon and Sockeye salmon but fails to cause Heart and Skeletal Muscle Inflammation. *PLoS ONE* **2016**, *11*, e0146229. [CrossRef]

12. Wessel, Ø.; Hansen, E.F.; Dahle, M.K.; Alarcon, M.; Vatne, N.A.; Nyman, I.B.; Soleim, K.B.; Dhamotharan, K.; Timmerhaus, G.; Markussen, T.; et al. Piscine Orthoreovirus-1 Isolates Differ in Their Ability to Induce Heart and Skeletal Muscle Inflammation in Atlantic Salmon (*Salmo salar*). *Pathogens* **2020**, *9*, 1050. [CrossRef]
13. Dhamotharan, K.; Tengs, T.; Wessel, Ø.; Braaen, S.; Nyman, I.B.; Hansen, E.F.; Christiansen, D.H.; Dahle, M.K.; Rimstad, E.; Markussen, T. Evolution of the Piscine orthoreovirus Genome Linked to Emergence of Heart and Skeletal Muscle Inflammation in Farmed Atlantic Salmon (*Salmo salar*). *Viruses* **2019**, *11*, 465. [CrossRef] [PubMed]
14. Solarte-Murillo, L.; Reyes, H.; Ojeda, L.; Cárcamo, J.G.; Pontigo, J.P.; Loncoman, C.A. Analyses and Insights into Genetic Reassortment and Natural Selection as Key Drivers of Piscine orthoreovirus Evolution. *Viruses* **2024**, *16*, 556. [CrossRef] [PubMed]
15. Vatne, N.A.; Stormoen, M.; Lund, M.; Devold, M.; Rimstad, E.; Wessel, Ø. Genetic grouping and geographic distribution of Piscine orthoreovirus-1 (PRV-1) in farmed Atlantic salmon in Norway. *Vet. Res.* **2021**, *52*, 131. [CrossRef]
16. Løvoll, M.; Alarcón, M.; Jensen, B.B.; Taksdal, T.; Kristoffersen, A.B.; Tengs, T.; Lovoll, M.; Alarcon, M.; Jensen, B.B.; Taksdal, T.; et al. Quantification of piscine reovirus (PRV) at different stages of Atlantic salmon *Salmo salar* production. *Dis. Aquat. Org.* **2012**, *99*, 7–12. [CrossRef]
17. Grammes, F.; Rørvik, K.A.; Takle, H. Tetradecylthioacetic acid modulates cardiac transcription in Atlantic salmon, *Salmo salar* L., suffering heart and skeletal muscle inflammation. *J. Fish. Dis.* **2012**, *35*, 109–117. [CrossRef]
18. Rennemo, J.; Myrvold, S.; Berge, K.; Kileng, Ø.; Pedersen, B.; Aksberg, D.S.; Lisik, P.; Crappe, D.; McGurk, C.; Rimstad, E. In-depth health surveillance and clinical nutrition in farmed Atlantic salmon: A strategic attempt to detect and mitigate an HSMI outbreak. *Vet. Res.* **2023**, *54*, 3. [CrossRef]
19. Martinez-Rubio, L.; Morais, S.; Evensen, Ø.; Wadsworth, S.; Ruohonen, K.; Vecino, J.L.G.; Bell, J.G.; Tocher, D.R. Functional feeds reduce heart inflammation and pathology in Atlantic salmon (*Salmo salar* L.) following experimental challenge with Atlantic salmon reovirus (ASRV). *PLoS ONE* **2012**, *7*, e40266. [CrossRef] [PubMed]
20. AquaGen. Resistance against HSMI. In *AquaGen*; AquaGen: Trondheim, Norway, 2017; Volume 2018.
21. Emilsen, V.; Bruheim, T.; Moen, T.; Kjoglum, S.; Korsvoll, S.; Santi, N. Marker assisted selection for improved HSMI-resistance in Atlantic salmon. In Proceedings of the 18th International Conference on the Diseases of Fish and Shellfish, Belfast, Northern Ireland, 4–8 September 2017.
22. Polinski, M.P.; Bradshaw, J.C.; Inkpen, S.M.; Richard, J.; Fritsvold, C.; Poppe, T.T.; Rise, M.L.; Garver, K.A.; Johnson, S.C. De novo assembly of Sockeye salmon kidney transcriptomes reveal a limited early response to piscine reovirus with or without infectious hematopoietic necrosis virus superinfection. *BMC Genom.* **2016**, *17*, 848. [CrossRef]
23. Zhang, Y.; Polinski, M.P.; Morrison, P.R.; Brauner, C.J.; Farrell, A.A.P.; Garver, K.A. High-load reovirus infections do not imply physiological impairment in salmon. *Front. Physiol.* **2019**, *10*, 114. [CrossRef] [PubMed]
24. Scholthof, K.-B.G. The disease triangle: Pathogens, the environment and society. *Nat. Rev. Microbiol.* **2007**, *5*, 152–156. [CrossRef]
25. Withler, R.E.; Supernault, K.J.; Miller, K.M. Genetic variation within and among domesticated Atlantic salmon broodstocks in British Columbia, Canada. *Anim. Genet.* **2005**, *36*, 43–50. [CrossRef]
26. Peterson, B.C.; Burr, G.S.; Pietrak, M.R.; Proestou, D.A. Genetic improvement of north American atlantic salmon and the eastern oyster Crassostrea virginica at the US Department of agriculture–agricultural research service national cold water marine aquaculture center. *N. Am. J. Aquac.* **2020**, *82*, 321–330. [CrossRef]
27. Liu, L.; Ang, K.P.; Elliott, J.A.; Kent, M.P.; Lien, S.; MacDonald, D.; Boulding, E.G. A genome scan for selection signatures comparing farmed Atlantic salmon with two wild populations: Testing colocalization among outlier markers, candidate genes, and quantitative trait loci for production traits. *Evol. Appl.* **2017**, *10*, 276–296. [CrossRef]
28. Yousaf, M.N.; Koppang, E.O.; Skjødt, K.; Hordvik, I.; Zou, J.; Secombes, C.; Powell, M.D. Comparative cardiac pathological changes of Atlantic salmon (*Salmo salar* L.) affected with heart and skeletal muscle inflammation (HSMI), cardiomyopathy syndrome (CMS) and pancreas disease (PD). *Vet. Immunol. Immunopathol.* **2013**, *151*, 49–62. [CrossRef] [PubMed]
29. Wiik-Nielsen, J.; Alarcón, M.; Jensen, B.B.; Haugland, Ø.; Mikalsen, A.B. Viral co-infections in farmed Atlantic salmon, *Salmo salar* L., displaying myocarditis. *J. Fish. Dis.* **2016**, *39*, 1495–1507. [CrossRef]
30. Quick, J.; Grubaugh, N.D.; Pullan, S.T.; Claro, I.M.; Smith, A.D.; Gangavarapu, K.; Oliveira, G.; Robles-Sikisaka, R.; Rogers, T.F.; Beutler, N.A. Multiplex PCR method for MinION and Illumina sequencing of Zika and other virus genomes directly from clinical samples. *Nat. Protoc.* **2017**, *12*, 1261–1276. [CrossRef]
31. Kibenge, M.J.T.; Iwamoto, T.; Wang, Y.; Morton, A.; Godoy, M.G.; Kibenge, F.S. Whole-genome analysis of piscine reovirus (PRV) shows PRV represents a new genus in family Reoviridae and its genome segment S1 sequences group it into two separate sub-genotypes. *Virol. J.* **2013**, *10*, 10–230. [CrossRef] [PubMed]
32. Shabardina, V.; Kischka, T.; Manske, F.; Grundmann, N.; Frith, M.C.; Suzuki, Y.; Makałowski, W. NanoPipe—A web server for nanopore MinION sequencing data analysis. *GigaScience* **2019**, *8*, giy169. [CrossRef] [PubMed]

33. Gao, G.; Waldbieser, G.C.; Youngblood, R.C.; Zhao, D.; Pietrak, M.R.; Allen, M.S.; Stannard, J.A.; Buchanan, J.T.; Long, R.L.; Milligan, M. The generation of the first chromosome-level de novo genome assembly and the development and validation of a 50K SNP array for the St. John River aquaculture strain of North American Atlantic salmon. *G3 Genes Genomes Genet.* **2023**, *13*, jkad138. [CrossRef] [PubMed]
34. Jombart, T. adegenet: A R package for the multivariate analysis of genetic markers. *Bioinformatics* **2008**, *24*, 1403–1405. [CrossRef]
35. Paradis, E. pegas: An R package for population genetics with an integrated–modular approach. *Bioinformatics* **2010**, *26*, 419–420. [CrossRef]
36. Siah, A.; Breyta, R.; Warheit, K.; Gagne, N.; Purcell, M.K.; Morrison, D.; Powell, J.; Johnson, S. Genomes reveal genetic diversity of Piscine orthoreovirus in farmed and free-ranging salmonids from Canada and USA. *Virus Evol.* **2020**, *6*, veaa054. [CrossRef] [PubMed]
37. Goudet, J. Hierfstat, a package for R to compute and test hierarchical F-statistics. *Mol. Ecol. Notes* **2005**, *5*, 184–186. [CrossRef]
38. Liao, Z.; Wan, Q.; Shang, X.; Su, J. Large-scale SNP screenings identify markers linked with GCRV resistant traits through transcriptomes of individuals and cell lines in Ctenopharyngodon idella. *Sci. Rep.* **2017**, *7*, 1184. [CrossRef] [PubMed]
39. Godoy, M.; Medina, D.A.; Suarez, R.; Valenzuela, S.; Romero, J.; Kibenge, M.; Wang, Y.; Kibenge, F. Extensive phylogenetic analysis of Piscine orthoreovirus genomic sequences shows the robustness of subgenotype classification. *Pathogens* **2021**, *10*, 41. [CrossRef] [PubMed]
40. Kibenge, M.J.T.; Wang, Y.; Gayeski, N.; Morton, A.; Beardslee, K.; McMillan, B.; Kibenge, F.S.B. Piscine orthoreovirus sequences in escaped farmed Atlantic salmon in Washington and British Columbia. *Virol. J.* **2019**, *16*, 41. [CrossRef] [PubMed]
41. Marty, G.D.; Morrison, D.B.; Bidulka, J.; Joseph, T.; Siah, A. Piscine reovirus in wild and farmed salmonids in British Columbia, Canada: 1974–2013. *J. Fish Dis.* **2015**, *38*, 713–728. [CrossRef]
42. Kongtorp, R.T.; Halse, M.; Taksdal, T.; Falk, K. Longitudinal study of a natural outbreak of heart and skeletal muscle inflammation in Atlantic salmon, *Salmo salar* L. *J. Fish Dis.* **2006**, *29*, 233–244. [CrossRef] [PubMed]
43. Mikalsen, A.B.; Haugland, O.; Rode, M.; Solbakk, I.T.; Evensen, O. Atlantic Salmon Reovirus Infection Causes a CD8 T Cell Myocarditis in Atlantic Salmon (*Salmo salar* L.). *PLoS ONE* **2012**, *7*, e37269. [CrossRef]
44. Muhamed, B.; Parks, T.; Sliwa, K. Genetics of rheumatic fever and rheumatic heart disease. *Nat. Rev. Cardiol.* **2020**, *17*, 145–154. [CrossRef] [PubMed]
45. Manríquez, R.A.; Sandoval, M.; Loncoman, C.; Tafalla, C.; Avendaño-Herrera, R.; Cárcamo, J.G. Epigenetic reprogramming around IFN1 and IFNy2 promoters in rainbow trout cells inoculated with infectious pancreatic necrosis virus (IPNV). *Fish Shellfish Immunol.* **2023**, *140*, 108947. [CrossRef] [PubMed]
46. Shang, X.; Wan, Q.; Su, J.; Su, J. DNA methylation of CiRIG-I gene notably relates to the resistance against GCRV and negatively-regulates mRNA expression in grass carp, Ctenopharyngodon idella. *Immunobiology* **2016**, *221*, 23–30. [CrossRef]

viruses

MDPI

Article

Influence of Viral Re-Infection on Head Kidney Transcriptome of Nervous Necrosis Virus-Resistant and -Susceptible European Sea Bass (*Dicentrarchus labrax*, L.)

Dimitra K. Toubanaki [1,*], Odysseas-Panagiotis Tzortzatos [1], Antonia Efstathiou [1], Vasileios Bakopoulos [2] and Evdokia Karagouni [1,*]

1 Immunology of Infection Group, Department of Microbiology, Hellenic Pasteur Institute, 11521 Athens, Greece; ptzortzatos@pasteur.gr (O.-P.T.); toniaef@pasteur.gr (A.E.)

2 Department of Marine Sciences, School of The Environment, University of the Aegean, University Hill, Lesvos, 81100 Mytilene, Greece; v.bakopoulos@aegean.gr

* Correspondence: dtouban@pasteur.gr (D.K.T.); ekaragouni@pasteur.gr (E.K.); Tel.: +30-210-647-8828 (D.K.T.); +30-210-647-8826 (E.K.)

Abstract: Fish viral infections have great environmental and economic implications in aquaculture. Nervous necrosis virus (NNV) is a pathogen affecting more than 120 different species, causing high mortality and morbidity. Herein, we study how NNV re-infection affects the European sea bass (*Dicentrarchus labrax*, L.) head kidney transcriptome in disease-resistant and -susceptible sea bass families. To determine how each family responds to re-infection, we performed the RNA-sequencing analysis of experimentally NNV-infected *D. labrax*. Fish were experimentally infected in a long-term study, and one month after the last recorded death, all surviving fish were re-infected by the same NNV strain. Fish tissues were sampled 7 days upon re-infection. The transcriptome profiles of infected vs. non-infected fish revealed 103 differentially expressed genes (DEGs) for the resistant family and 336 DEGs for the susceptible family. Only a few pathways were commonly enriched in the two families, further indicating that the resistant and susceptible families utilize completely different mechanisms to fight the NNV re-infection. Protein–protein interaction analysis identified a variety of hub genes for the resistant and the susceptible families, quite distinct in their function on NNV resistance. In conclusion, NNV-resistant and -sensitive sea bass transcriptomes were analyzed following NNV survivors' viral re-infection, offering a glimpse into how host attempts to control the infection depending on its genetic background in relation with virus resistance.

Keywords: viral nervous necrosis; nervous necrosis virus; nodavirus; European sea bass; disease resistance; transcriptome; viral re-infection; host–pathogen interaction

1. Introduction

Nervous necrosis virus (NNV), or nodavirus, is the causative agent of viral encephalopathy and retinopathy (VER) disease, which can cause high mortality and morbidity in more than 120 different species from marine and freshwater environments, and it is responsible for great economic losses in the aquaculture industry, affecting human nutrition and the environment [1–4]. NNV belongs to the genus Betanodavirus and is a member of the Nodaviridae family, characterized by single-stranded RNA genome with two positive-sense molecules (RNA1 and RNA2) in a non-enveloped icosahedral capsid. Betanodaviruses can be classified into four different genotypes: red-spotted grouper nervous necrosis virus (RGNNV), striped jack nervous necrosis virus (SJNNV), tiger puffer nervous necrosis virus (TPNNV), and barfin flounder nervous necrosis virus (BFNNV), based on the T4 region of RNA2 [5]. The disease is characterized by severe damage of the fish nervous system (e.g., brain, retina, spinal cord), and the clinical symptoms include abnormal swimming behavior, loss of appetite, swim bladder hyperinflation, coloration abnormalities, eventually resulting in the potential death of infected hosts [2,3,6].

European sea bass (*Dicentrarchus labrax*) is one of the most susceptible hosts of NNV. It is a teleost fish found in the Mediterranean, north-eastern Atlantic Ocean, and Black Sea. It is well characterized [7] and economically important, since it is the main marine fish farmed in the Mediterranean basin, along with gilthead seabream [8,9], which is considered an asymptomatic carrier of NNV [10]. In general, the onset of a disease outbreak depends both on pathogen virulence and the host immune response [11], and vaccination may be used to protect fish from NNV [12]. However, the use of commercial vaccines to protect sea bass against VNN in the Mediterranean is still limited due to the cost, technical, and logistic problems presented when combined with other vaccines [6,13]. As an alternative solution, selective breeding in order to have traits such as disease resistance has gained attention in recent years [14].

The molecular bases of disease resistance mechanisms are often complex, and resistance is rarely attributable to a single immune function characteristic [15]. The development of fish genetic maps and -omics technologies allowed for the identification of several genes and pathways which are related to host resistance against several pathogens in a range of fish hosts [15–21]. For example, disease-resistant fish seem to have significantly higher expressions of pro-inflammatory genes and transcription factors, like those in the salmon head kidney following viral infection [18]. Resistant fish appear to have a limited and prolonged immune response to the virus, while susceptible fish have an acute short response to viral infection [19]. In another study, genes related to cytokine activity and inflammatory response were up-regulated in susceptible fish, but that mechanism failed to protect the salmon against the virus, whereas the resistant fish had a milder immune response, including the up-regulation of genes relating to the M2 macrophage system, which seems to be a more effective way of surviving a viral infection [20]. A transcriptome profile comparison of Japanese flounder both resistant to and susceptible against bacterial infection showed that genes related to hematopoietic cell lineage, innate immune-related inflammatory factors, antigen processing and presentation, and T/B cell receptor signaling pathways were significantly enriched in the resistant family. These results indicate that the resistance molecular mechanism is controlled by multiple genes in the immune signaling pathway [15]. In *Aeromonas salmonicida* bacterium-resistant/susceptible turbot, a transcriptome analysis revealed that the resistant family displayed a more controlled inflammatory response to infection, and several up-regulated genes in the head kidney were related to antigen presentation and T-cell activity [21].

The genetic assessments of European sea bass resistance to NNV infection suggests moderate heritability for resistance [13,22,23]. A strong quantitative trait locus (QTL) at the LG12 chromosome was reported [13,23], along with putative traits at LG8, LG15, and LG19. Significant SNPs, associated with the putative QTL genotype at LG12, included ITPK1 (inositol tetrakisphosphate 1-kinase 1), PLK4 (polo-like kinase 4), REEP1 (receptor expression-enhancing protein 1), CHMP2 (charged multivesicular body protein 2), MRPL35 (mitochondrial ribosomal protein L35), and SCUBE1 (signal peptide-CUB-EGF (epidermal growth factor) domain-containing protein 1). The highest significant SNP was located within the intron of the HSPA4L gene (heat shock protein family A, member four) which belongs to the heat shock proteins (HSPs) family. HSPs are implicated in cell responses to harmful circumstances and protect cells from stress [13]. Genetic analysis provides valuable insights for selective breeding. However, in order to understand host resistance mechanisms, more parameters should be analyzed, including transcriptome, proteome, and metabolic profile studies of resistant and susceptible fish.

In the present study, in order to investigate the VNN resistance mechanism of European sea bass, two genetically distinct families were chosen for head kidney transcriptome analysis; i.e., one NNV-resistant and one susceptible family originated from a well-studied family-based breeding program with 100 sea bass families [13]. Transcriptome analysis was performed to compare NNV-resistant and NNV-susceptible families' responses to viral re-infection in the fish head kidney because it is one of the most important lymphoid tissues in teleosts and plays a crucial role in the immune responses following an infection [24]. Based on the transcriptome data, the resistant and susceptible families' responses were assessed in three complementary ways: gene ontology (GO) enrichment analysis, pathway (KEGG and reactome) analysis, and protein–protein interaction (PPI) analysis, which were used to reveal the key genes and pathways important for disease resistance. The obtained data are critical to understanding the functional basis of genetic resistance to NNV in European sea bass. Overall, such studies could provide general perspectives on host resistance mechanisms against viral pathogens in teleosts, with implications on rational prognostic and therapeutics strategies for sustainable aquaculture.

2. Materials and Methods

2.1. Experimental Fish

The present experimentation was performed in the frame of a larger project involving two families of European seabass with different levels of resistance to NNV infection (i.e., an NNV-resistant family (R) and an NNV-susceptible family (S)), as determined on a family-based breeding program [13]. Briefly, healthy non-vaccinated fish were used and transported by Nireus Aquaculture S.A. at the Laboratory of Ichthyology-Aquaculture and Aquatic Animal Health (ICHTHYAI), Department of Marine Sciences, University of The Aegean, Greece. A total of 300 randomly selected fish from both families (weight: 125.85 ± 29.8 g) were acclimatized for 3–4 days in 1 m^3 cylindroconical fiberglass tanks connected to a closed recirculated sea water system, with a 20 m^3 total volume capacity. Water was recirculated via a 14 m^3 h^{-1} sea water pump, filtered via a sand filter, disinfected via 5×39 W UV-C lamps, treated in a biological filter, and aerated via air stones connected to five 150 L h^{-1} air pumps. Seawater in the system was renewed by 1/3 every 1.5 months. The seawater temperature was maintained at 22–23 °C during the acclimatization period. Salinity was 3.8–3.9%, dissolved O_2 was maintained above 4.8 mg L^{-1}, total ammonia nitrogen and nitrite was kept below 0.05 ppm and 0.5 ppm, respectively, and nitrate levels were maintained below 40 mg L^{-1}. pH ranged between 7.9–8.1. Temperature, dissolved oxygen, and nitrogen metabolites were measured daily, while salinity and pH were measured on a weekly basis. Fish were reared in a 12 h light–12 h dark photoperiod and were fed with 1–2% of their biomass commercial (Feedus, Blueline) diet 3 times a day with 6 h intervals.

2.2. Nervous Necrosis Virus Challenge

NNV was originally isolated from naturally infected *D. labrax* (genotype: RGNNV [25]) and was propagated as previously described [26]. Briefly, fish brains were homogenized in EMEM (Eagle Minimum Essential Medium; Sigma-Aldrich, Steinheim, Germany) or Leibovitz L15c medium (Biochrom, Berlin, Germany) containing 10% FBS (Fetal bovine serum; Biochrom, Berlin, Germany). Following inoculation, the SSN-1 cells (The European Collection of Animal Cell Cultures, Salisbury, UK) were grown at 26 °C in Falcon Primaria cell culture flasks (Becton Dickinson Labware, Franklin Lakes, NJ, USA) containing Leibovitz's L15 medium, supplemented with 10% FBS, 100 μ mL^{-1} penicillin, 100 μg mL^{-1}

streptomycin, and 2 mM glutamine (Gibco, Paisley, UK). Virus titration was performed on monolayers of SSN-1 cells grown in a 96-well plate. Viral suspensions were prepared with 12-fold serial dilutions in EMEM supplemented with 10% FBS. Quadruplicates of 50 μL of each dilution were added in a 96-well plate seeded with SSN-1 cells. Cultures were incubated at 26 °C for 6 days. During this period, the cell monolayers were observed for the appearance of a cytopathic effect (CPE), and the final titer, expressed as $TCID_{50}$ mL^{-1}, was estimated by the end-point titration method [27].

A summary of the experimental setup is given in Figure 1. Prior to infection, groups of 20 and 50 fish were transported to separate tanks (in triplicates), for family R and family S, respectively; the temperature was gradually raised to 27.0–27.2 °C and the fish were acclimatized for 7 more days. The temperature remained constant to 27.0–27.2 °C throughout the experiment. Before the infection experiment, total RNA was extracted from three randomly selected individuals' brains and amplified with an NNV RT-qPCR assay using the Quantitect Probe RT-PCR master mix (Qiagen) to ensure that those specimens were not infected. Sea bass were challenged by intramuscular injection in the dorsal muscle with 7×10^6 $TCID_{50}$ mL^{-1} of nodavirus-containing supernatant (200 μL; serially diluted with PBS from a 1×10^{11} $TCID_{50}$ mL^{-1} nodavirus-containing supernatant stock). As the negative control group, uninfected sea bass from the same families were mock-challenged with 100 μL PBS. Fish were monitored twice a day, and mortalities were recorded up to the end of the experimentation. The first stage of experimentation was up to 28 dpi and was previously described in [28]. After an additional period of 2 weeks (42 days post infection in total), the fish that had survived were re-infected by intramuscular injection in the dorsal muscle with 1×10^5 $TCID_{50}$ mL^{-1} of nodavirus-containing supernatant (100 μL). All deceased fish were removed, and one week following re-infection, 9 fish of the remaining live population (3 samples from each tank) were randomly selected. Before sampling, the specimens were anesthetized with 0.2% phenoxyethanol and weighed. Following, head kidney and brain tissues were removed aseptically and were either subjected in RNA isolation immediately or stored in RNAlater (Qiagen, Hilden, Germany) at −80 °C. The same procedure was followed for 9 fish belonging to the negative control groups for each family. For the determination of the brain viral load, a quantitative RT-qPCR assay was used (n = 5) [29].

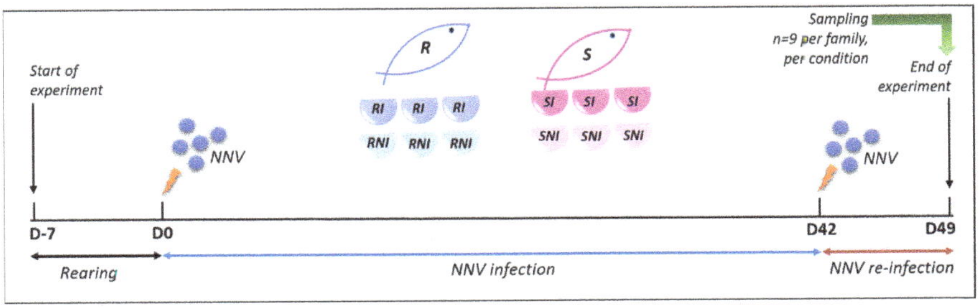

Figure 1. Experimental setup (NNV: nervous necrosis virus; R: resistant; RI: resistant infected; RNI: resistant non-infected; S: susceptible; SI: susceptible infected; SNI: susceptible non-infected; D: days post infection).

2.3. Ethics Statement

Experimentation was performed in the Laboratory of Ichthyology-Aquaculture and Aquatic Animal Health (ICHTHYAI) (Government Issue 1255/28-4-2016). ICHTHYAI has been granted all the required permits for producing (EL 83 BIObr 01), supplying (EL 83 BIOsup 01), and experimenting on aquatic organisms (EL83 BioExp 01), according to the Presidential Decree 56/2013 conforming to Directive 2010/63/EE (Decision

No 4053/14-3-2017 of the competent Regional Veterinary Authority). Fish were euthanized using a procedure listed on the appropriate license, and the protocol for the experimental infection performed in this study has been approved by Decision No 5379/4-4-2017 of the competent Regional Veterinary Authority.

2.4. RNA Extraction

Total RNA from the head kidney was extracted using a Trizol-based protocol to perform total RNA sequencing. Tissues (~75 mg) from NNV-infected (n = 9) and non-challenged (n = 5) fish were homogenized using the TissueLyzer mechanical homogenizer with 5 mm steel beads (Qiagen). The TRIzol® Reagent solution (Invitrogen, Carlsbad, CA, USA) was used for total RNA extraction, following the manufacturer's instructions. RNA quantity and quality were assessed by spectrophotometry (NanoDrop 2000), a Qubit 2.0 fluorometer (Thermo Fisher Scientific, Waltham, MA, USA) and the RNA integrity number (RIN) was measured using a 2100 Bioanalyzer instrument (Agilent Technologies, Santa Clara, CA, USA). Only RNA samples of high quality (RIN $\geq$ 7) were used for constructing the cDNA libraries, and each samples' transcriptome was analyzed individually.

2.5. cDNA Library Construction, Sequencing, and Transcriptome Mapping

RNA samples were used to create the sequencing libraries by using the Ion Total RNA-Seq Kit v2 kit (Thermo Fisher Scientific, Waltham, MA, USA). The analysis was comprised of 16 individual samples: 2 genotypes (resistant, susceptible) $\times$ 2 challenge states (challenged, control) $\times$ 3–5 biological replicates. The 16 libraries were prepared according to the manufacturer's instructions and then sequenced using Ion Torrent technology (Ion S5XL platform, (Thermo Fisher Scientific Inc., Waltham, MA, USA)). Raw data were processed to remove adapters and were normalized for inherent systematic or experimental biases using the Bioconductor package DESeq2, v. 1.46.0. The reads were aligned to the *Dicentrarchus labrax* (seabass_V1.0-GCA_000689215.1) reference genome [7] with hisat2 (v. 2.2.1) and bowtie2 (v. 2.4.5). Post-mapping quality control was assessed with the Bioconductor package metaseqR2, v. 1.18.0 [28,29]. All raw data reads are available in the NCBI database (Accession No. PRJNA1030357).

2.6. Differential Expression and Enrichment Analyses

Basic differential expression analysis was performed with the Bioconductor packages metaseqR2 and/or edgeR [30–32]. A transcript was considered as a differentially expressed transcript if the adjusted p-value (FDR) threshold was less than 0.05 (significance level) and the log2-transformed fold change (log2FC) was more than 1.5. The function of the identified differentially expressed transcripts was analyzed in OmicsBox software (Version 3.2.4) by first using BLASTX against an NCBI non-redundant (NR) database to search for the possible top hit proteins (accessed on 6 June 2024). To obtain high-quality results, the '*Actinopterygii*' (Taxid:7898) taxonomy filter was applied. Thereafter, blasted sequences were subjected to gene ontology (GO) mapping and annotation with default parameters. GO functional enrichment and pathway analysis were carried out by BLAST2GO (BioBam Bioinformatics, Valencia, Spain) using the total transcripts dataset as the reference background. The annotated DEGs were subjected to Fisher's exact test and were considered significantly enriched in GO terms when their Bonferroni adjusted p-value was less than 0.05. The results were reduced to the most specific terms. Enriched KEGG pathways were determined by Gene Set Enrichment Analysis (GSEA), with adjusted p-value less than 0.05. All bubble plots and pathway summaries were generated using the SRplot [33].

2.7. Construction and Analysis of PPI Networks and Functional Annotation

To further investigate the relationship of resistance-related genes following virus re-infection, the Search Tool for the Retrieval of Interacting Genes/Proteins database (STRING v11.5; http://www.string-db.org, accessed on 27 November 2024) was used to construct their PPI network [34]. All blasted differentially expressed transcripts were manually curated and translated to reference gene names orthologous to the model organism, zebrafish (*Danio rerio*; Taxid: 7955), in the Uniprot database (accessed on 27 November 2024) [https://www.uniprot.org/].

2.8. Real-Time Quantitative Polymerase Chain Reaction (qPCR) Validation

Real-time PCR assays were carried out to confirm the differential expression data of transcriptome analysis. Specific primers used for the expression analysis of genes from both families (Table S1) were designed utilizing Primer-Blast (https://www.ncbi.nlm.nih.gov/tools/primer-blast/, accessed on 12 July 2024), IDT_Primer Quest (https://www.idtdna.com/pages/tools/primerquest, accessed on 12 July 2024), and Primer3Plus (https://www.primer3plus.com/, accessed on 12 July 2024). All qRT-PCR primers were designed according to the minimum information needed for the publication of qRT-PCR experiment (MIQE) guidelines [35]. The potential primer secondary structures (homo- or cross-dimers and hairpin structures) and primer specificity were checked with IDT_Primer Quest and Primer-Blast, respectively. The elongation factor gene was chosen in our study as the reference gene. Before sample quantification experiments, the specificity of each primer pair was studied using positive and negative samples.

Total RNA was used as a template to synthesize cDNA using the QuantiNova Reverse Transcription kit (Qiagen) according to the manufacturer's instructions. Approximately 5 μg RNA were used as the input material. Real-time PCR reactions were performed with QuantiNova SYBR Green PCR (Qiagen) using 1 μL of a 1:10 dilution of cDNA. Primers for all genes were used at 500 nM. The thermal conditions used were as follows: 2 min at 95 °C of pre-incubation followed by 40 cycles at 95 °C for 10 s and 60 °C for 30 s. An additional temperature ramping step was utilized to produce melting curves from 62 to 95 °C to verify the amplification of a unique single product on all samples. All reactions were performed in technical triplicate using a RotorGene Q PCR Detection System (Qiagen). The quantification was performed according to the comparative CT method [36]. The value for each experimental condition was expressed as normalized relative expression, calculated in relation to the values of the control group and normalized against those of the reference gene (by its geometric average).

3. Results

3.1. Experimental Challenge of NNV-Resistant and -Susceptible D. Labrax Families

The experimental NNV challenging of the *D. labrax* fish of both resistant (R) and susceptible (S) families was utilized to study the NNV infection progress, as depicted in Figure 1.

Following the first injection challenge, in NNV-resistant fish (R), typical signs of VNN were observed from day 3 and the first death occurred on day 4, while the NNV-susceptible fish (S) had apparent VNN symptoms and the first death occurred on day 2. The R family appeared to experience a decline in the survival of infected fish from day 4 to day 15 post infection (86.3% survival/ 13.7% mortality), which ceased until the end of the experiment. On the contrary, the S family had a sharp decline in fish survival from day 2 to day 11 post infection, with a maximum mortality on day 5 (50.5% survival/ 49.5% mortality) (Figure 2). No mortalities were recorded following the re-infection of both families. No mortalities were registered in non-infected sea bass, whose survival remained 100% up until the end of the experiment.

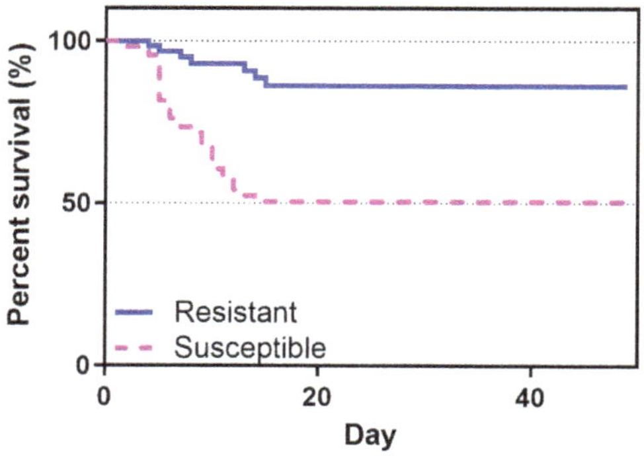

Figure 2. Cumulative survival of NNV-challenged resistant (continuous blue line) and susceptible (dashed magenta line) sea bass families. The mean mortality is represented in each time point (days post infection).

The sea bass brain was tested for viral NNV RNA1 presence in both families at the final time point, i.e., 7 days post re-infection (dpri). The viral load of the surviving fish 7 dpri was $20.1 \pm 1.6 \times 10^{11}$ $TCID_{50}$ of the NNV per µg of total RNA for the R family and $26.0 \pm 8.9 \times 10^{11}$ $TCID_{50}$ of the NNV per µg of total RNA for the S family. Despite the presence of the virus in the brain, no phenotypic or behavioral changes were observed in infected fish and no natural fish loss (i.e., not related to NNV infection) occurred during the experiment following re-infection. No viral load was observed in non-infected groups.

3.2. Summary and Assessment of NNV-Challenged D. Labrax Head Kidney RNA Sequencing (RNA-Seq) Data

Head kidney total RNA isolated from both resistant and susceptible NNV-infected ($n = 10$) and non-infected ($n = 6$) fish were subjected to RNA-seq. The sequencing of the 16 libraries yielded a total of 224,837,867 and 123,140,003 sequences for NNV-infected and non-infected head kidney samples, respectively. Raw data reads are available in the NCBI database (Accession No. PRJNA1030357).

3.3. Differential Expression Analysis Between Infected and Non-Infected Groups

Differentially expressed transcripts were identified according to $\log_2FC \geq 1.5$, and the adjusted *p*-value (FDR) threshold was set as less than 0.05. The transcriptome data were compared among the four groups: the resistant infected fish (RI), the susceptible infected fish (SI), the resistant non-infected fish (RNI), and the susceptible non-infected fish (SNI). The pairwise comparison of RI vs. RNI identified 57 significantly up-regulated genes and 46 significantly down-regulated genes, while the SI vs. SNI comparison resulted in 219 significantly up-regulated genes and 117 significantly down-regulated genes. The pairwise comparisons were also performed between the resistant and susceptible families; the RI vs. SI comparison resulted in 215 significantly up-regulated genes and 116 significantly down-regulated genes. The RNI vs. SNI comparison identified 57 significantly up-regulated genes and 62 significantly down-regulated genes.

The up-regulated and down-regulated DEGs corresponding to the described comparisons are depicted in Figure 3A. Most DEGs were overexpressed after infection for both families (RI vs. RNI and SI vs. SNI comparisons). Also, more up-regulated DEGs were found in the RI vs. SI comparison. The number of common de-regulated DEGs between RI vs. RNI and SI vs. SNI was relatively low (20 common genes), reflecting a totally different head kidney response upon re-infection between the resistant and the susceptible family (Figure 3B). A differential gene expression clustering analysis indicated high intragroup similarity and intergroup differences between all compared groups (Figures 3C and S1).

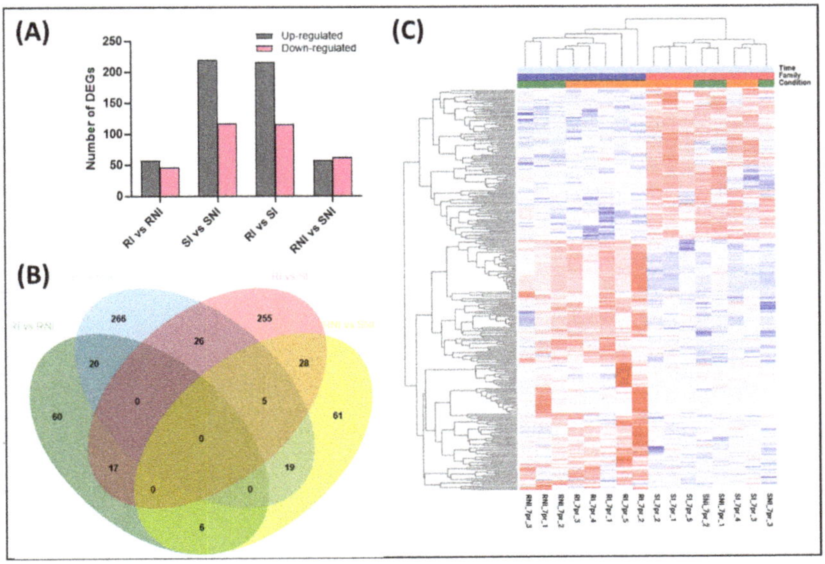

Figure 3. (**A**) Statistics of differentially expressed genes (DEGs) in RI vs. RNI, SI vs. SNI, SI vs. RI, and SNI vs. RNI groups. The X axis and Y axis represent a comparison of samples and the number of DEGs. The first column at each time-point graph corresponds to up-regulated genes, and the second column corresponds to down-regulated genes. (**B**) Venn diagram showing the number of common and exclusive DEGs in RI vs. RNI, SI vs. SNI, SI vs. RI, and SNI vs. RNI groups. (**C**) Cluster analysis of DEGs. Heat map of the genes showed a more than 1.5-fold difference between the resistant (R: blue bar) and susceptible (S: magenta bar) families in NNV-infected (I: orange bar) and non-infected (NI: green bar) experimental groups. Red and blue colors indicate up- and down-regulation in Log2 cpm values, respectively. Each column in the graph represents a sample, each row represents a gene, and the expression of genes in different samples is represented by different colors, with redder colors indicating higher expression and bluer colors indicating lower expression.

3.4. *Resistant Infected (RI) Versus Resistant Non-Infected NNV (RNI) Experimental Group Analysis Results*

3.4.1. GO Enrichment Analysis of DEGs Between Resistant Infected (RI) and Non-Infected NNV (RNI) Experimental Groups

In order to identify the potential molecular mechanisms needed to overcome NNV infection in the resistant family, the functional enrichment analyses were performed between the resistant infected (RI) and non-infected NNV (RNI) experimental groups. For the up-regulated DEGs, the GOs result from the enrichment analysis (adj. $p < 0.05$) are listed in Table S2 and the top 30 GOs are shown in Figure 4A. The most enriched category was the biological process (25/43), followed by the cellular component (9/43) and the molecular function (9/43) categories. For the down-regulated DEGs, the GOs resulting from enrichment analysis are listed in Table S3 and the top 30 GOs are shown in Figure 4B. The most enriched category was the biological process (25/47), followed by the cellular component (11/47) and the molecular function (11/4) categories.

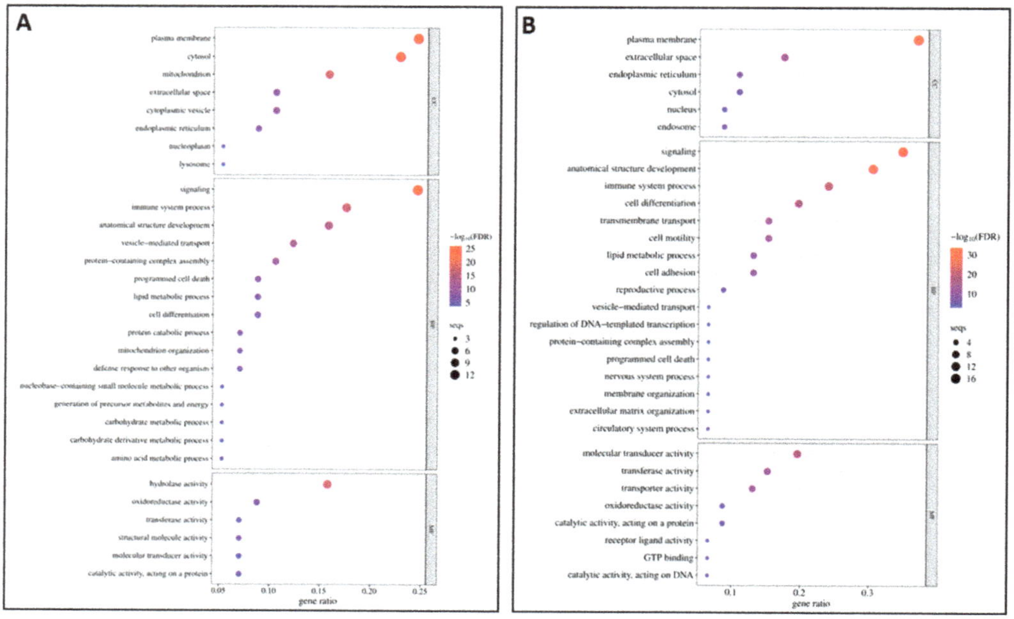

Figure 4. The top 30 gene ontology (GO)-enriched terms in RI vs. RNI groups: (**A**) up-regulated and (**B**) down-regulated DEGs. BP: biological process, MF: molecular function, CC: cellular component.

Among the top enriched categories for both the up- and down-regulated genes were (1) signaling, immune system process, and vesicle-mediated transport; (2) hydrolase activity, molecular transducer activity, and transferase activity; (3) cytosol, plasma membranes, and mitochondrion. The categories uniquely enriched in the up-regulated genes were cytoplasmic vesicle, structural molecule activity, mitochondrion organization, protein catabolic process, catalytic activity acting on a protein, carbohydrate metabolic process, amino acid metabolic process, the generation of precursor metabolites and energy, nucleoplasm, nucleobase containing small molecule metabolic process, molecular function regulator activity, chromatin organization, nucleocytoplasmic transport, autophagy, DNA metabolic process, and cytoskeletal protein binding. The categories uniquely enriched in the down-regulated genes were transmembrane transport, transporter activity, cell adhesion, endosome, catalytic activity acting on a protein, nucleus, GTP binding, catalytic activity acting on DNA, receptor ligand activity, circulatory system process, nervous system process, extracellular matrix organization, muscle system process, endocrine process, cell junction organization, inflammatory response, Golgi apparatus, vitamin metabolic process, extracellular matrix, and lipid binding.

3.4.2. Pathway Analysis of DEGs Between Resistant Infected (RI) and Non-Infected NNV (RNI) Experimental Groups

Pathway analysis was applied between the resistant infected (RI) and non-infected NNV (RNI) experimental groups. All annotated transcripts were run against the KEGG and Reactome databases. It was found that 97 pathways were enriched in the KEGG database, while 19 pathways were enriched in the Reactome database. The enriched pathway categories for both databases are summarized in Table 1 and analyzed in Table S4 and the top 20 pathways are shown in Figure 5. These pathways are involved in (1) immune system: antigen processing and presentation, class I MHC-mediated antigen processing and presentation, NOD-like receptor signaling, C-type lectin receptor signaling, cytosolic DNA-sensing, neutrophil extracellular trap formation, and IL-17 signaling; (2) signaling molecules and interaction: cytokine–cytokine receptor interaction, viral protein interaction with cytokine and cytokine receptors, and cell adhesion molecules; (3) signal transduction: Rap1 signaling, HIF-1 signaling, Ras signaling, signaling by PTK6, and RHO GTPase cycle; (4) cell growth and death: apoptosis, necroptosis, p53 signaling, and the apoptotic cleavage of cellular and cell adhesion proteins; (5) genetic information processing: proteasome, RNA degradation, RNA polymerase, mRNA surveillance, and spliceosome; (6) metabolism: amino acid metabolism, carbohydrate metabolism, oxidative phosphorylation, and fatty acid degradation; and (7) viral infectious diseases, immune diseases, and neurodegenerative diseases.

Table 1. Enriched pathway categories on KEGG and Reactome databases for RI vs. RNI groups.

Database	Pathway Category	Found Pathways	Enriched Pathways
KEGG	Human Diseases	96	34
	Genetic Information Processing	31	15
	Metabolism	115	14
	Organismal Systems	91	15
	Cellular Processes	31	10
	Environmental Information Processing	39	9
Reactome	Signal Transduction	141	10
	Programmed Cell Death	13	4
	Immune System	131	**3**
	Developmental Biology	30	1
	Hemostasis	26	1

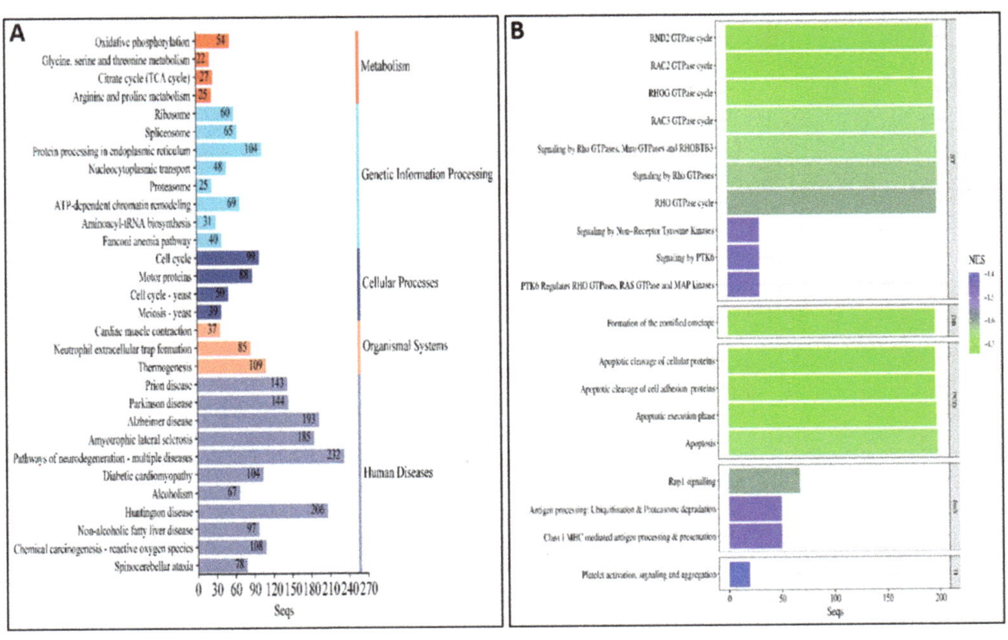

Figure 5. (**A**) The top 30 enriched KEGG pathways in RI vs. RNI groups. (**B**) The enriched Reactome pathways in RI vs. RNI groups. ST: signal transduction; DB: developmental biology; PCD: programmed cell death; ImS: immune system; H: hemostasis; NES: normalized enrichment score.

3.4.3. Protein-Protein Interaction (PPI) Analysis of DEGs Between Resistant Infected (RI) and Non-Infected NNV (RNI) Experimental Groups

To gain further insights into the disease resistance mechanisms, the zebrafish orthologs of all differentially expressed genes in between the resistant infected (RI) and non-infected NNV (RNI) experimental groups were manually determined. Of the 140 proteins, 105 were found in the Search Tool for the Retrieval of Interacting Genes/Proteins database of the STRING database and analyzed in terms of gene clustering (Figure 6). PPI analysis with k-means clustering resulted in five distinct clusters containing more than three genes. The hub genes include vegfc (vascular endothelial growth factor C); aldh18a1 (delta-1-pyrroline-5-carboxylate synthetase); canx (calnexin/ MHC class I antigen, partial); mxf (interferon-induced GTP-binding protein Mx) and alox12 (arachidonate 12-lipoxygenase); and ddit4 (DNA damage-inducible transcript 4 protein).

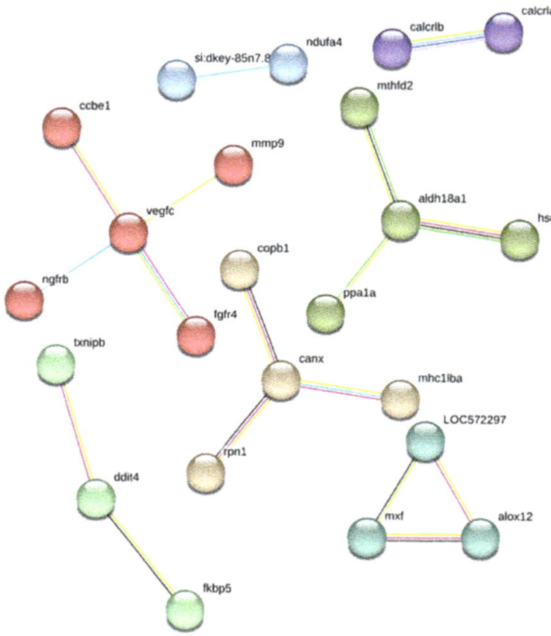

Figure 6. Protein–protein interaction (PPI) networks, with k-means clustering for the zebrafish orthologs of the differentially expressed genes in *D. labrax* RI vs. RNI groups. It is retrieved via API access to the STRING database (https://string-db.org) (accessed on 27 November 2024) and was performed based on the *Danio rerio* protein database. Each colored group represents a different cluster. The edges represent protein–protein interactions.

3.5. Susceptible Infected (SI) Versus Susceptible Non-Infected NNV (SNI) Experimental Groups Analysis Results

3.5.1. GO Enrichment Analysis of DEGs Between Susceptible Infected (SI) and Non-Infected NNV (SNI) Experimental Groups

In order to identify potential molecular mechanisms to overcome NNV infection by the susceptible family, the functional enrichment analyses were performed between the susceptible infected (SI) and non-infected NNV (SNI) experimental groups. For the up-regulated DEGs, the GOs resulting from the enrichment analysis (adj. $p < 0.05$) are listed in Table S5 and are shown in Figure 7A. The most enriched category was molecular function (8/16), followed by the biological process (6/16) and the cellular component (2/16) categories. For the down-regulated DEGs, the GOs resulting from the enrichment analysis are listed in Table S6 and are shown in Figure 7B. Only four GO categories were enriched belonging to the cellular component (2/4), the biological process (1/4), and the molecular function (1/4) categories.

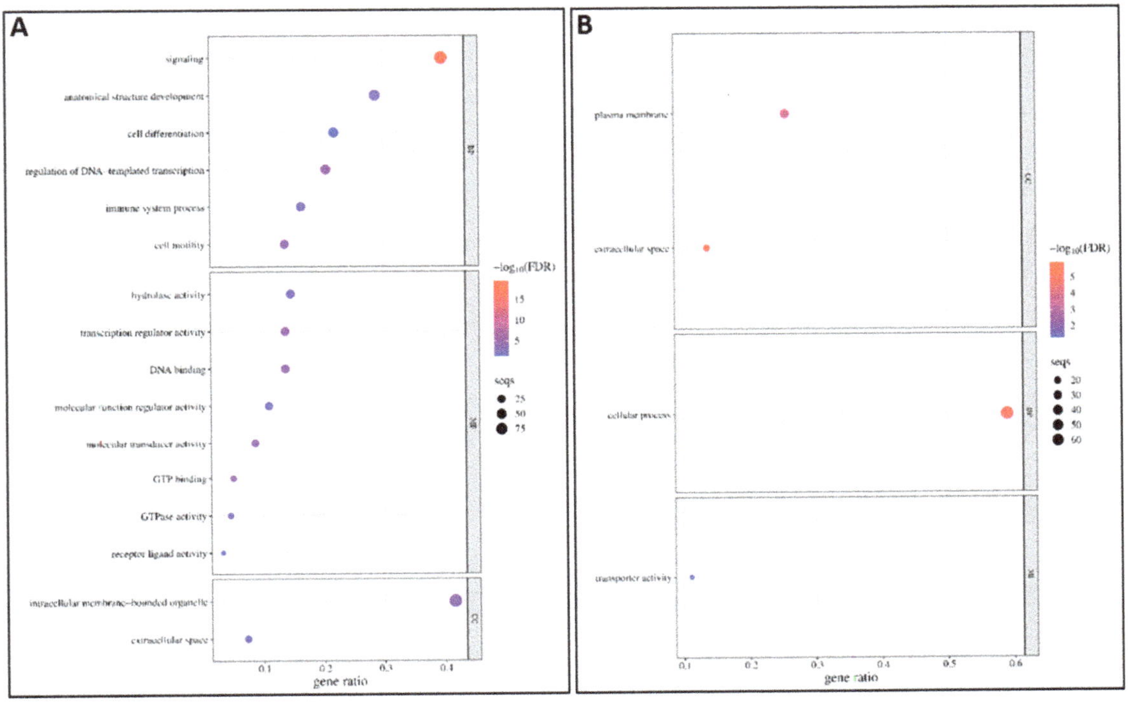

Figure 7. The gene ontology (GO)-enriched terms in SI vs. SNI groups: (**A**) up-regulated and (**B**) down-regulated DEGs. BP: biological process, MF: molecular function, CC: cellular component.

The categories uniquely enriched in the up-regulated genes are signaling, the regulation of DNA-templated transcription, transcription regulator activity, DNA binding, GTP binding, molecular transducer activity, cell motility, intracellular membrane-bounded organelle, hydrolase activity, anatomical structure development, GTPase activity, immune system process, molecular function regulator activity, cell differentiation, and receptor ligand activity. The categories uniquely enriched in the down-regulated genes are cellular process, plasma membrane, and transporter activity. The only category enriched for both up- and down-regulated genes was the extracellular space.

3.5.2. Pathway Analysis of DEGs Between Susceptible Infected (SI) and Non-Infected NNV (SNI) Experimental Groups

Pathway analysis was applied between the susceptible infected (SI) and non-infected NNV (SNI) experimental groups. All annotated transcripts were run against the KEGG and Reactome databases. It was found that 36 pathways were enriched in the KEGG database, while 35 pathways were enriched on the Reactome database. The enriched pathway categories for both databases are summarized in Table 2 and analyzed in Table S7, and the top 20 pathways are shown in Figure 8. These pathways are involved in (1) the immune system: NOD-like receptor signaling, C-type lectin receptor signaling, cytosolic DNA-sensing, T cell receptor signaling, IL-17 signaling, cytokine signaling in immune system, toll-like receptor cascades, MyD88-dependent cascades, and the innate immune system;

(2) signaling molecules and interaction: interleukin-1 family signaling and the regulation of NF-kappa B signaling; (3) signal transduction: NRIF signals cell death from the nucleus; (4) cell growth and death: cell cycle, necroptosis, and the regulation of actin cytoskeleton; (5) genetic information processing: proteasome, RNA polymerase, and nucleotide excision repair; (6) metabolism: carbohydrate metabolism, oxidative phosphorylation, and nicotinate and nicotinamide metabolism; (7) viral infectious diseases and immune diseases, but no neurodegenerative diseases.

Table 2. Enriched pathway categories on KEGG and Reactome databases for SI vs. SNI groups.

Database	Pathway Category	Found Pathways	Enriched Pathways
KEGG	Human diseases	96	13
	Genetic information processing	31	4
	Metabolism	129	4
	Organismal systems	91	9
	Cellular processes	31	6
	Environmental information processing	39	-
Reactome	Immune system	500	23
	Metabolism	2206	5
	Signal transduction	728	4
	Metabolism of proteins	367	1
	Drug ADME	61	1
	Gene expression (Transcription)	252	1

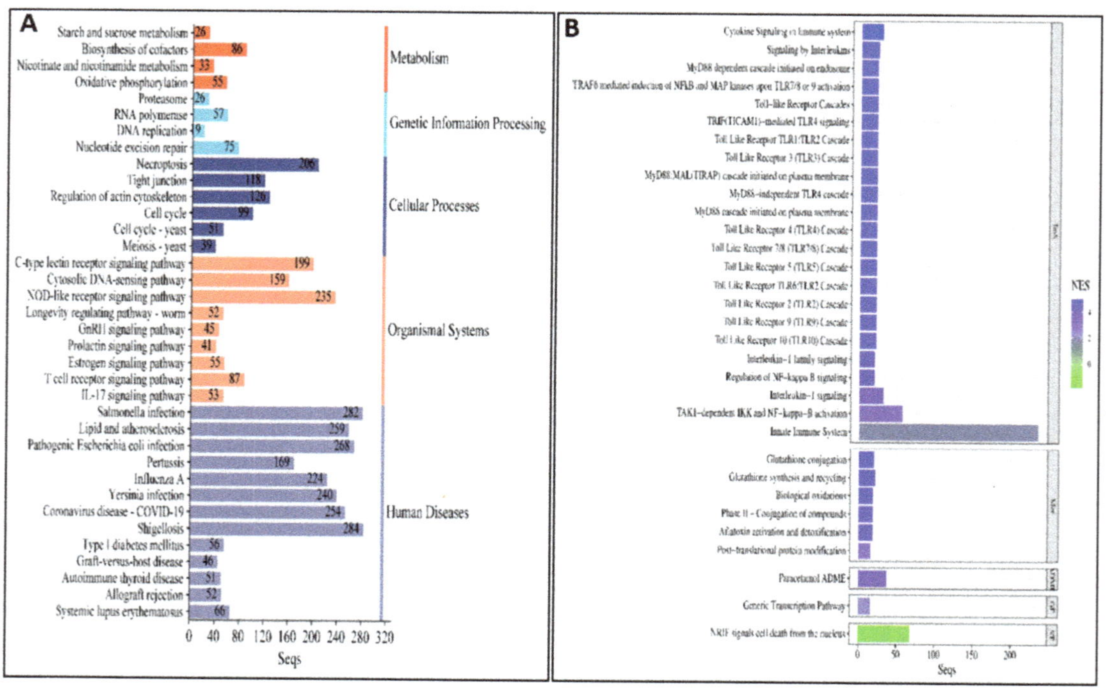

Figure 8. (**A**) The top enriched KEGG pathways in SI vs. SNI groups. (**B**) The enriched Reactome pathways in SI vs. SNI groups. ImS: immune system, Met: metabolism, ADME: drug ADME, GE: gene expression (transcription), ST: signal transduction, NES: normalized enrichment score.

3.5.3. Protein–Protein Interaction Analysis of DEGs Between Susceptible Infected (SI) and Non-Infected NNV (SNI) Experimental Groups

The zebrafish orthologs of all differentially expressed genes in between the susceptible infected (SI) and non-infected NNV (SNI) experimental groups were manually determined. Of the 321 proteins, 217 were found in the Search Tool for the Retrieval of Interacting Genes/Proteins database of the STRING database and analyzed in terms of gene clustering (Figure 9). PPI analysis with k-means clustering, resulted in four distinct clusters containing more than three genes. The hub genes include egr1 (early growth response protein 1), jun (jun proto-oncogene, AP-1 transcription factor subunit), fosb (FBJ murine osteosarcoma viral oncogene homolog B); mmp13b and mmp19 (matrix metallopeptidases); and galr1b (galanin receptor 1b), a member of the G-protein-coupled receptor 1 family.

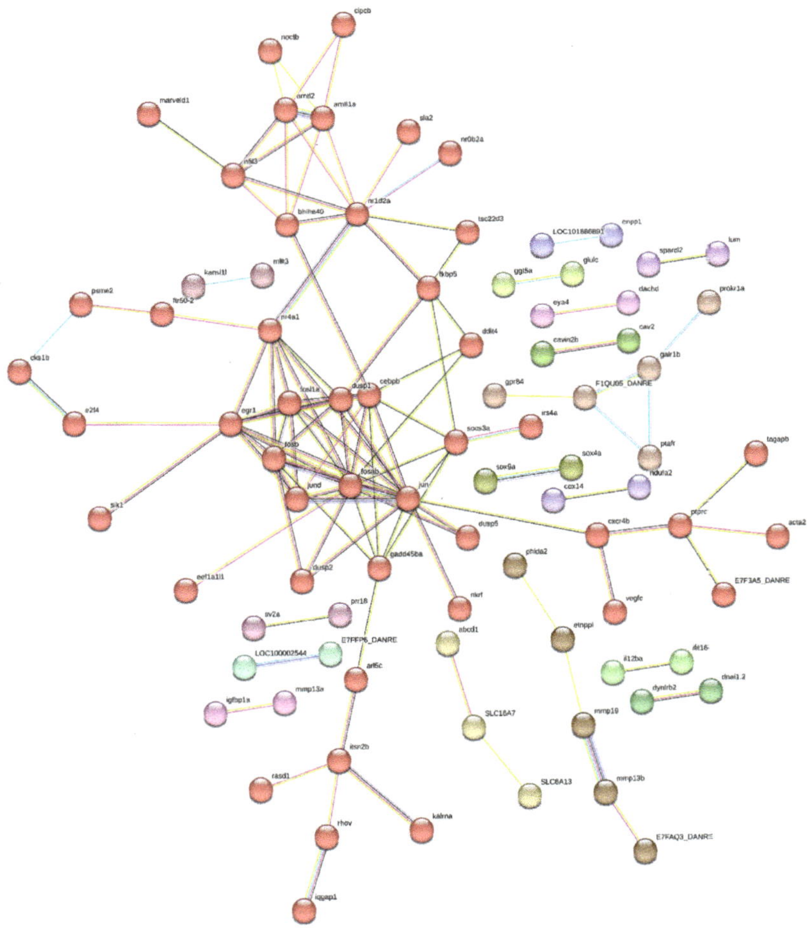

Figure 9. Protein–protein interaction (PPI) networks, with k-means clustering for the zebrafish orthologs of the differentially expressed genes in *D. labrax* SI vs. SNI groups. It is retrieved via API access to the STRING database (https://string-db.org) (accessed on 27 November 2024) and was performed based on the *Danio rerio* protein database. Each colored group represents a different cluster. The edges represent protein-protein interactions.

3.6. *qRT-PCR Assay of Selected Genes in the Head Kidney of NNV-Resistant and NNV-Susceptible D. labrax Families*

The expression levels of six selected genes were analyzed by quantitative RT-PCR to validate the DEGs identified by RNA-seq. The analyzed genes were the arylsulfatase g (arsg), g-protein-coupled receptor 84 (GPR84), and transmembrane protein 120a (TACAN), which were highly expressed by the resistant family, and zinc finger bed domain-containing protein 1 (SUMO) and hydroxycarboxylic acid receptor 2-like (12S-HETE) and collagenase 3-like (mmp13), which were highly expressed by the susceptible family. As shown in Figure S2, the expression of the genes for both families when the infected fish were compared with non-infected fish were consistent with the expression results obtained from the transcriptome analysis for each family.

4. Discussion

European sea bass is an economically important marine fish species widely farmed in the Mediterranean, and nervous necrosis virus is causing significant economic losses to the aquaculture industry [3]. It has been reported that betanodavirus survivors from natural infections are resistant to the disease recurrence [37], suggesting a protective immune response of the fish to infection. In another study, it has been reported that the surviving fish were found to produce neutralizing antibodies, which possibly explains their resistance to natural re-infection [6]; however, in a recent study on convalescent sevenband groupers, almost no NNV-neutralizing antibodies were detected, even though they were strongly protected against re-infection with NNV [38]. Thus, the potential molecular mechanisms of immune responses against NNV, especially following survivors' re-infection, remains poorly understood.

The viral load in resistant sea bass brain was found to be slightly lower than the susceptible sea bass brain; however, both families had a considerable amount of the replicated virus in the target organ. This observation is in agreement with previous studies [20], confirming that fish with high levels of genetic resistance to the virus are actively infected; thus, genetic resistance cannot be entirely due to the inability of the virus to infect the fish. Although VNN infection is primarily localized to the brain, systemic immune responses are anticipated to manifest in the head kidney. To this end, numerous differentially expressed genes (DEGs) were identified in the head kidneys of the resistant infected vs. non-infected (RI vs. RNI) and the susceptible infected vs. non-infected (SI vs. SNI) experimental groups. The resistant and susceptible families' responses were studied utilizing different approaches, i.e., gene ontology (GO) enrichment, pathway (KEGG and Reactome) enrichment, and protein–protein interaction (PPI) analysis, in order to examine the expression patterns and obtain a more comprehensive understanding of the biological functions of DEGs from each family. The 'uniquely' implicated components of each family response resulting from these analyses are summarized in Table 3.

Table 3. The 'uniquely' implicated components of each family response, resulting from gene ontology (GO), pathway enrichment, and protein–protein (PPI) analysis.

Analysis	Resistant Family		Susceptible Family
GO enrichment	vesicle-mediated transport	cytoskeletal protein binding	transcription regulator activity
	transferase activity	transmembrane transport	intracellular membrane-bounded organelle
	cytosol	cell adhesion	GTPase activity

Table 3. *Cont.*

Analysis	Resistant Family		Susceptible Family
	mitochondrion	endosome	cellular process
	cytoplasmic vesicle	catalytic activity acting on a protein	
	structural molecule activity	nucleus	
	mitochondrion organization	catalytic activity acting on DNA	
	protein catabolic process	circulatory system process	
	catalytic activity acting on a protein	nervous system process	
	carbohydrate metabolic process	extracellular matrix organization	
	amino acid metabolic process	muscle system process	
	generation of precursor metabolites and energy	endocrine process	
	nucleoplasm	cell junction organization	
	nucleobase-containing small molecule metabolic process	inflammatory response	
	chromatin organization	Golgi apparatus	
	nucleocytoplasmic transport	vitamin metabolic process	
	autophagy	extracellular matrix	
	DNA metabolic process	lipid binding	
Pathway enrichment	antigen processing and presentation	apoptotic cleavage of cellular and cell adhesion proteins	T cell receptor signaling
	class I MHC-mediated antigen processing and presentation	RNA degradation	cytokine signaling in immune system
	neutrophil extracellular trap formation	mRNA surveillance	toll-like receptor cascades
	cytokine–cytokine receptor interaction	spliceosome	MyD88 dependent cascades
	viral protein interaction with cytokine and cytokine receptor	amino acid metabolism	innate immune system
	cell adhesion molecules	neurodegenerative diseases	interleukin-1 family signaling
	Rap1 signaling		regulation of NF-kappa B signaling
	HIF-1 signaling		NRIF signals cell death from the nucleus
	Ras signaling		cell cycle
	signaling by PTK6, RHO GTPase cycle		regulation of actin cytoskeleton
	apoptosis		nucleotide excision repair
	p53 signaling		nicotinate and nicotinamide metabolism

Table 3. *Cont.*

Analysis	Resistant Family	Susceptible Family
PPI	vegfc	egr1
	aldh18a1	jun
	canx	fosb
	mxf	mmp13b and mmp19
	alox12	galr1b
	ddit4	

The DEGs between RI vs. RNI were significantly enriched in genetic information processing, organismal systems (including immune system), and viral infection-related pathways, while the DEGs between SI vs. SNI were significantly enriched in immune-related and molecular function activity pathways. The GO terms 'immune system process' and 'signaling' were enriched in the up-regulated transcripts of both families, as expected, but we also found enrichment in the down-regulated transcripts of the resistant family.

In the up-regulated genes of the resistant family, many enriched GO terms were related to metabolism (e.g., 'carbohydrate metabolic process', 'amino acid metabolic process', 'generation of precursor metabolites and energy', and 'nucleobase containing small molecule metabolic process'), which may be attributed to the host's attempt to overcome lower energy availability as an effect of appetite loss, as described for other fish viruses [20]. Other metabolism-related GO terms were enriched in the down-regulated transcripts of the resistant family, including 'vitamin metabolic process', 'lipid metabolic process', and 'lipid binding'. In vitro studies of RGNNV infection mechanisms proved that the virus requires lipogenesis for infection and induced the formation of lipid droplets [39]; therefore, the down-regulation of the related genes may reflect an attempt of the resistant sea bass to control the infection.

In addition, GO terms related to 'nervous system process' and 'muscle system process' were found to be down-regulated in the resistant family, which possibly makes an effort to control the disease symptoms (i.e., nervous system damage and abnormal swimming).

'Vesicle-mediated transport' and 'cytoplasmic vesicle' terms were also enriched exclusively in the resistant family, implying a crucial role of vesicle-mediated transport in response to the NNV infection. It has been suggested that the betanodavirus enters host cells mainly via clathrin-mediated endocytosis in a cholesterol-, pH-, and cytoskeleton-dependent manner, involving clathrin-coated vesicles containing internalized viruses [40]. In general, viruses hijack such cellular pathways to promote their propagation; however, the vesicle-related term enrichment in the resistant family may indicate the fish's effort to modify its cells normal vesicular transport pathways to defend itself from the infection. For example, the cells could recognize and degrade viral components either via pattern recognition receptor detection, compartmentalization, and phagocytosis stimulation or by aggresome formation and autophagy induction [41]. On the other hand, the susceptible sea bass attempts to control infection utilizing specific molecules' activity by enriching genes associated more with molecular function terms. The 'transcription regulator activity' is highly enriched in the up-regulated transcripts, since viral transcription regulators of gene expression are central to disease pathogenesis due to their ability to control the expression of both viral and host genes [42]; therefore, sea bass tries to effectively control their activity.

The pathway enrichment analysis was performed using both the KEGG and Reactome pathway databases. It was found that some immune-related pathways ('NOD-like receptor signaling', 'C-type lectin receptor signaling', 'cytosolic DNA sensing', and 'IL-17 signaling') were enriched in both resistant and susceptible fish, while different immune system components seem to be involved in each family response. The resistant family utilizes pathways related to 'class I MHC-mediated antigen processing and presentation', 'neutrophil extracellular trap formation', 'viral protein interaction with cytokine and cytokine receptor', as well as 'cytokine–cytokine receptor interaction' and 'cell adhesion molecules'. The 'cytokine–cytokine receptor interaction' pathway was also found to be enriched in Atlantic salmon fry challenged with the infectious pancreatic necrosis virus in resistant fish at a late time point (20 dpi) [20]. The response in the susceptible family depends mostly on 'T cell receptor signaling', 'cytokine signaling in immune system', 'toll-like receptor cascades', 'MyD88-dependent cascades', the 'innate immune system', 'interleukin-1 family signaling', and the 'regulation of NF-kappa B signaling'.

Apoptosis related-pathways (i.e., 'apoptosis', 'p53 signaling', and 'apoptotic cleavage of cellular and cell adhesion proteins') are enriched only in the resistant family, while the 'necroptosis' pathway is enriched by both families. When an organism is infected, its cells can regulate (or "program") their death to tailor immune responses, thereby changing the impact that their death will have on the surroundings [43]. Necroptosis is a non-apoptotic form of cell death which has evolved to detect pathogens and promote tissue repair. The process culminates with the loss of membrane integrity and passive leakage of intracellular contents (e.g., cytokines, DAMPs, and PAMPs), which provide pro-inflammatory cues that recruit immune cells but also induces inflammation, which can be detrimental to the host. Apoptosis induces the breakdown of the nuclear membrane, the cleavage of many intracellular proteins, membrane blebbing, the breakdown of genomic DNA into nucleosomal structures, and the release of cytochrome c from mitochondria [43]. Apoptotic death usually leads to immunologically silent responses, and its activation does not promote a significant inflammatory response, thereby preserving homeostatic integrity [44]. Its utilization by the resistant family in combination with necroptosis, in a balanced way, is possibly responsible for better outcomes of host health.

Generic viral disease ('Influenza A' and 'Coronavirus disease—COVID-19') pathways were enriched in both families, in accordance with similar studies [20], as well as 'RNA polymerase' and 'proteasome' in the genetic information processing KEGG category. However, the 'RNA degradation', 'mRNA surveillance', and 'spliceosome' pathways were only enriched in the resistant family, indicating a more intense effort to control the host cell machinery response to the RNA virus.

Protein–protein interaction analysis of DEGs from each family resulted in discrete clustering with a variety of hub genes. Interactions of the resistant family are linked by vegfc, aldh18a1, canx, mxf, alox12, and ddit4. The **vascular endothelial growth factor (VEGF) C** (vegfc) is a member of the VEGF family, which are secreted polypeptides acting through a family of cognate receptor kinases in endothelial cells to stimulate blood vessel formation in vertebrates [45]. In humans, many viruses seek the up-regulation of VEGF, either by utilizing HIF-1α, COX-2 and AP1 target pathways for the virus-mediated upregulation of VEGF or by the activation of inflammatory mediators [46]. Moreover, VEGFs were found to participate in the immune response of invertebrates [47]. Thus, VGEFs seem to play major role in infection by pathogens, but their role in sea bass is still unclear. On the contrary, the **interferon-induced GTP-binding protein Mx** have a well-characterized antiviral role and show a strict dependence on type-I interferons (IFN α/β) for its expression in different vertebrates. Mx proteins survey exocytic events, target viral nucleocapsid-like structures, and mediate vesicle trafficking to trap essential viral components in order to prevent viral replication at early time points [48]. European sea bass expresses two different Mx genes (MxA and MxB) [49] and the MxA profile in the head kidney during the VNN infection has been studied extensively [11,28,50–52]. Different levels of Mx expression have been observed in viral disease-resistant/susceptible fish hosts. An Atlantic salmon family resistant to IPNV demonstrated higher Mx expression in the head kidney compared to susceptible fish, while in Atlantic salmon fry, Mx was moderately up-regulated in resistant fish, but highly up-regulated in susceptible fish at various time points [20]. **Arachidonate 12-lipoxygenase** (alox12) is a lipid mediator which plays a vital role in the innate immune responses of teleost fish following pathogen infection by encoding enzymes that act on different polyunsaturated fatty acid substrates to generate bioactive lipid mediators (e.g., eicosanoids) and mediate inflammatory responses [53,54].

Calnexin (canx) is an endoplasmic reticulum (ER) membrane-bound lectin chaperone, which comprises a specialized maturation system combined with the lectin chaperone calreticulin. Its main functions are glycoprotein folding, the quality control of ER protein synthesis, and Ca^{2+} storage, but canx also plays roles in phagocytosis and dendritic cell immunity [55]. Canx is a highly conserved chaperone widely distributed in eukaryotic organisms, and it has been identified in channel catfish [56], zebrafish [57], rainbow trout [58], pufferfish [59], shrimp [55], and crab [60]. Calnexin is highly associated with the correct folding and assembly of MHC molecules in ER [56] and is a potent modulator of antibacterial immune responses [55,59,60]. **Aldehyde dehydrogenase 18A1** (aldh18a1) belongs to the aldehyde dehydrogenase (ALDH) superfamily of enzymes. Human ALDH18A1 encodes a bifunctional enzyme, designated as delta 1-pyrroline-5-carboxylate synthase, which catalyzes the first two steps in proline, ornithine, and arginine biosynthesis [61]. However, its presence and function in teleost fish is not well studied, with the exception of Atlantic salmon and zebrafish [62]. The **DNA damage-inducible transcript 4 protein** (ddit4) is expressed under stress situations, turning off the metabolic activity triggered by the mammalian targeting of rapamycin (mTOR) in humans, and plays a crucial role in cancer [63], but is not well studied in fish. In a recent report on turbot resistance to bacterial infection, ddit4 was expressed at higher levels in the resistant family head kidney and liver, in accordance with our study [21]. Interestingly, in a study on the administration of an aldehyde dehydrogenase (recombinant ALDH7A1) in Atlantic salmon infected with *Aeromonas salmonicida,* it was found that ddit4 was among the differentially expressed transcripts [64].

On the other hand, interactions of the susceptible family are linked by egr1, jun, fosb, mmp13b, mmp19, galr1b, and ptafr. **Early growth response 1** (egr1) is a multifunctional transcription factor capable of both enhancing and/or inhibiting gene expression, highly conserved between numerous species including zebrafish and human. EGR1 can be activated by a wide array of stimuli (e.g., growth factors, cytokines) and various cellular stress states (e.g., viral infections (HSV-1 (herpes simplex 1), HIV (human immunodeficiency virus), EBV (Epstein–Barr virus)), acting either by the activation or suppression of virus infectivity [65]. **Jun** and **Fos** proteins are DNA-binding proteins that are involved in gene expression through transcriptional regulation, including modulating cell proliferation or cell death, in response to different biological stress signals. Both transcription factors are members of the AP1 (activator protein one) complex [66], which in mammals, also mediates gene regulation in response to various stimuli, including cytokines, growth factors, stress signals, and bacterial and viral infections [67].

The matrix metalloproteinase (MMP) gene family is responsible for regulating the degradation of extra cellular matrix (ECM) proteins, which are important for physiological processes, such as wound healing, tissue remodeling, and stress response [68], and they are actively involved in the regulation of the host immune system following infection [69]. The matrix metalloprotease **mmp13** is a collagenase actively involved in the process of bone formation. In teleosts, MMP-13 is required for normal embryogenesis (zebrafish) and is up-regulated following infection (Japanese flounder, channel catfish, and rainbow trout) [17,69,70]. **MMP19** was originally isolated from a rheumatoid arthritis patient. In mammals, MMP19 is able to cleave typical ECM components and the insulin-like growth factor binding protein-3 (IGFBP-3), and it is expressed in macrophages and up-regulated under inflammatory conditions [71]. In teleosts, however, its role is not well understood, with the exception of studies focused on reproduction mechanisms [72]. Interestingly, the **galanin receptor 1b** (galr1b) hub gene is, also, related with the galinergic system, which is mainly involved in reproductive functions and feeding regulation in fish [73].

Overall, the NNV-resistant fish seem to better control their immune responses to the virus compared to the susceptible fish, which appears to have higher inflammatory response following survivors' re-infection. For example, the same trend has been observed in Atlantic salmon fry infected with IPNV [20] or infectious salmon anemia virus (ISAV) [74], among others. In similar studies, the susceptible fish appeared to have high immune responses, which includes the high representation and expression of inflammatory pathway members, IFN-responsive elements, and cytokines, leading to eventual apoptosis but failing to stop the infection.

5. Conclusions

The host immune response to an infectious pathogen is a systematic and complex biological process. The present study offered glimpses of the mechanisms underlying a disseminated systemic response of sea bass belonging to families with varying VVN disease resistance, following survivors' re-infection. To the best of our knowledge, the present study is the first report of a transcriptome profile comparison of NNV-resistant and -susceptible families of European sea bass following survivors' re-infection. The obtained transcriptome data showed significant differences between the resistant and susceptible families before and after viral infection. In conclusion, the transcriptome profiles revealed 103 DEGs for the resistant family (RI vs. RNI) and 336 DEGs for the susceptible family (SI vs. SNI). The pathway analysis indicated that immune-related pathways were enriched in both resistant and susceptible fish, but different immune system components were involved in each family response. Apoptosis-related pathways were enriched only in the resistant family, while the 'necroptosis' pathway was enriched in both families. Furthermore, protein–protein interaction analysis identified a variety of hub genes for the resistant and the susceptible families, quite distinct in their function on NNV resistance. The presented results offer glimpses of the mechanisms of European sea bass responses to nervous necrosis virus survivors' re-infection, depending on different host genetic backgrounds, providing valuable datasets for further research on viral disease resistance in teleost fish.

Supplementary Materials: The following supporting information can be downloaded at https://www.mdpi.com/article/10.3390/v17020230/s1, Figure S1: Cluster analysis of DEGs. Each column in the graph represents a sample, each row represents a gene, and the expression of genes in different samples is represented by different colors, with redder colors indicating higher expression and bluer colors indicating lower expression. Heat map of the genes showed more than a 1.5-fold difference between the resistant (R: blue bar) and susceptible (S: magenta bar) families or between the NNV-infected (I: orange bar) and non-infected (NI: green bar) experimental groups. Red and blue colors indicate up- and down-regulation in Log2 cpm values, respectively. Each column in the graph represents a sample, each row represents a gene, and the expression of genes in different samples is represented by different colors, with redder colors indicating higher expression and bluer colors indicating lower expression; Figure S2: Validation of the RNA-Seq results by qPCR. Arsg: arylsulfatase g, GPR84: g-protein coupled receptor 84, TACAN: transmembrane protein 120a; SUMO: zinc finger bed domain-containing protein 1, 12S-HETE: hydroxycarboxylic acid receptor 2-like, mmp13: collagenase 3-like. Table S1: Primers and probes used in the present work; Table S2: The gene ontology (GO)-enriched terms in RI vs. RNI group up-regulated DEGs. BP: Biological process,

MF: molecular function, CC: cellular component; Table S3: The gene ontology (GO)-enriched terms in RI vs. RNI group down-regulated DEGs. BP: biological process, MF: molecular function, CC: cellular component; Table S4: enriched pathway categories on KEGG and Reactome databases for RI vs. RNI groups; Table S5: The gene ontology (GO)-enriched terms in SI vs. SNI groups up-regulated DEGs. BP: biological process, MF: molecular function, CC: cellular component; Table S6: The gene ontology (GO)-enriched terms in SI vs. SNI groups down-regulated DEGs. BP: biological process, MF: molecular function, CC: cellular component; Table S7: enriched pathway categories on KEGG and Reactome databases for SI vs. SNI groups.

Author Contributions: Conceptualization, V.B. and E.K.; methodology, D.K.T., V.B. and E.K.; validation, D.K.T., O.-P.T. and A.E.; formal analysis, D.K.T.; investigation, D.K.T., O.-P.T., A.E., V.B. and E.K.; resources, V.B. and E.K.; data curation, D.K.T.; writing—original draft preparation, D.K.T.; writing—review and editing, D.K.T., O.-P.T., A.E., V.B. and E.K.; visualization, D.K.T.; supervision, E.K.; project administration, D.K.T., V.B. and E.K.; funding acquisition, D.K.T., V.B. and E.K. All authors have read and agreed to the published version of the manuscript.

Funding: This research was funded by European Maritime and Fisheries Fund and Hellenic Ministry of Rural Development and Food in the frame of O.P. Maritime & Fisheries 2014–2020, grant number MIS: 5010925. The APC was kindly waived by MDPI.

Institutional Review Board Statement: The animal study protocol was approved by the committee on Bioethics of University of the Aegean, on the Laboratory of Ichthyology-Aquaculture and Aquatic Animal Health (ICHTHYAI) (Government Issue 1255/28-4-2016). ICHTHYAI has been granted all the required permits for producing (EL 83 BIObr 01), supplying (EL 83 BIOsup 01) and experimenting on aquatic organisms (EL83 BioExp 01), according to the Presidential Decree 56/2013 con-forming to Directive 2010/63/EE (Decision No 4053/14-3-2017 of the competent Regional Veterinary Authority). The protocol for the experimental infection performed in this study has been approved by Decision No 5379/4-4-2017 of the competent Regional Veterinary Authority.

Informed Consent Statement: Not applicable.

Data Availability Statement: The raw sequencing data are available in the NCBI database (accession no. PRJNA1030357).

Acknowledgments: We would like to thank Daniella White, Michail-Aggelos Valsamidis, and Akindynos Palaiologos for their support during collection of samples and Panagiotis Moulos for assistance with bioinformatics analysis.

Conflicts of Interest: The authors declare no conflict of interest. The funders had no role in the design of the study; in the collection, analyses, or interpretation of data; in the writing of the manuscript; or in the decision to publish the results.

References

1. Munday, B.L.; Kwang, J.; Moody, N. Betanodavirus infections of teleost fish: A review. *J. Fish Dis.* **2002**, *25*, 127–142. [CrossRef]
2. Doan, Q.K.; Vandeputte, M.; Chatain, B.; Morin, T.; Allal, F. Viral encephalopathy and retinopathy in aquaculture: A review. *J. Fish Dis.* **2017**, *40*, 717–742. [CrossRef] [PubMed]
3. Bandín, I.; Souto, S. Betanodavirus and VER Disease: A 30-year Research Review. *Pathogens* **2020**, *9*, 106. [CrossRef] [PubMed]
4. Ahmed, N.; Thompson, S. The blue dimensions of aquaculture: A global synthesis. *Sci. Total. Environ.* **2019**, *652*, 851–861. [CrossRef]
5. Nishizawa, T.; Furuhashi, M.; Nagai, T.; Nakai, T.; Muroga, K. Genomic classification of fish nodaviruses by molecular phylogenetic analysis of the coat protein gene. *Appl. Environ. Microbiol.* **1997**, *63*, 1633–1636. [CrossRef]
6. Costa, J.Z.; Thompson, K.D. Understanding the interaction between Betanodavirus and its host for the development of prophylactic measures for viral encephalopathy and retinopathy. *Fish Shellfish Immunol.* **2016**, *53*, 35–49. [CrossRef]
7. Tine, M.; Kuhl, H.; Gagnaire, P.A.; Louro, B.; Desmarais, E.; Martins, R.S.; Hecht, J.; Knaust, F.; Belkhir, K.; Klages, S.; et al. European sea bass genome and its variation provide insights into adaptation to euryhalinity and speciation. *Nat. Commun.* **2014**, *5*, 5770. [CrossRef] [PubMed]

8. Hough, C. *Regional Review on Status and Trends in Aquaculture Development in Europe—2020*; FAO Fisheries and Aquaculture, FAO: Rome, Italy, 2022. [CrossRef]
9. FAO. *The State of World Fisheries and Aquaculture 2022. Towards Blue Transformation*; FAO: Rome, Italy, 2022. [CrossRef]
10. Castri, J.; Thiéry, R.; Jeffroy, J.; de Kinkelin, P.; Raymond, J.C. Sea bream *Sparus aurata*, an asymptomatic contagious fish host for nodavirus. *Dis. Aquat. Org.* **2001**, *47*, 33–38. [CrossRef] [PubMed]
11. Moreno, P.; Gemez-Mata, J.; Garcia-Rosado, E.; Bejar, J.; Labella, A.M.; Souto, S.; Alonso, M.C. Differential immunogene expression profile of European sea bass (*Dicentrarchus labrax*, L.) in response to highly and low virulent NNV. *Fish Shellfish Immunol.* **2020**, *106*, 56–70. [CrossRef]
12. Gonzalez-Silvera, D.; Guardiola, F.A.; Espinosa, C.; Chaves-Pozo, E.; Esteban, M.Á.; Cuesta, A. Recombinant nodavirus vaccine produced in bacteria and administered without purification elicits humoral immunity and protects European sea bass against infection. *Fish Shellfish Immunol.* **2019**, *88*, 458–463. [CrossRef] [PubMed]
13. Vela-Avitúa, S.; Thorland, I.; Bakopoulos, V.; Papanna, K.; Dimitroglou, A.; Kottaras, E.; Papaharisis, L.; Guinand, B.; Tsigenopoulos, C.S.; Aslam, M.L. Genetic Basis for Resistance against Viral Nervous Necrosis: GWAS and Potential of Genomic Prediction Explored in Farmed European Sea Bass (*Dicentrarchus labrax*). *Front. Genet.* **2022**, *13*, 804584. [CrossRef] [PubMed]
14. Huang, Y.; Li, Z.; Li, M.; Zhang, X.; Shi, Q.; Xu, Z. Fish Genomics and Its Application in Disease-Resistance Breeding. *Rev. Aquac.* **2025**, *17*, e12973. [CrossRef]
15. Wang, L.; Xu, X.; Zhang, Z.; Li, K.; Yang, Y.; Zheng, W.; Sun, H.; Chen, S. Transcriptome analysis and protein-protein interaction in resistant and susceptible families of Japanese flounder (*Paralichthys olivaceus*) to understand the mechanism against *Edwardsiella tarda*. *Fish Shellfish Immunol.* **2022**, *123*, 265–281. [CrossRef]
16. Camp, K.L.; Wolters, W.R.; Rice, C.D. Survivability and immune responses after challenge with *Edwardsiella ictaluri* in susceptible and resistant families of channel catfish, *Ictalurus punctatus*. *Fish Shellfish Immunol.* **2000**, *10*, 475–487. [CrossRef] [PubMed]
17. Langevin, C.; Blanco, M.; Martin, S.A.; Jouneau, L.; Bernardet, J.F.; Houel, A.; Lunazzi, A.; Duchaud, E.; Michel, C.; Quillet, E.; et al. Transcriptional responses of resistant and susceptible fish clones to the bacterial pathogen *Flavobacterium psychrophilum*. *PLoS ONE* **2012**, *7*, e39126. [CrossRef]
18. Cofre, C.; Gonzalez, R.; Moya, J.; Vidal, R. Phenotype gene expression differences between resistant and susceptible salmon families to IPNV. *Fish Physiol. Biochem.* **2014**, *40*, 887–896. [CrossRef]
19. Reyes-López, F.E.; Romeo, J.S.; Vallejos-Vidal, E.; Reyes-Cerpa, S.; Sandino, A.M.; Tort, L.; Mackenzie, S.; Imarai, M. Differential immune gene expression profiles in susceptible and resistant full-sibling families of Atlantic salmon (*Salmo salar*) challenged with infectious pancreatic necrosis virus (IPNV). *Dev. Comp. Immunol.* **2015**, *53*, 210–221. [CrossRef]
20. Robledo, D.; Taggart, J.B.; Ireland, J.H.; McAndrew, B.J.; Starkey, W.G.; Haley, C.S.; Hamilton, A.; Guy, D.R.; Mota-Velasco, J.C.; Gheyas, A.A.; et al. Gene expression comparison of resistant and susceptible Atlantic salmon fry challenged with Infectious Pancreatic Necrosis virus reveals a marked contrast in immune response. *BMC Genom.* **2016**, *17*, 279. [CrossRef] [PubMed]
21. Pereiro, P.; Tur, R.; García, M.; Figueras, A.; Novoa, B. Unravelling turbot (*Scophthalmus maximus*) resistance to *Aeromonas salmonicida*: Transcriptomic insights from two full-sibling families with divergent susceptibility. *Front. Immunol.* **2024**, *15*, 1522666. [CrossRef] [PubMed]
22. Palaiokostas, C.; Cariou, S.; Bestin, A.; Bruant, J.S.; Haffray, P.; Morin, T.; Cabon, J.; Allal, F.; Vandeputte, M.; Houston, R.D. Genome-wide association and genomic prediction of resistance to viral nervous necrosis in European sea bass (*Dicentrarchus labrax*) using RAD sequencing. *Genet. Sel. Evol.* **2018**, *50*, 30. [CrossRef] [PubMed]
23. Griot, R.; Allal, F.; Phocas, F.; Brard-Fudulea, S.; Morvezen, R.; Bestin, A. Genome-wide Association Studies for Resistance to Viral Nervous Necrosis in Three Populations of European Sea Bass (*Dicentrarchus labrax*) Using a Novel 57k SNP Array DlabChip. *Aquaculture* **2021**, *530*, 735930. [CrossRef]
24. Bjørgen, H.; Koppang, E.O. Anatomy of Teleost Fish Immune Structures and Organs. In *Principles of Fish Immunology*; Buchmann, K., Secombes, C.J., Eds.; Springer: Cham, Switzerland, 2022.
25. Toubanaki, D.K.; Margaroni, M.; Karagouni, E. Development of a Novel Allele-Specific PCR Method for Rapid Assessment of Nervous Necrosis Virus Genotypes. *Curr. Microbiol.* **2015**, *71*, 529–539. [CrossRef] [PubMed]
26. Toubanaki, D.K.; Margaroni, M.; Prapas, A.; Karagouni, E. Development of a Nanoparticle-based Lateral Flow Strip Biosensor for Visual Detection of Whole Nervous Necrosis Virus Particles. *Sci. Rep.* **2020**, *10*, 6529. [CrossRef] [PubMed]
27. Reed, L.J.; Muench, H. A Simple Method of Estimating Fifty Per Cent Endpoints. *Am. J. Epidemiol.* **1938**, *27*, 493–497. [CrossRef]
28. Toubanaki, D.K.; Efstathiou, A.; Tzortzatos, O.P.; Valsamidis, M.A.; Papaharisis, L.; Bakopoulos, V.; Karagouni, E. Nervous Necrosis Virus Modulation of European Sea Bass (*Dicentrarchus labrax*, L.) Immune Genes and Transcriptome towards Establishment of Virus Carrier State. *Int. J. Mol. Sci.* **2023**, *24*, 16613. [CrossRef]
29. Baud, M.; Cabon, J.; Salomoni, A.; Toffan, A.; Panzarin, V.; Bigarré, L. First generic one step real-time Taqman RT-PCR targeting the RNA1 of betanodaviruses. *J. Virol. Methods* **2015**, *211*, 1–7. [CrossRef] [PubMed]
30. Moulos, P.; Hatzis, P. Systematic integration of RNA-Seq statistical algorithms for accurate detection of differential gene expression patterns. *Nucleic Acids Res.* **2015**, *43*, e25. [CrossRef]

31. Fanidis, D.; Moulos, P. Integrative, normalization-insusceptible statistical analysis of RNA-Seq data, with improved differential expression and unbiased downstream functional analysis. *Brief. Bioinform.* **2021**, *22*, bbaa156. [CrossRef] [PubMed]
32. Robinson, M.D.; McCarthy, D.J.; Smyth, G.K. edgeR: A Bioconductor package for differential expression analysis of digital gene expression data. *Bioinformatics* **2010**, *26*, 139–140. [CrossRef]
33. Tang, D.; Chen, M.; Huang, X.; Zhang, G.; Zeng, L.; Zhang, G.; Wu, S.; Wang, Y. SRplot: A free online platform for data visualization and graphing. *PLoS ONE* **2023**, *18*, e0294236. [CrossRef]
34. Szklarczyk, D.; Kirsch, R.; Koutrouli, M.; Nastou, K.; Mehryary, F.; Hachilif, R.; Gable, A.L.; Fang, T.; Doncheva, N.T.; Pyysalo, S.; et al. The STRING database in 2023: Protein-protein association networks and functional enrichment analyses for any sequenced genome of interest. *Nucleic Acids Res.* **2023**, *51*, D638–D646. [CrossRef] [PubMed]
35. Bustin, S.A.; Benes, V.; Garson, J.A.; Hellemans, J.; Huggett, J.; Kubista, M.; Mueller, R.; Nolan, T.; Pfaffl, M.W.; Shipley, G.L.; et al. The MIQE guidelines: Minimum information for publication of quantitative real-time PCR experiments. *Clin. Chem.* **2009**, *55*, 611–622. [CrossRef]
36. Schmittgen, T.D.; Livak, K.J. Analyzing real-time PCR data by the comparative C(T) method. *Nat. Protoc.* **2008**, *3*, 1101–1108. [CrossRef]
37. Tanaka, S.; Mori, K.; Arimoto, M.; Iwamoto, T.; Nakai, T. Protective immunity of sevenband grouper, *Epinephelus septemfasciatus Thunberg*, against experimental viral nervous necrosis. *J. Fish Dis.* **2001**, *24*, 15–22. [CrossRef]
38. Gye, H.J.; Oh, M.J.; Nishizawa, T. Lack of nervous necrosis virus (NNV) neutralizing antibodies in convalescent sevenband grouper *Hyporthodus septemfasciatus* after NNV infection. *Vaccine* **2018**, *36*, 1863–1870. [CrossRef]
39. Huang, Y.; Zhang, Y.; Zheng, J.; Wang, L.; Qin, Q.; Huang, X. Metabolic profiles of fish nodavirus infection in vitro: RGNNV induced and exploited cellular fatty acid synthesis for virus infection. *Cell. Microbiol.* **2020**, *22*, e13216. [CrossRef]
40. Huang, R.; Zhu, G.; Zhang, J.; Lai, Y.; Xu, Y.; He, J.; Xie, J. Betanodavirus-like particles enter host cells via clathrin-mediated endocytosis in a cholesterol-, pH- and cytoskeleton-dependent manner. *Vet. Res.* **2017**, *48*, 8. [CrossRef]
41. Hassan, Z.; Kumar, N.D.; Reggiori, F.; Khan, G. How Viruses Hijack and Modify the Secretory Transport Pathway. *Cells* **2021**, *10*, 2535. [CrossRef] [PubMed]
42. Liu, X.; Hong, T.; Parameswaran, S.; Ernst, K.; Marazzi, I.; Weirauch, M.T.; Fuxman Bass, J.I. Human virus transcriptional regulators. *Cell* **2020**, *182*, 24–37. [CrossRef]
43. Bertheloot, D.; Latz, E.; Franklin, B.S. Necroptosis, pyroptosis and apoptosis: An intricate game of cell death. *Cell. Mol. Immunol.* **2021**, *18*, 1106–1121. [CrossRef]
44. Heckmann, B.L.; Tummers, B.; Green, D.R. Crashing the computer: Apoptosis vs. necroptosis in neuroinflammation. *Cell Death Differ.* **2019**, *26*, 41–52. [CrossRef] [PubMed]
45. Holmes, D.I.; Zachary, I. The vascular endothelial growth factor (VEGF) family: Angiogenic factors in health and disease. *Genome Biol.* **2005**, *6*, 209. [CrossRef] [PubMed]
46. Alkharsah, K.R. VEGF Upregulation in Viral Infections and Its Possible Therapeutic Implications. *Int. J. Mol. Sci.* **2018**, *19*, 1642. [CrossRef] [PubMed]
47. Zhou, B.; Zhang, Y.; Ni, M.; Bai, Y.; Shi, Q.; Zheng, J.; Cui, Z. The involvement of VEGF and VEGFR in bacterial recognition and regulation of antimicrobial peptides in *Eriocheir sinensis*. *Int. J. Biol. Macromol.* **2024**, *270*, 132242. [CrossRef]
48. Sadler, A.J.; Williams, B.R. Interferon-inducible antiviral effectors. *Nat. Rev. Immunol.* **2008**, *8*, 559–568. [CrossRef] [PubMed]
49. Novel, P.; Fernández-Trujillo, M.A.; Gallardo-Gálvez, J.B.; Cano, I.; Manchado, M.; Buonocore, F.; Randelli, E.; Scapigliati, G.; Alvarez, M.C.; Béjar, J. Two Mx genes identified in European sea bass (*Dicentrarchus labrax*) respond differently to VNNV infection. *Vet. Immunol. Immunopathol.* **2013**, *153*, 240–248. [CrossRef]
50. Poisa-Beiro, L.; Dios, S.; Montes, A.; Aranguren, R.; Figueras, A.; Novoa, B. Nodavirus increases the expression of Mx and inflammatory cytokines in fish brain. *Mol. Immunol.* **2008**, *45*, 218–225. [CrossRef]
51. Scapigliati, G.; Buonocore, F.; Randelli, E.; Casani, D.; Meloni, S.; Zarletti, G.; Tiberi, M.; Pietretti, D.; Boschi, I.; Manchado, M.; et al. Cellular and molecular immune responses of the sea bass (*Dicentrarchus labrax*) experimentally infected with betanodavirus. *Fish Shellfish Immunol.* **2010**, *28*, 303–311. [CrossRef]
52. Chaves-Pozo, E.; Guardiola, F.A.; Meseguer, J.; Esteban, M.A.; Cuesta, A. Nodavirus infection induces a great innate cell-mediated cytotoxic activity in resistant, gilthead seabream, and susceptible, European sea bass, teleost fish. *Fish Shellfish Immunol.* **2012**, *33*, 1159–1166. [CrossRef]
53. Gómez-Abellán, V.; Sepulcre, M.P. The role of prostaglandins in the regulation of fish immunity. *Mol. Immunol.* **2016**, *69*, 139–145. [CrossRef]
54. Xue, X.; Caballero-Solares, A.; Hall, J.R.; Umasuthan, N.; Kumar, S.; Jakob, E.; Skugor, S.; Hawes, C.; Santander, J.; Taylor, R.G.; et al. Transcriptome Profiling of Atlantic Salmon (*Salmo salar*) Parr With Higher and Lower Pathogen Loads Following *Piscirickettsia salmonis* Infection. *Front. Immunol.* **2021**, *12*, 789465. [CrossRef]
55. Zhang, Q.; Wang, X.Q.; Jiang, H.S.; Jia, W.M.; Zhao, X.F.; Wang, J.X. Calnexin functions in antibacterial immunity of *Marsupenaeus japonicus*. *Dev. Comp. Immunol.* **2014**, *46*, 356–363. [CrossRef] [PubMed]

56. Fuller, J.R.; Pitzer, J.E.; Godwin, U.; Albertino, M.; Machon, B.D.; Kearse, K.P.; McConnell, T.J. Characterization of the molecular chaperone calnexin in the channel catfish, *Ictalurus punctatus*, and its association with MHC class II molecules. *Dev. Comp. Immunol.* **2004**, *28*, 603–617. [CrossRef] [PubMed]
57. Hung, I.C.; Cherng, B.W.; Hsu, W.M.; Lee, S.J. Calnexin is required for zebrafish posterior lateral line development. *Int. J. Dev. Biol.* **2013**, *57*, 427–438. [CrossRef]
58. Sever, L.; Vo, N.T.; Bols, N.C.; Dixon, B. Rainbow trout (*Oncorhynchus mykiss*) contain two calnexin genes which encode distinct proteins. *Dev. Comp. Immunol.* **2014**, *42*, 211–219. [CrossRef]
59. Huang, Y.; Wang, R.X.; Jiang, F.H.; Xu, X.T.; Shi, Y.; Zhao, Z. A new calnexin modulates antibacterial immune response in obscure puffer *Takifugu obscurus*. *Dev. Comp. Immunol.* **2022**, *127*, 104288. [CrossRef] [PubMed]
60. Huang, Y.; Hui, K.; Jin, M.; Yin, S.; Wang, W.; Ren, Q. Two endoplasmic reticulum proteins (calnexin and calreticulin) are involved in innate immunity in Chinese mitten crab (*Eriocheir sinensis*). *Sci. Rep.* **2016**, *6*, 27578. [CrossRef] [PubMed]
61. Holmes, R.S. Comparative and evolutionary studies of ALDH18A1 genes and proteins. *Chem. Biol. Interact.* **2017**, *276*, 2–8. [CrossRef] [PubMed]
62. Holmes, R.S. Polyploidy among salmonid aldehyde dehydrogenase genes and proteins. *Chem. Biol. Interact.* **2019**, *303*, 22–26. [CrossRef] [PubMed]
63. Tirado-Hurtado, I.; Fajardo, W.; Pinto, J.A. DNA Damage Inducible Transcript 4 Gene: The Switch of the Metabolism as Potential Target in Cancer. *Front. Oncol.* **2018**, *8*, 106. [CrossRef] [PubMed]
64. Li, X.; Fan, K.; Liu, Y.; Liu, Y.; Liu, P.F. Administration of a recombinant ALDH7A1 (rA7) indicates potential regulation of the metabolite and immunology pathways in Atlantic salmon infected with *Aeromonas salmonicida*. *J. Fish Dis.* **2021**, *44*, 961–977. [CrossRef]
65. Woodson, C.M.; Kehn-Hall, K. Examining the role of EGR1 during viral infections. *Front. Microbiol.* **2022**, *13*, 1020220. [CrossRef]
66. Wei, S.; Huang, Y.; Huang, X.; Qin, Q. Characterization of c-Jun from orange-spotted grouper, *Epinephelus coioides* involved in SGIV infection. *Fish Shellfish Immunol.* **2015**, *431*, 230–240. [CrossRef] [PubMed]
67. Foletta, V.C.; Segal, D.H.; Cohen, D.R. Transcriptional regulation in the immune system: All roads lead to AP-1. *J. Leukoc. Biol.* **1998**, *63*, 139–152. [CrossRef] [PubMed]
68. Xie, H.; Hu, J.; Wang, Y.; Wang, X. Identification of the matrix metalloproteinase (MMP) gene family in Japanese flounder (*Paralichthys olivaceus*): Involved in immune response regulation to temperature stress and *Edwardsiella tarda* infection. *Fish Shellfish Immunol.* **2023**, *139*, 108878. [CrossRef]
69. Jiang, Y.; Abernathy, J.W.; Peatman, E.; Liu, H.; Wang, S.; Xu, D.H.; Kucuktas, H.; Klesius, P.; Liu, Z. Identification and characterization of matrix metalloproteinase-13 sequence structure and expression during embryogenesis and infection in channel catfish (*Ictalurus punctatus*). *Dev. Comp. Immunol.* **2010**, *34*, 590–597. [CrossRef] [PubMed]
70. Matsuyama, T.; Fujiwara, A.; Nakayasu, C.; Kamaishi, T.; Oseko, N.; Hirono, I.; Aoki, T. Gene expression of leucocytes in vaccinated Japanese flounder (*Paralichthys olivaceus*) during the course of experimental infection with *Edwardsiella tarda*. *Fish Shellfish Immunol.* **2007**, *22*, 598–607. [CrossRef]
71. Beck, I.M.; Rückert, R.; Brandt, K.; Mueller, M.S.; Sadowski, T.; Brauer, R.; Schirmacher, P.; Mentlein, R.; Sedlacek, R. MMP19 is essential for T cell development and T cell-mediated cutaneous immune responses. *PLoS ONE* **2008**, *3*, e2343. [CrossRef] [PubMed]
72. Lin, H.; Zhou, Z.; Zhao, J.; Zhou, T.; Bai, H.; Ke, Q.; Pu, F.; Zheng, W.; Xu, P. Genome-Wide Association Study Identifies Genomic Loci of Sex Determination and Gonadosomatic Index Traits in Large Yellow Croaker (*Larimichthys crocea*). *Mar. Biotechnol.* **2021**, *23*, 127–139. [CrossRef]
73. Martins, R.S.; Pinto, P.I.; Guerreiro, P.M.; Zanuy, S.; Carrillo, M.; Canário, A.V. Novel galanin receptors in teleost fish: Identification, expression and regulation by sex steroids. *Gen. Comp. Endocrinol.* **2014**, *205*, 109–120. [CrossRef]
74. Jørgensen, S.M.; Afanasyev, S.; Krasnov, A. Gene expression analyses in Atlantic salmon challenged with infectious salmon anemia virus reveal differences between individuals with early, intermediate and late mortality. *BMC Genom.* **2008**, *9*, 179. [CrossRef]

viruses

MDPI

Article

scTRIM44 Positively Regulated Siniperca Chuatsi Rhabdovirus Through RIG-I- and MDA5-Mediated Interferon Signaling

Yinjie Niu, Xinmei Yang, Hongru Liang, Xia Luo, Baofu Ma, Qiang Lin, Xiaozhe Fu and Ningqiu Li *

Ministry of Agriculture and Rural Affairs, Guangdong Provincial Key Laboratory of Aquatic Animal Immunology and Sustainable Aquaculture, Key Laboratory of Fishery Drug Development, Pearl River Fisheries Research Institute, Chinese Academy of Fishery Sciences, Guangzhou 510380, China; niuyinjie0530@163.com (Y.N.); yxm122623@163.com (X.Y.); hrliang13@126.com (H.L.); lxwenhao@163.com (X.L.); mabf@prfri.ac.cn (B.M.); lin9902057@163.com (Q.L.); fuxiaozhe-1998@163.com (X.F.)

* Correspondence: liningq@126.com

Abstract: Tripartite Motif-Containing 44 (TRIM44) is responsible for cancers, neurodegenerative diseases, and viral infections. However, the role of *Siniperca chuatsi* TRIM44 (scTRIM44) during viral infection remains unclear. In the present study, we analyzed the molecular characteristics of scTRIM44 and its role in infectious spleen and kidney necrosis virus (ISKNV), largemouth bass virus (LMBV), and Siniperca chuatsi rhabdovirus (SCRV) infection. ScTRIM44 contained one B-box domain (B, 166–207 aa) and a coiled-coil domain (CC, 279–309 aa), but lacked the canonical RING domain of E3 ubiquitin ligases. The scTRIM44 mRNA was expressed relatively high in immune-related tissues. The mRNA expression of scTRIM44 significantly decreased in vivo and vitro post-ISKNV and -LMBV infection. However, the expression of scTRIM44 mRNA showed significant up-regulation post-SCRV infection. ScTRIM44 positively regulated SCRV infection in CPB cells, but copies of ISKNV and LMBV showed no significant alteration in over-expressed or knocked-down scTRIM44 cells. Moreover, scTRIM44 positively regulated RIG-I- and MDA5-mediated interferon molecule signaling. These data suggested that scTRIM44 promoted SCRV infection by positively regulating RIG-I- and MDA5-mediated interferon molecule signaling, but didn't regulate ISKNV and LMBV infection. This research provided a comprehensive insight into the antiviral activity of scTRIM44.

Keywords: scTRIM44; LMBV; ISKNV; SCRV; mandarin fish

1. Introduction

The mandarin fish (*Siniperca chuatsi*) is a universally cultured species in Asian countries with delicious taste, abundant nutrition, and high market value [1]. Infectious spleen and kidney necrosis virus disease (ISKNVD), Siniperca chuatsi rhabdovirus disease (SCRVD), and largemouth bass virus disease (LMBVD) are major viral diseases; they have caused significant economic losses and hindered the sustainable development of the mandarin fish industry [2–5]. ISKNV and LMBV are both enveloped, icosahedral double-stranded DNA viruses, which belong to the *Iridoviridae* family [6,7]. SCRV is a negative-sense single-stranded RNA, which belongs to the *Rhabdoviridae* family [8]. The tripartite motif (TRIM) proteins induced by IFNs play an important role in the mammalian immune defense against microbial pathogens infection [9].

The TRIM protein family is characterized by three conserved N-terminal domains (named RBCC domains), including a canonical RING domain (R), one or two B-box domains (B), and one coiled-coil domain (CC) [10,11]. TRIM proteins containing the RING-finger domain are defined as the E3 ubiquitase. Recently, the expression of TRIM genes in many fish has been shown to become induced by viral infection or poly I:C in vitro and in vivo, including TRIM8, TRIM13, TRIM16L, TRIM32, TRIM35, TRIM39, TRIM62, and TRIM47 [12,13]. TRIM23 regulated RIG-I/MDA5-mediated antiviral innate response through ubiquitination. *Siniperca chuatsi* TRIM59 (scTRIM59) negatively regulated ISKNV,

SCRV infection, and IRF3/IRF7-mediated signal genes [14]. The tripartite motif-containing 44 (TRIM44) protein is a new member of the TRIM family, which was cloned from a mouse brain cDNA library in 2001 [15]. TRIM44 has a zinc finger domain in the N-terminal region, but lacks the RING finger domain, it is an atypical TRIM family protein with the activity of ubiquitin-specific proteases (USps) [16]. Many studies reported that TRIM44 is involved in multiple disorders such as neurodegenerative diseases, cancers, and viral infections [17,18]. Yang et al. demonstrated that TRIM44 promoted virus-triggered IFN production through increasing the RIG-mediated signaling pathway [19]. When a viral pathogen entered the body, both RIG-1 and MDA5 caused the conformational change after recognizing the released RNA. Subsequently, the RIG-1 and MDA5 interacted with the Virus-Induced Signaling Adapter (VISA), which can activate IRF3 and IRF7 to promoting the production of type I interferons [20]. Zheng et al. reported that the expression of Singapore grouper iridovirus (SGIV) MCP and VP19 had no significant changes, but the CP and RdRp expression of red-spotted grouper nervous necrosis virus (RGNNV) was significantly up-regulated in EcTRIM44-over-expressing cells [21]. However, the molecular characteristics and antiviral activities of scTRIM44 are not fully understood.

In this study, we analyzed the molecular characteristics and structures of scTRIM44, and explored the changes in scTRIM44 mRNA post-ISKNV, -LMBV, and -SCRV infection. Moreover, we elucidated the antiviral activity of scTRIM44 in ISKNV, LMBV, and SCRV infection, and the regulation of scTRIM44 on interferon-mediated signaling molecules. These data provided a novel insight into the antiviral activity of scTRIM44 through interferon response signaling.

2. Materials and Methods

2.1. Cells and Virus Strains

We constructed the CPB (Chinese perch brain cell line) cells and cultured them in L-15 medium with 8% FBS [22]. The LMBV, ISKNV, and SCRV virus strains were isolated in our laboratory [23,24].

2.2. Molecular Analysis of scTRIM44

The nucleotide sequence of the scTRIM44 coding region was translated into an amino acid (aa) sequence with MEGA7.0.26. The structural characteristics of scTRIM44 amino acids were predicated by TMHMM (http://www.cbs.dtu.dk/services, accessed on 9 August 2023).

The aa sequences of scTRIM44 and 17 other fish species were aligned by using Clustal W. Then, we constructed the phylogenetic tree of 18 scTRIM44 protein sequences by using the neighbor-joining algorithm of MEGA 7.0.26 (bootstrap replications: 1000).

2.3. The Distribution of scTRIM44 mRNA in Mandarin Fish

To investigate the expression of scTRIM44 mRNA in healthy fish, the brain, heart, liver, stomach, hemocyte, intestine, head kidney, trunk kidney, and spleen were collected from three healthy fish (9 ± 0.5 cm). These organs were chosen based on immunological functions and the targets for replication by the three chosen viruses [24]. The scTRIM44 mRNA was detected with the SYBR Premix Ex Taq kit (Takara Bio, Ostsu, Japan).

2.4. The Changes in scTRIM44 mRNA Post-SCRV, -ISKNV, and -LMBV Infection

Ninety mandarin fish (9 ± 1 cm) were divided into three groups (A, B, C). First, three fish of each group were taken out as the negative sample at 0 h; then the remaining fish in the A, B, and C groups were injected with SCRV (10^4 TCID50/fish), ISKNV (10^5 TCID50/fish), or LMBV (10^4 TCID50/fish), respectively. The spleens of fish infected with SCRV were collected at 0, 3, 6, 12, 24, 48, and 72 h post-infection. The spleens of fish infected with ISKNV were collected at 0, 3, 6, 12, 24, 48, 72, 96, and 168 h post-infection. The spleens of fish infected with LMBV were collected at 0, 3, 6, 12, 24, 48, 72, 96, and 120 h post-infection. Three fishes were dissected at each indicated time.

The CPB cells were incubated with SCRV (MO1 = 0.1), ISKNV (MO1 = 1), and LMBV (MOI = 0.1). Three parallel replicate cultures of CPB cells were collected at 0, 1, 2, 4, 6, and 8 h post-SCRV infection, respectively. Three parallel wells were collected at 0, 2, 4, 6, 8, 12, 48, and 72 h post-ISKNV infection, respectively. The CPB cells infected with LMBV were gathered at 0, 3, 6, 12, 24, 48, and 72 h post-infection.

2.5. The Effect of scTRIM44 on SCRV, ISKNV, and LMBV Infection

The siRNAs were synthesized at Genepharma. scTRIM44 was knocked down by transfecting 25 nM siRNA with TransIntro™ EL transfection reagent according to the manufacturer's instructions. The RNA of cells transfected with siRNAs were extracted and reversed, and gene targets amplified and quantified by qPCR. Then, the cells transfected with siRNA were infected with ISKNV (MOI = 0.1), SCRV (MOI = 0.001), and LMBV (MOI = 0.01), respectively.

The CDs (coding sequences, CDs) of scTRIM44 were amplified by PCR, then ligated to the parallel eukaryotic expression vector pCMV-EGFR-scTRIM44. The over-expressing scTRIM44 cells were constructed by transfecting pCMV-EGFP-scTRIM44 with FuGENE®6 and selecting with 0.1 mg/mL geneticin. Then, over-expressed scTRIM44 cells were infected with ISKNV (MOI = 0.1), SCRV (MOI = 0.001), and LMBV (MOI = 0.01), respectively.

2.6. TaqMan Real-Time PCR and Quantitative RT-PCR

We detected the copies of SCRV RNA, ISKNV DNA, and LMBV DNA with TaqMan PCR methods. TaqMan-PCR amplification was performed as follows: 2 × Premix (Takara, Japan) Ex Taq™ 10 μL, forward primer 0.4 μL, and reverse primer 0.4 μL, probe 0.4 μL, 50 × Rox Reference Dye II 0.4 μL, DNA 2 μL, and ddH_2O 6.4 μL.

The relative expression level of scTRIM44 mRNA was detected by quantitative RT-PCR, then was calculated by using a $2^{-\Delta\Delta CT}$ method relative to mean expression in uninfected fish or uninfected CPB cells (time point 0); the 18S rRNA was used as the reference gene. The used primers are listed as Table 1.

Table 1. The primers used in this study.

Primer Name	Sequence	
TRIM44-F	5′ ATGGACCACAAAGGGGAAC 3′	
TRIM44-R	5′ TCAGGGGGCGTGGTCCATG 3′	
pCMV-TRIM44-F	5′ CG**GAATTC**ATCGACCACAAAGGGGAAC 3′	EcoRI
pCMV-TRIM44-R	5′ GG**GGTACC**TCAGGGGGCGTGGTCCATC 3′	KpnI
ScTRIM44-RT-F	ACTTGGCACCAAAAGAGACTCC	
ScTRIM44-RT-R	TCTCACTGTGTCCCTCTTCCCA	
TRIM44-1-sense	GCUGAAACAAGAGGAACUUTT	Trim44-siRNA
TRIM44-1-antisense	UAGUGCCAAGUCCAUUAGCTT	
TRIM44-2-sense	GCUAAUGGACUUGGCACUATT	
TRIM44-2-antisense	UAGUGCCAAGUCCAUUAGCTT	
TRIM44-3-sense	GGAGGAGAAGAGGACCCUUTT	
TRIM44-3-antisense	AAGGGUCCUCUUCUCCUCCTT	
NC-sense	UUCUCCGAACGUGUCACGUTT	
NC-antisense	ACGUGACACGUUCGGAGAATT	
IRF3-RT-F	GTCTACAGCCCTGAACTCAACGG	RT-PCR
IRF3-RT-R	AAATCTCTTGGGGCTGTGTGGTC	
IRF7-RT-F	AGTTCACCTCTGCAGCCATGTAT	
IRF7-RT-R	GTTAAGGACGCGGTTGGTGAAAT	
RIG-I-RT-F	AAGTGCAAGATGTTTGCGTGTC	
RIG-I-RT-R	GAAGTTGATGGGCTTTCTGTGAG	
MAD5-RT-F	CTCCCGACAGGAAGTGGTAAA	
MAD5-RT-R	GCGGAATAATGCTGCTCAAC	
IFN-α-RT-F	TGAGGATGCTGGAGTGACC	
IFN-α-RT-R	GCCTGCCGAGTAACATTGAC	
IFN-β-RT-F	ACGGATCTCAAGTCAGGGTC	
IFN-β-RT-R	TGAGTAGGGTATGAGGGCATT	
18S-F	CATTCGTATTGTGCCGCTAGA	
18S-R	CAAATGCTTTCGCTTTGGTC	

2.7. Statistical Analysis

The fish and cell experiments were performed in triplicate. Statistical data were analyzed by one-way analysis of variance (ANOVA) (expressed as mean ± SD). All data were compared using SPSS 13.0 (SPSS, Chicago, IL, USA). The relative expression of scTRIM44 expression was relative to time-zero. $p < 0.05$ represented a statistical difference, $p < 0.01$ represented a significant difference, $p < 0.001$ represented an extremely significant difference.

3. Results

3.1. Molecular Organization and Characteristics of ScTRIM44

Multiple sequence alignment indicated that TRIM44 differed considerably among different species (Figure 1). Phylogenetic analysis suggested that the relative of scTRIM44 was close to *Sebastes umbrosus* and *Micropterus dolomieu* and belonged to the *Perciformes* branch (Figure 2). The TRIM44 identity in the Perciformes branch was 44.23%, and the conservative regions were predominately located in the C-terminal (1–83 aa) and 157–328 aa. The identity was 90.90%, 42.89%, 26%, and 23% between scTRIM44 and TRIM44 of *Micropterus dolomieu, Danio rerio, Rana temporaria,* and *Homo sapiens*, respectively (Figure 1).

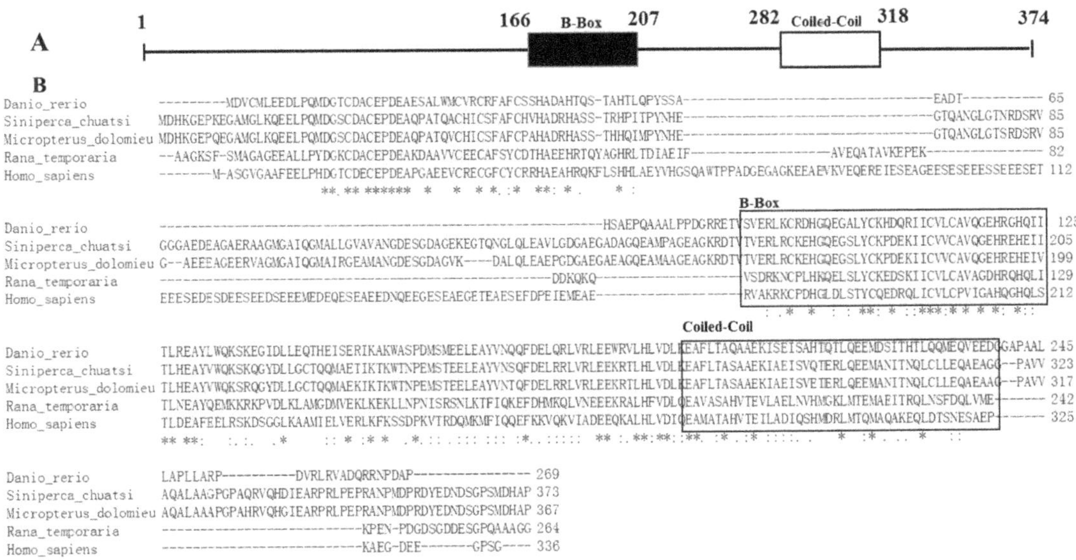

Figure 1. The molecular characteristics of scTRIM44. A: The structure diagram of scTRIM44. B: The aa sequence and structure component of scTRIM44. * represented the identity; . and : represented the similarity.

The trans-membrane analysis indicated that scTRIM44 was an intracellular protein. It contained a B-box-type zinc finger superfamily (Bbox_SF), one B-box domain (B, 166–207 aa), and a coiled-coil domain (CC, 279–309 aa), but lacked the canonical RING domain. The domains (B and CC) of scTRIM44 were 95.77%, 70.43, 56.34%, and 33.81% identical with *Micropterus dolomieu, Danio rerio, Rana temporaria* and *Homo sapiens*, respectively (Figure 1). The structure of scTRIM44 was similar to mammalian TRIM44, which indicated that scTRIM44 had similar antiviral function of mammalian TRIM44.

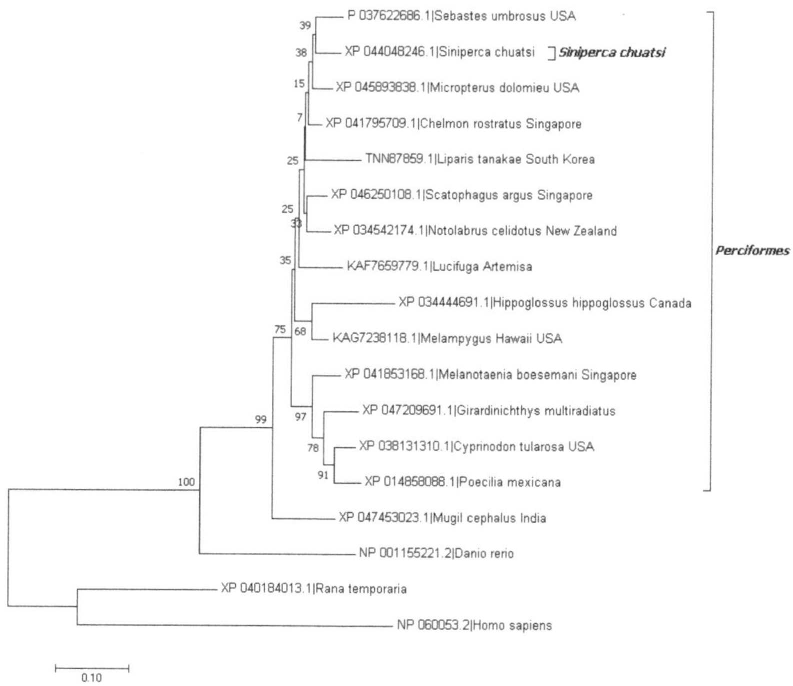

Figure 2. The phylogenetic tree of 18 TRIM44 aa sequences constructed by using MEGA 7.0.

3.2. Expression of scTRIM44 in *Siniperca chuatsi*

The expression of scTRIM44 mRNA can be detected in the heart, liver, stomach, brain, hemocyte, intestine, trunk kidney, head kidney, and spleen. Especially, the expression of scTRIM44 mRNA was relatively abundant in immune organs, including the head kidney, trunk kidney, and spleen. This result indicated that scTRIM44 plays an important role in the immune function of Mandarin fish (Figure 3).

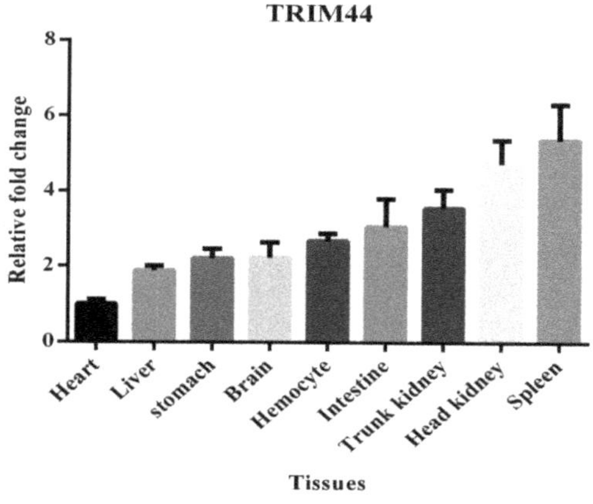

Figure 3. The scTRIM44 mRNA expression in different tissues of mandarin fish. The relative fold change of scTRIM44 was calculated relative to that of the heart.

3.3. The Effect of SCRV Infection on scTRIM44

The changes of scTRIM44 mRNA in spleen and CPB cells were detected by qRT-PCR. The expression of scTRIM44 mRNA was rapidly up-regulated at 1 and 2 h, and it increased slightly at 6 and 8 h in CPB cells post-SCRV infection (Figure 4A); The scTRIM44 mRNA significantly increased at 6 h, and it showed a slight up-regulation at 3, 12, 48, and 72 h in spleen post-SCRV infection (Figure 4B). Those results indicated that the expression was up-regulated in vivo and vitro post-SCRV infection.

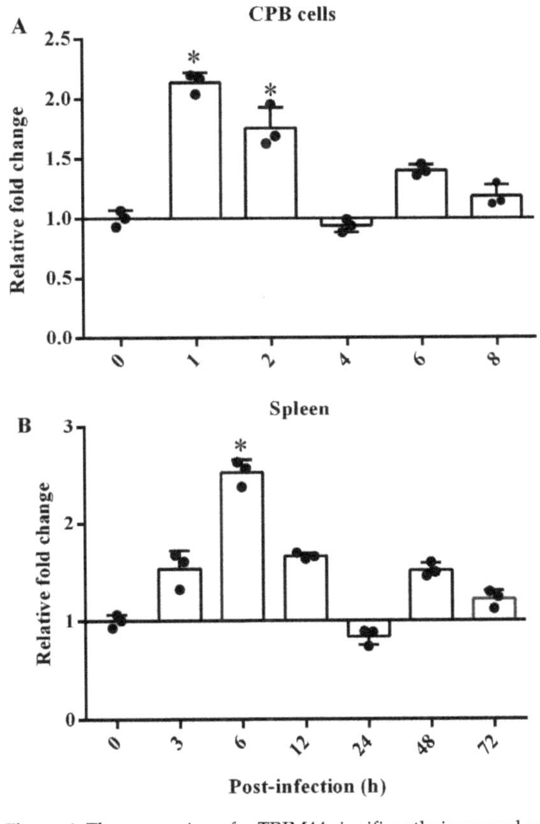

Figure 4. The expression of scTRIM44 significantly increased post-SCRV infection. (**A**): scTRIM44 in cells infected with SCRV was detected by using qRT-PCR; the expression of scTRIM44 represented up-regulation at 1, 2, 6, and 8 h in CPB cells post-SCRV infection. (**B**): The scTRIM44 mRNA also was up-regulated at 3, 6, 12, 48, and 72 h in spleen post-SCRV infection. *: $p < 0.05$.

3.4. The Changes in scTRIM44 Post-ISKNV or -LMBV Infection

To determine the changes of scTRIM44 mRNA induced by ISKNV or LMBV infection, we analyzed the expression of scTRIM44 mRNA in CPB cells and spleen post-ISKNV or LMBV infection, respectively. In CPB cells, the expression of scTRIM44 mRNA was significantly reduced at 4, 8, 48, and 72 h and the expression of scTRIM44 mRNA showed a slightly down-regulation at 6 h, but scTRIM44 mRNA was significantly up-regulated at 12 h post-ISKNV infection (Figure 5A). In spleen, the expression of scTRIM44 mRNA was markedly down-regulated at 3, 6, 12, 48, 96, and 168 h, and the scTRIM44 mRNA slightly increased at 72 h post-ISKNV infection (Figure 5B). Overall, the expression of scTRIM44 mRNA was significantly down-regulated in vivo and in vitro post-ISKNV infection.

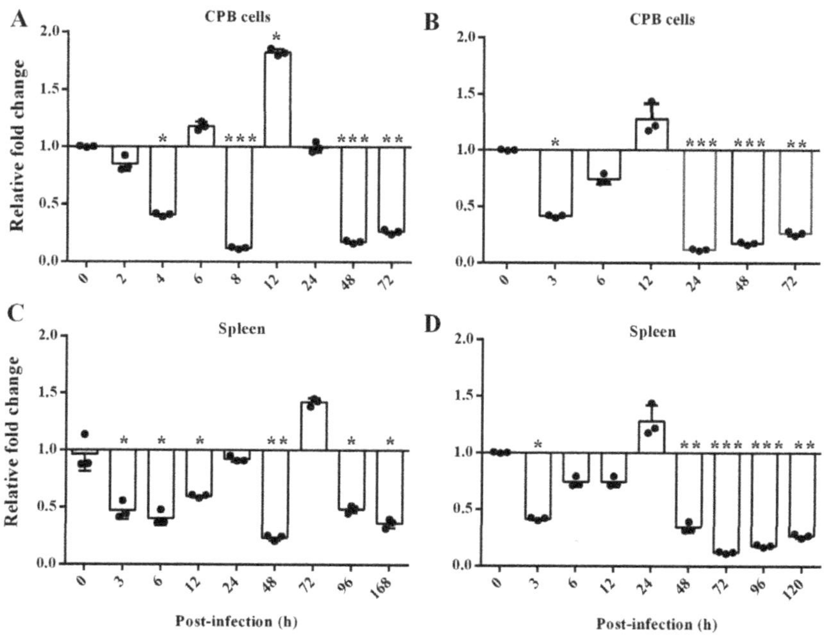

Figure 5. The response of scTRIM44 mRNA was significantly down-regulated post-ISKNV and LMBV infection. (**A**): The expression of scTRIM44 mRNA was significantly down-regulated at 4, 8, 48, and 72 h in CPB cells; (**C**): The scTRIM44 mRNA significantly decreased at 3, 6, 12, 48, 96, and 168 h in spleen after ISKNV infection; (**B**): The expression of scTRIM44 mRNA markedly decreased in CPB cells at 3, 24, 48, and 72 h post-LMBV infection; (**D**): The scTRIM44 mRNA significantly decreased in spleen at 3, 48, 72, 96, and 120 h post-LMBV infection. The changes in scTRIM44 were detected by using qRT-PCR post-ISKNV infection, *: $p < 0.05$; **: $p < 0.01$; ***: $p < 0.001$.

In cells, the expression of scTRIM44 mRNA was significantly decreased at 3, 24, 48, and 72 h, and the scTRIM44 mRNA was slightly down-regulated at 6 h post-LMBV infection. In spleen, the expression of scTRIM44 mRNA was markedly down-regulated at 3, 48, 72, 96, and 120 h, and the expression of scTRIM44 mRNA showed a slight down-regulation at 6 and 12 h post-LMBV infection. This result indicated that expression of scTRIM44 mRNA presented a significantly down-regulation in cells and spleen post-LMBV infection.

3.5. scTRIM44 Positively Regulated SCRV Infection but Did Not Regulate ISKNV and LMBV Infection

The copies of the SCRV genome were significantly up-regulated in over-expressing scTRIM44 cells; however, the copies of the SCRV genome were markedly decreased in knocked-down scTRIM44 cells by transfecting siRNA (Figure 6A,B). The copies of the ISKNV and LMBV genomes represented no significant change in over-expressed scTRIM44 cells (Figure 6A,B). After knocking down scTRIM44, there was also no significant change of the copies of the ISKNV and LMBV genomes (Figure 6A,B). These results indicated that scTRIM44 positively regulated SCRV infection but was not involved in ISKNV and LMBV infection.

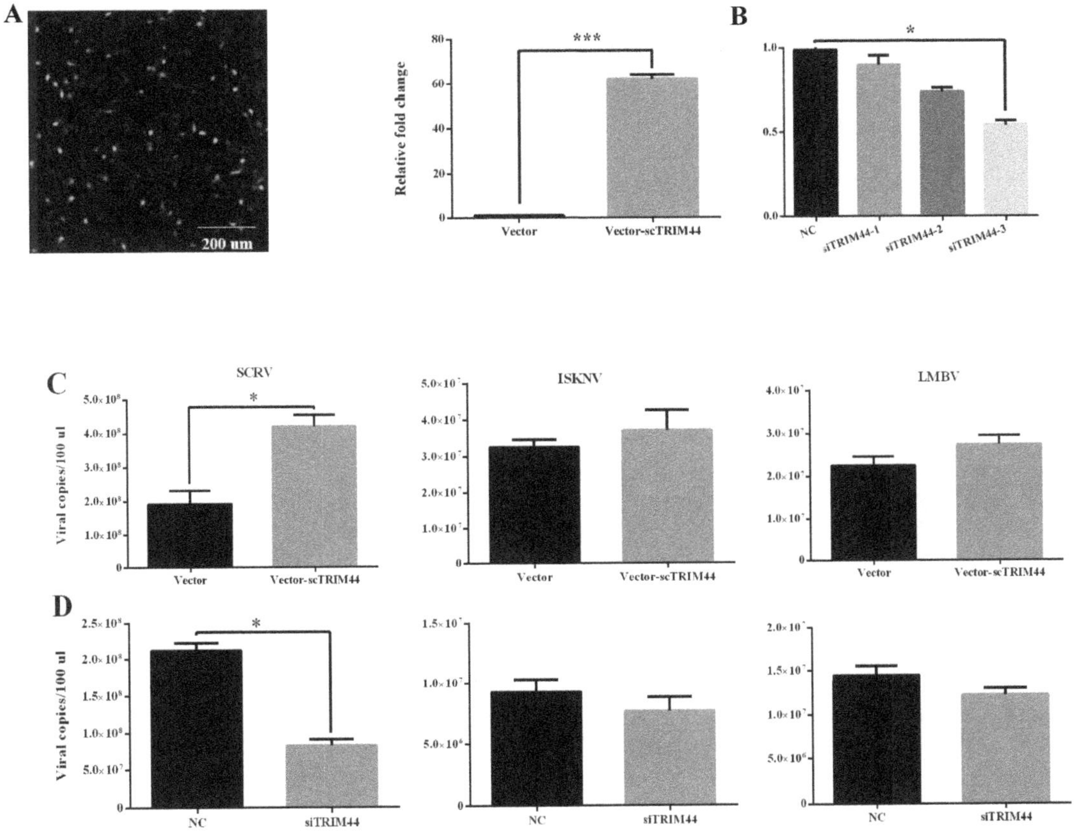

Figure 6. scTRIM44 positively regulated SCRV infection but didn't regulate ISKNV and LMBV infection. (**A**): The green fluorescence showed that in over-expressing TRIM44 cells, the scTRIM44 mRNA in over-expressing scTRIM44 cells was significantly up-regulated compared to the vector group. (**B**): The expression of TRIM44 mRNA was markedly down-regulated in cells transfected with siTRIM44-3. (**C**): The viral copies of the SCRV genome significantly decreased in over-expressing scTRIM44 cells, and the viral copies of the ISKNV and LMBV genome did not change in over-expressing scTRIM44 cells. (**D**): The viral copies of the SCRV genome were markedly down-regulated in cells transfected with siTRIM44-3, but the viral copies of ISKNV and LMBV genome were no significant differences compared with NC group, NC: negative control. *: $p < 0.05$; ***: $p < 0.001$.

3.6. scTRIM44 Positively Regulated the RIG-I- or MDA5-Mediated Interferon Pathway

We detected the mRNA expression of RIG-I, MDA5, IRF3, IRF7, IFN-α, and IFN-β in over-expressing scTRIM44 or knocked-down scTRIM44 cells. The expression of RIG-I, IRF3, IRF7, IFN-α, and IFN-β in the mRNA level was significantly increased in over-expressed scTRIM44 cells (Figure 7). After knocking down scTRIM44, the expression of RIG-I, IRF3, IRF7, IFN-α, and IFN-β mRNA were significantly down-regulated (Figure 7). These results suggested that scTRIM44 may regulate the interferon pathway induced by RIG-I or MDA5.

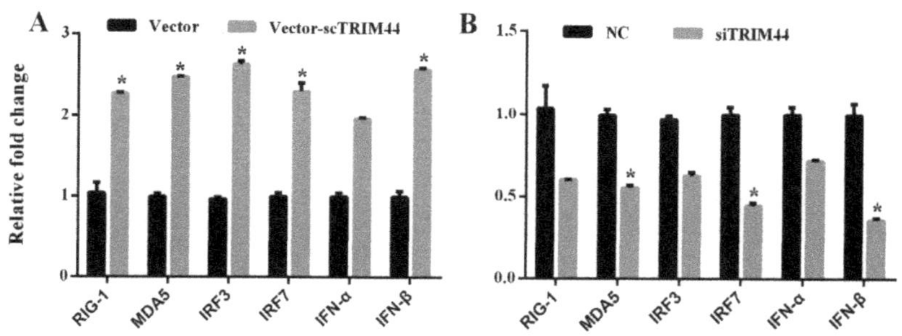

Figure 7. scTRIM44 positively regulated RIG-I- and MDA5-mediated interferon signaling molecules. (**A**): Compared to the cells transfected with the empty vector, the expression of RIG-I, MDA5, IRF3, IRF7, IFN-α, and IFN-β mRNA were up-regulated in over-expressiing scTRIM44 cells. (**B**): The expression of RIG-I, MDA5, IRF3, IRF7, IFN-α, and IFN-β mRNA expression were down-regulated in cells transfected with siTRIM44. NC: negative control. *: $p < 0.5$.

4. Discussion

Many studies have indicated that TRIM family proteins possessed antiviral activity by directly effecting viral replication or via regulation of innate immune pathways [9,25]. Several fish TRIM proteins have proven to regulate fish viral infection through interferon-related signaling [26,27]. Here, the expression of scTRIM44 mRNA in many mandarin fish tissues can be detected, and it showed relatively high expression in immune-related organs. After ISKNV and LMBV infection, mRNA of scTRIM44 in spleen and cells appeared down-regulated, but the expression of scTRIM44 mRNA was up-regulated in CPB cells and spleen infected with SCRV. ScTRIM44 positively regulated SCRV infection but didn't regulate ISKNV and SCRV infection. Meanwhile, scTRIM44 positively regulated the RIG-I- and MDA-5-mediated interferon pathways. These results suggested that scTRIM44 positively regulated SCRV infection through RIG-I- and MDA-5-mediated interferon signaling.

The C-terminal of the TRIM family protein consists of a RING zinc finger domain, one or two B-Box zinc finger structures, and a coiled-coil region, and the RING domain possesses E3 ubiquitin ligase activity [28]. TRIM44 is a member of the TRIM family, but it lacks the RING domain. The TRIM44 of orange spotted grouper contained the conserved B-Box domain and coiled-coil domain, but not a RING domain [21]. scTRIM44 also contained B-box-type zinc finger superfamily (Bbox_SF), one B-box domain (B, 166–207 aa), and the coiled-coil domain (CC, 279–309 aa), but lacked the canonical RING domain. The domains (B and CC) of scTRIM44 were 95.77%, 70.43, 56.34%, and 33.81% identical with *Micropterus dolomieu*, *Danio rerio*, *Rana temporaria*, and *Homo sapiens*, respectively. These results indicated that the variations in TRIM44 aa sequences were large among different species; however, the motif characteristics of fish TRIM44 structure was similar to that of mammal TRIM44, which indicated that scTRIM44 possessed a similar function to mammalian TRIM44.

The up-regulation of TRIM44 can be detected in various kinds of cancer, such as osteosarcoma, gastric cancer, lung cancer, breast cancer, papillary thyroid cancer, testicular germ cell tumor, gastric cancer, and prostate cancer, which suggested that TRIM44 presented a broad spectrum of expression in tissues [29,30]. ScTRIM44 mRNA can also be detected in different tissues, with especially high expression in immune organs; this phenomenon is similar to the expression of TRIM59 mRNA in mandarin fish [14]. The expression of scTRIM44 mRNA was significantly down-regulated in cells and spleen infected with ISKNV and LMBV, but the expression of scTRIM44 mRNA increased in cells and spleen post-SCRV infection. The changing trends of scTRIM44 mRNA was consistent with scTRIM59 infected with ISKNV [14]. These results indicated that scTRIM44 played different roles in ISKNV, LMBV, and SCRV infection.

Studies have suggested that TRIM44 played important regulatory roles in type I IFN signaling and antiviral immunity [16,31]. Yang et al. demonstrated that TRIM44 can positively regulate the virus-triggered immune response by enhancing the stability of VISA [19]. However, grouper EcTRIM44L was proven to increase virus replication by inhibiting the MDA5- or MAVS-induced interferon immune response [21]. The TRIM protein usually acts as ubiquitinase to regulating RIG-I and MDA5, which can recruite IRF3/IRF7 to promoting the production of type I interferons. In this study, scTRIM44 promoted the expression of RIG-I, MDA5, IRF3, IRF7, IFN-α, and IFN-β mRNA. This result indicated that scTRIM44 positively regulated RIG-I- or MDA5-mediated interferon immune response, which was consistent with Yang's report. The expression of RGNNV CP and RdRp were significantly up-regulated in EcTRIM44L-over-expressing cells; however, the expression of SGIV MCP and VP19 showed no significant changes in EcTRIM44L-over-expressing cells [21]. In the present study, the copies of the SCRV genome were significantly increased in scTRIM44-over-expressing cells, but the copies of ISKNV and LMBV genome showed no significant alterations in over-expressing scTRIM44 cells. RGNNV and SCRV are both RNA viruses, while SGIV, ISKNV, and LMBV are double-stranded DNA viruses belonging to iridoviridae family [32]. These results showed that fish TRIM44 positively regulated fish RNA virus infection, but it showed no effect on fish double-stranded DNA viruses. The reason for this different antiviral activity of scTRIM44 against fish RNA viruses and double-stranded DNA viruses will need further studies.

Overall, scTRIM44 aa was 90.90%, 42.89%, 26%, and 23% identical with *Micropterus dolomieu*, *Danio rerio*, *Rana temporaria*, and *Homo sapiens*, respectively. The scTRIM44 consisted of one B-box domain, and a coiled-coil domain, but not the RING domain. The expression of scTRIM44 mRNA was significantly down-regulated in cells and spleen infected with ISKNV and LMBV, but the expression of scTRIM44 mRNA increased in cells and spleen post-SCRV infection. scTRIM44 negatively regulated RIG-I- or MDA5-mediated interferon immune response. The viral copies of the SCRV genome were significantly increased in scTRIM44-over-expressing cells, but there was no significant change of the expression of the ISKNV and LMBV genome in over-expressing scTRIM44 cells. These results suggested that fish TRIM44 positively regulated fish RNA virus infection, but there was no influence on fish double-stranded DNA virus infection. This data provided the basis for further studying the antiviral mechanism of scTRIM44 and a novel insight into the antiviral activity of TRIM44 through the interferon response signaling.

Author Contributions: Conceptualization, B.M.; Methodology, X.Y.; Software, H.L.; Validation, X.L.; Formal analysis, Q.L. and X.F.; Writing—original draft, Y.N.; Writing—review & editing, N.L. All authors have read and agreed to the published version of the manuscript.

Funding: This study was supported by the National Key Research and Development Program of China, (2023YFD2400701) and Central Public-interest Scientific Institution Basal Research Fund, CAFS (NO. 2023TD48).

Institutional Review Board Statement: The animal study protocol was approved by the Laboratory Animal Ethics Committee of Pearl River Fisheries Research Institute, CAFS (LACE-PRFRI-2024-03-45 and 2023.03.24).

Informed Consent Statement: Not applicable.

Data Availability Statement: The datasets used and/or analyzed during the current study are available from the corresponding author upon reasonable request.

Conflicts of Interest: The authors declare no conflicts of interest.

References

1. Bao, Y.; Zhang, Y.; Xu, W. Effects of Different Freezing Rate and Frozen Storage Temperature on the Quality of Large-Mouth Bass (*Micropterus salmoides*). *Molecules* **2023**, *28*, 5432. [CrossRef] [PubMed]
2. Zhao, X.; Petrashen, A.P.; Sanders, J.A.; Peterson, A.L.; Sedivy, J.M. SLC1A5 glutamine transporter is a target of MYC and mediates reduced mTORC1 signaling and increased fatty acid oxidation in long-lived Myc hypomorphic mice. *Aging Cell* **2019**, *18*, e12947. [CrossRef]

3. Wang, Y.Q.; Lü, L.; Weng, S.P.; Huang, J.N.; Chan, S.-M.; He, J.G. Molecular epidemiology and phylogenetic analysis of a marine fish infectious spleen and kidney necrosis virus-like (ISKNV-like) virus. *Arch. Virol.* **2007**, *152*, 763–773. [CrossRef]
4. Xu, Z.; Liao, J.; Zhang, D.; Liu, S.; Zhang, L.; Kang, S.; Xu, L.; Chen, H.; Peng, W.; Zhou, S.; et al. Isolation, Characterization, and Transcriptome Analysis of an ISKNV-Like Virus from Largemouth Bass. *Viruses* **2023**, *15*, 398. [CrossRef]
5. Fu, X.; Ming, Y.; Li, C.; Niu, Y.; Lin, Q.; Liu, L.; Liang, H.; Huang, Z.; Li, N. *Siniperca chuatsi* rhabdovirus (SCRV) induces antophagy via PI3K/Akt-mTOR pathway in CPB cells. *Fish Shellfish Immunol.* **2020**, *2020*, 381–388. [CrossRef] [PubMed]
6. He, J.G.; Denga, M.; Weng, S.P.; Lia, Z.; Zhou, S.Y.; Long, Q.X.; Wang, X.Z.; Chanb, S.M. Complete Genome Analysis of the Mandarin Fish Infectious Spleen and Kidney Necrosis Iridovirus. *Virology* **2001**, *291*, 126–139. [CrossRef]
7. Yu, X.-D.; Ke, F.; Zhang, Q.-Y.; Gui, J.-F. Genome Characteristics of Two Ranavirus Isolates from Mandarin Fish and Largemouth Bass. *Pathogens* **2023**, *12*, 730. [CrossRef] [PubMed]
8. Tao, J.-J.; Zhou, G.-Z.; Gui, J.-F.; Zhang, Q.-Y. Genomic sequence of mandarin fish rhabdovirus with an unusual small non-transcriptional ORF. *Virus Res.* **2008**, *132*, 86–96. [CrossRef]
9. Koepke, L.; Gack, M.U.; Sparrer, K.M. The antiviral activities of TRIM proteins. *Curr. Opin. Microbiol.* **2021**, *59*, 50–57. [CrossRef]
10. Short, K.M.; Cox, T.C. Subclassification of the RBCC/TRIM Superfamily Reveals a Novel Motif Necessary for Microtubule Binding. *J. Biol. Chem.* **2006**, *281*, 8970–8980. [CrossRef]
11. Uchil, P.D.; Hinz, A.; Siegel, S.; Coenen-Stass, A.; Pertel, T.; Luban, J.; Mothes, W. TRIM Protein-Mediated Regulation of Inflammatory and Innate Immune Signaling and Its Association with Antiretroviral Activity. *J. Virol.* **2012**, *87*, 257–272. [CrossRef] [PubMed]
12. Yu, Y.; Liu, J.; Zhang, J.; Hu, Y.; Yang, Y.; Huang, Y.; Qin, Q. Fish TRIM32 functions as a critical antiviral molecule against iridovirus and nodavirus. *Fish Shellfish Immunol.* **2017**, *60*, 33–43. [CrossRef]
13. Yu, Y.; Huang, X.; Liu, J.; Zhang, J.; Hu, Y.; Yang, Y.; Huang, Y.; Qin, Q. Fish TRIM35 negatively regulates the interferon signaling pathway in response to grouper nodavirus infection. *Fish Shellfish Immunol.* **2017**, *69*, 142–152.
14. Niu, Y.; Fu, X.; Lin, Q.; Liang, H.; Luo, X.; Zuo, S.; Liu, L.; Li, N. The composition and antiviral activity of scTRIM59 in Mandarin fish. *Fish Shellfish Immunol.* **2022**, *130*, 86–92. [CrossRef] [PubMed]
15. Kawabata, H.; Azuma, K.; Ikeda, K.; Sugitani, I.; Kinowaki, K.; Fujii, T.; Osaki, A.; Saeki, T.; Horie-Inoue, K.; Inoue, S. TRIM44 Is a Poor Prognostic Factor for Breast Cancer Patients as a Modulator of NF-κB Signaling. *Int. J. Mol. Sci.* **2017**, *18*, 1931. [CrossRef]
16. Luo, Q.; Lin, H.; Ye, X.; Huang, J.; Lu, S.; Xu, L. Trim44 facilitates the migration and invasion of human lung cancer cells via the NF-κB signaling pathway. *Int. J. Clin. Oncol.* **2014**, *20*, 508–517. [CrossRef]
17. Zhu, H.; Wang, G.; Sun, Q.; Zhu, H.; Xu, A. Elevation of TRIM44 potentiates propagation of gastric cancer stem cells. *Genes Dis.* **2022**, *9*, 1156–1159. [CrossRef]
18. Liu, S.; Meng, F.; Ding, J.; Ji, H.; Lin, M.; Zhu, J.; Ma, R. High TRIM44 expression as a valuable biomarker for diagnosis and prognosis in cervical cancer. *Biosci. Rep.* **2019**, *39*, BSR20181639. [CrossRef]
19. Yang, B.; Wang, J.; Wang, Y.; Zhou, H.; Wu, X.; Tian, Z.; Sun, B. Novel Function of Trim44 Promotes an Antiviral Response by Stabilizing VISA. *J. Immunol.* **2013**, *190*, 3613–3619. [CrossRef]
20. Hou, F.; Sun, L.; Zheng, H.; Skaug, B.; Jiang, Q.-X.; Chen, Z.J. MAVS Forms Functional Prion-like Aggregates to Activate and Propagate Antiviral Innate Immune Response. *Cell* **2011**, *146*, 448–461. [CrossRef]
21. Zheng, J.; Zhang, Y.; Zhi, L.; Lv, S.; Xiao, L.; Huang, X.; Huang, Y.; Qin, Q.; Huang, Y.; Qin, Q. The novel gene TRIM44L from orange-spotted grouper negatively regulates the interferon response. *Fish Shellfish Immunol.* **2019**, *92*, 746–755. [CrossRef] [PubMed]
22. Fu, X.; Li, N.; Lai, Y.; Luo, X.; Wang, Y.; Shi, C.; Huang, Z.; Wu, S.; Su, J. A novel fish cell line derived from the brain of Chinese perch *Siniperca chuatsi*: Development and characterization. *J. Fish Biol.* **2015**, *86*, 32–45. [CrossRef] [PubMed]
23. Fu, X.; Lin, Q.; Liang, H.; Liu, L.; Huang, Z.; Li, N.; Su, J. The biological features and genetic diversity of novel fish rhabdovirus isolates in China. *Arch. Virol.* **2017**, *162*, 2829–2834. [CrossRef]
24. Fu, X.; Li, W.; Liu, C.; Luo, X.; Lin, Q.; Niu, Y.; Liang, H.; Ma, B.; Li, N. A naturaly attenuated largemouth bass ranavirus strain provided protection for Micropterus salmoides by immersion immunization. *Fish Shellfish Immunol.* **2024**, *153*, 109871. [CrossRef]
25. Van Tol, S.; Hage, A.; Giraldo, M.I.; Bharaj, P.; Rajsbaum, R. The TRIMendous Role of TRIMs in Virus–Host Interactions. *Vaccines* **2017**, *5*, 23. [CrossRef] [PubMed]
26. Jin, Y.; Jia, K.; Zhang, W.; Xiang, Y.; Jia, P.; Liu, W.; Yi, M. Zebrafish TRIM25 Promotes Innate Immune Response to RGNNV Infection by Targeting 2CARD and RD Regions of RIG-I for K63-Linked Ubiquitination. *Front. Immunol.* **2019**, *10*, 2805. [CrossRef]
27. Yu, T.; Kuang, H.; Chen, J.; Lin, X.; Wu, Y.; Chen, K.; Zhang, M.; Zhang, W.; Wen, Z. Tripartite-motif family protein 35–28 regulates microglia development by preventing necrotic death of microglial precursors in zebrafish. *J. Biol. Chem.* **2020**, *295*, 8846–8856. [CrossRef]
28. Cai, C.; Tang, Y.-D.; Zhai, J.; Zheng, C. The RING finger protein family in health and disease. *Signal Transduct. Target. Ther.* **2022**, *7*, 300. [CrossRef]
29. Yamada, Y.; Takayama, K.; Fujimura, T.; Ashikari, D.; Obinata, D.; Takahashi, S.; Ikeda, K.; Kakutani, S.; Urano, T.; Fukuhara, H.; et al. A novel prognostic factor TRIM44 promotes cell proliferation and migration, and inhibits apoptosis in testicular germ cell tumor. *Cancer Sci.* **2017**, *108*, 32–41. [CrossRef]

30. Wu, J.; Guo, N.-Z.; Cui, L.-L.; Wang, W.; Xiong, C.-Q.; Zhang, X.-Y. Correlation between tripartite motif-containing protein 44 protein expression and the prognosis of postoperative patients exhibiting skin squamous cell carcinoma. *Medicine* **2018**, *97*, e13021. [CrossRef]
31. He, H.; Cai, T.; Chen, Q.; Chen, Z.; Zhang, B.; Chen, C.; Wang, Y.; Liu, Y.; Wang, Y.; Luo, Y.; et al. TRIM44 Promotes Rabies Virus Replication by Autophagy-Dependent Mechanism. *Int. J. Mol. Sci.* **2024**, *25*, 4616. [CrossRef] [PubMed]
32. Leiva-Rebollo, R.; Labella, A.M.; Gémez-Mata, J.; Castro, D.; Borrego, J.J. Fish Iridoviridae: Infection, vaccination and immune response. *Vet. Res.* **2024**, *55*, 88. [CrossRef] [PubMed]

viruses

MDPI

Systematic Review

Viral Threats to Australian Fish and Prawns: Economic Impacts and Biosecurity Solutions—A Systematic Review

Md. Mizanur Rahaman [1], Bhavya Sharma [2], Saranika Talukder [3], Muhammad Jasim Uddin [4,5], Muhammad A. B. Siddik [6] and Subir Sarker [1,*]

1 Biomedical Sciences and Molecular Biology, College of Medicine and Dentistry, James Cook University, Townsville, QLD 4811, Australia; mdmizanur.rahaman@my.jcu.edu.au
2 School of Health & Biomedical Sciences, STEM College, RMIT University, Bundoora, VIC 3083, Australia; bhavyasharma731@gmail.com
3 College of Science and Engineering, James Cook University, Townsville, QLD 4811, Australia; saranika.talukder@jcu.edu.au
4 School of Veterinary Medicine, Murdoch University, Perth, WA 6150, Australia; jasim.uddin@murdoch.edu.au
5 Centre for Biosecurity and One Health, Harry Butler Institute, Murdoch University, Perth, WA 6150, Australia
6 Nutrition and Seafood Laboratory (NuSea.Lab), School of Life and Environmental Sciences, Deakin University, Queenscliff, VIC 3225, Australia; m.siddik@deakin.edu.au
* Correspondence: subir.sarker@jcu.edu.au

Abstract: Viral diseases pose significant threats to aquaculture industries worldwide, including the Australian fish and prawn farming sectors, which contribute over AUD 1.6 billion annually to the national economy. The Australian aquaculture industry relies heavily on wild-caught broodstock for seedstock production, introducing substantial and unprecedented biosecurity risks. This systematic review consolidates current knowledge on the viral pathogens affecting key Australian fish and prawn species, their economic impacts, and the biosecurity measures implemented for mitigation. Notably, viral outbreaks have led to losses exceeding AUD 100 million in some sectors, highlighting the urgent need for improved management. Existing biosecurity strategies, including surveillance systems, molecular diagnostics, and pathogen exclusion protocols, are critically assessed for their effectiveness. Emerging approaches such as genetic resistance breeding, advanced vaccination technologies, and integrated risk management frameworks are also explored. Key knowledge gaps, particularly in the context of emerging viral pathogens and their ecological interactions under changing environmental conditions, are identified as priority areas for future research. This review emphasises the necessity of adopting a multidisciplinary approach to enhance the resilience of Australian aquaculture, advocating for stronger biosecurity frameworks and innovative technologies to mitigate the escalating risks posed by viral diseases.

Keywords: viruses; fish; prawn; economic impact; biosecurity

1. Introduction

Australia is renowned for its diverse marine species, hosting approximately 33,000 species within its marine ecosystems [1]. Over the past 50 years, the production of aquatic animals has increased from 33.65 million tons to 214.92 million tons and is projected to increase further [2,3]. Aquaculture originates from ancient Chinese wisdom and has progressively advanced into one of the most efficient methods of food production [4]. It supplies nearly one-third of the animal protein consumed in China and contributes approximately 60% of global aquaculture production [5].

While these figures illustrate the global importance of aquaculture, Australia represents a unique and critical case due to its vast marine biodiversity, geographic isolation, and reliance on both wild-sourced and farmed aquatic species [6]. Understanding the local context is essential to addressing the challenges unique to Australian aquaculture systems.

Fish and prawns are recognised as critical and sustainable nutritional resources, particularly in the contexts of the growing global population, the impacts of climate change, and health concerns, as they provide essential protein and nutrients, which is especially important in regions where alternative food sources are limited [7]. However, the increasing demand for these resources has driven the rapid growth of aquaculture, which has, in turn, facilitated the spread and emergence of highly pathogenic viruses that are frequently transmitted globally through aquaculture practices [8].

While the demand for seafood continues to rise, the expansion of the aquaculture industry faces significant constraints, limiting its capacity to fully meet market demands. Various aquatic species are cultivated at high densities in diverse water aquaculture systems, such as ponds, tanks, and cages. Intensive farming exposes aquatic species to stressful conditions, increasing their vulnerability to diseases and mortality [9]. Despite technological advancements in aquaculture production, achieving profitability remains a big challenge. Aquaculture technologies require higher investments in maintaining water quality and have a strong focus on preventing viral diseases [7]. Improper farming practices, combined with the challenges inherent to aquaculture, contribute to the rapid spread of diseases and diminish the ability of aquatic species to resist infections [9].

Both long-standing and newly emerging viral pathogens pose significant challenges, as the development of effective treatments remains elusive [9]. Additionally, co-infections involving multiple viral pathogens are emerging as a significant concern in aquaculture systems. These co-infections—such as white spot syndrome virus (WSSV) with infectious hypodermal and haematopoietic necrosis virus (IHHNV) in prawns [10] or salmonid alphavirus (SAV) alongside piscine orthoreovirus (PRV) in fish—can exacerbate disease severity, complicate diagnostics, and hinder effective disease management [11]. Despite their critical impact, co-infections remain under-investigated in the context of Australian aquaculture, representing an important gap in current surveillance and biosecurity frameworks.

Nevertheless, significant progress has been made in areas such as genetic breeding, seed industry advancement, nutrition and feed development, disease diagnosis and prevention, environmental protection, ecological engineering, and modernisation of aquaculture infrastructure, all of which contribute to enhanced disease control and the overall sustainability of the industry [4,12,13].

Diagnostic methods such as real-time PCR (qPCR) are commonly employed to detect the specific pathogens associated with disease outbreaks [14]. qPCR provides high sensitivity and specificity, enabling the quantification and typing of pathogens that impact Australia's marine environment [15]. Emerging epizootics frequently cause significant losses in fish and prawn populations, leading to substantial economic impacts in commercial markets and posing serious threats to valuable aquatic animal stocks [16,17]. The severity of viral diseases is frequently exacerbated by delayed diagnoses, insufficient identification of causative agents, and inadequate understanding of critical epidemiological factors [18]. Although Australia is renowned for its pristine environment, strong biosecurity, and high terrestrial and aquatic animal production yields, implementing appropriate biosecurity in each industry remains a challenge. Thus, insufficient biosecurity is often a contributory factor to disease outbreaks and mass mortality events in aquatic populations [19].

Moreover, the Australian aquaculture industry heavily relies on wild-sourced broodstock for the black tiger prawn (*Penaeus monodon*), which presents considerable biosecurity risks. A recent study analysed 7472 pleopod samples from broodstock using TaqMan qPCR to detect viral pathogens. The study revealed a high prevalence of IHHNV (30%) and gill-associated virus (GAV) (28.1%) [20]. Reliance on wild-sourced broodstock for *P. monodon* farming exposes the industry to biosecurity risks, which can affect economic outcomes due to challenges related to domestication and genetic selection.

Despite ongoing advancements in disease diagnosis, genetics, and aquaculture infrastructure, significant knowledge gaps remain. In particular, the full economic and ecological implications of emerging and re-emerging viral threats in Australian aquaculture are not yet well understood. Moreover, the role of viral co-infections and their compounding effects on disease outcomes remain underexplored, complicating disease management and policy development. This review addresses existing knowledge gaps by systematically updating the current understanding of the viral diseases affecting Australian fish and prawns, with a focus on the economic impacts, co-infection dynamics, and biosecurity challenges. It also critically evaluates existing biosecurity policies and proposes innovative, evidence-based strategies for disease prevention, early detection, and sustainable aquaculture management. By doing so, this review aims to enhance the resilience of Australia's aquaculture sector against present and future threats.

2. Materials and Methods

This systematic review investigates the viral pathogens affecting fish and prawns in Australia, assessing their economic impacts and associated biosecurity measures. The review follows the PRISMA 2020 guidelines to ensure a standardised and transparent reporting process (Supplementary Table S1) [21].

2.1. Information Sources and Search Strategy

A comprehensive literature search was conducted across five major electronic databases: Google Scholar, MEDLINE, PubMed, Web of Science, and Scopus. The search included studies published up to January 2025. The keywords used were virus, fish, prawn, pathogen, biosecurity, and economic impacts of viral diseases in fish and prawn. Only articles published in English and focused on Australian aquaculture were included. Duplicate records were identified and removed using EndNote software, followed by manual verification.

2.2. Inclusion and Exclusion Criteria

2.2.1. Inclusion Criteria

The following inclusion criteria were used in this systematic review:

1. Investigated viral pathogens affecting fish and prawns in Australia.
2. Focused on biosecurity practices and/or the economic impacts of viral diseases in Australian aquaculture.
3. Included research conducted through in vivo, in vitro, or epidemiological studies, whether or not the mechanism of action was elucidated.

2.2.2. Exclusion Criteria

The exclusion criteria listed below were carefully implemented to ensure the thoroughness and applicability of this systematic review:

1. Were duplicated or did not meet inclusion criteria upon title/abstract screening.
2. Focused on non-viral pathogens or unrelated aquaculture animals.
3. Addressed topics outside the scope of biosecurity and economic impacts in terms of Australian fish and prawn farming.

2.3. Risk of Bias in Individual Studies

To assess the reliability of the results from the selected studies, we used the Cochrane risk-of-bias assessment criteria, specifically the RoB 2 tool version, dated 22 August 2019. This tool evaluates the risk across several domains: randomisation process, deviations from intended interventions, missing outcome data, outcome measurement, selection of reported results, and other potential sources of bias. Each category comprises questions across these domains, with the responses categorised as "YES" (indicating a low risk of bias, colour-coded green), "NO" (indicating a high risk of bias, colour-coded red), or "Some Concern" (indicating an uncertain risk of bias, colour-coded yellow) (Figure 1).

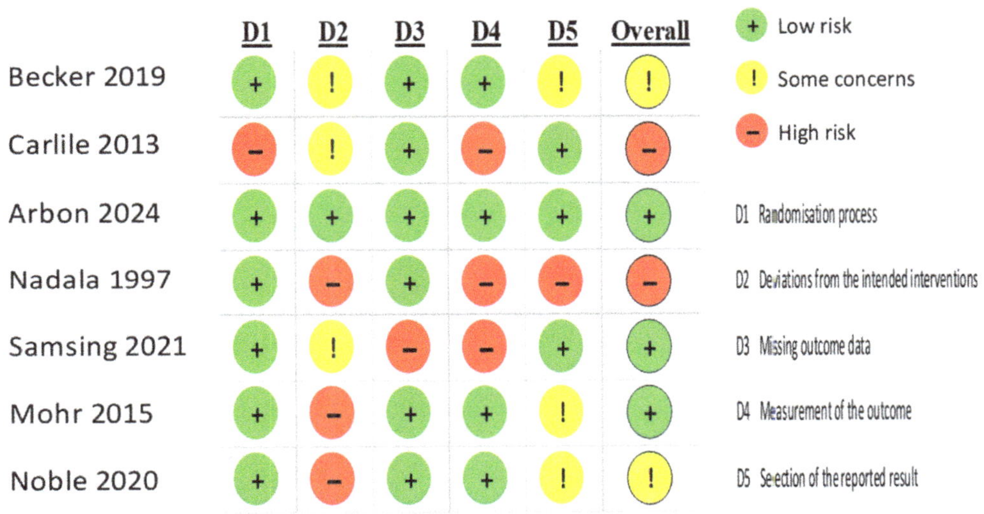

Figure 1. Risk of bias assessment. According to the Cochrane risk-of-bias tool for randomised trials (RoB 2): +, low risk of bias; −, high risk of bias; !, some risk of bias [22].

2.4. Study Selection Process

After the initial database searches, 5393 records were identified. Following duplicate removal, 2277 records were screened based on the titles and abstracts. Subsequently, 331 full-text articles were assessed for eligibility. After retrieved 265 removed and selected 66 for further study. Finally excluding 48 articles on the basis of qualitative synthesis and 18 studies were included in the final manuscript. The full study selection process is illustrated in a PRISMA flow diagram (Figure 2).

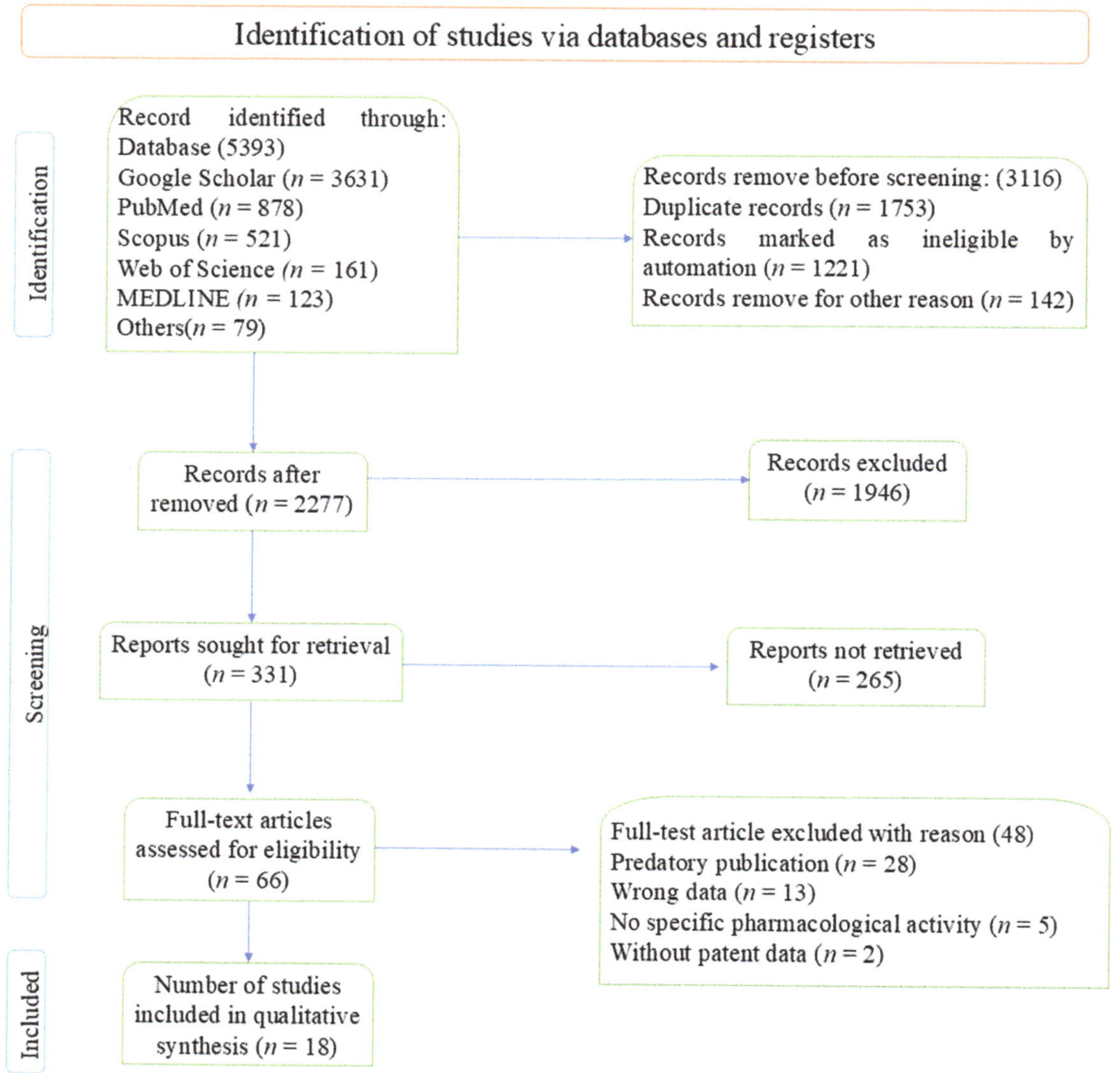

Figure 2. PRISMA flow diagram based on the data extraction.

2.5. Data Extraction and Characteristics of Sources

Data were extracted using a standardised form, focusing on the virus type, host species, disease mechanisms, environmental interactions, economic impacts, and biosecurity measures. Special attention was paid to emerging viral pathogens and mitigation strategies under changing environmental conditions.

3. Results and Discussion

3.1. Pathogenic Viruses in Fish

Eight fish viruses were identified via the search process. Table 1 lists the nomenclature, source, and year of publication for seven of these viruses. One virus has been excluded from Table 1 due to insufficient data. However, a description has been added for clarification.

Table 1. Pathogenic viruses reported in the literature from fish species found in Australia.

Pathogenic Virus	Viral Genus	Viral Family	Type	Susceptible Fish Species	Origin	Year of Study	Impact Level	References
EHNV	*Ranavirus*	*Iridoviridae*	dsDNA	Wild redfin (*Perca fluviatilis*)	Victoria, Australia	July 2007 and June 2011	High	[23]
CyHV2	*Cyvirus*	*Alloherpesviridae*	dsDNA	Goldfish (*Cyprinus carpio*)	Western Australia	*	Moderate	[24]
TSRV	*Aquareovirus*	*Reoviridae*	dsRNA	farmed Atlantic salmon (*Salmo salar*)	Tasmania	Since 2005	Moderate	[25,26]
POMV	Unclassified	*Orthomyxoviridae*	ssRNA	Atlantic salmon	Southern Tasmania	2012	High	[27]
NNV	*Betanodavirus*	*Nodaviridae*	ssRNA	Barramundi (*Lates calcarifer*)	New South Wales, Northern Territory, Queensland, South Australia, Tasmania, and Western Australia	*	High	[28]
ISKNV	*Megalocytivirus*	*Iridoviridae*	dsDNA	Ornamental fish	Exotic—not recorded in Australia	2008 and 2011	Potential threat	[29]
BIV	*Ranavirus*	*Iridoviridae*	dsDNA	Tilapia fry (*Oreochromis mossambicus*)	North Queensland, Australia	*	Moderate	[30]

* Year the study was performed was not mentioned. Table 1 is a summary of the major viral pathogens affecting fish species in Australia. The table highlights the pathogen type, host range, and geographic distribution, with a classification of the impact level based on the known or potential disease severity and economic consequences.

3.1.1. Epizootic Haematopoietic Necrosis Virus (EHNV)

Epizootic haematopoietic necrosis virus (EHNV), a member of the family *Iridoviridae* and genus *Ranavirus*, is a significant pathogen that causes severe necrosis of haematopoietic tissues in fish [17]. EHNV is a large icosahedral virus, approximately 175 nm in size, with a double-stranded DNA genome of approximately 127 kb [9]. In 1984, EHNV was first detected in Victoria. It has caused epidemic mortality in *P. fluviatis* and mild disease among farmed rainbow trout (*Oncorhynchus mykiss*) over the last two decades across New South Wales (NSW), the Australian Capital Territory (ACT), and South Australia [23]. EHNV primarily infects *P. fluviatilis*, though it is also capable of infecting a broad range of native and non-native fish species under experimental conditions [31]. Variable levels of infection and mortality were observed in mosquito fish (*Gambusia affinis*), Australian bass (*Macquaria novemaculeata*), Atlantic salmon (*S. salar*), and other perch species, including Macquarie perch (*Macquaria australis*), golden perch (*Macquaria ambigua*), and silver perch (*Bidyanus bidyanus*) under controlled laboratory conditions [32]. Infected fish displayed clinical symptoms including an enlarged abdomen, darkened skin, petechiae, and haemorrhages at the gill and fin bases [33]. Infection with EHNV also led to multifocal necrosis of haematopoietic tissue in the liver, spleen, and kidney of infected fish [34]. EHNV is known to be transmitted between susceptible hosts within a population via transfer through water or ingestion of tissues from infected fish. In the aquaculture industry, the primary mode of transmission was likely the movement of infected trout fingerlings, which may carry subclinical infections in a small proportion of individuals [35]. However, a recent report, including cases from 2021 at Lake Hume Resort, indicated that EHNV continues to be a concern, particularly with environmental changes facilitating the movement of carrier species like *P. fluviatilis* into new habitats [36]. The diagnostic methods for EHNV have evolved to include antibody detection via Western blot analyses, antigen-capture enzyme-linked immunosorbent assays (ELISAs), and polymerase chain reaction (PCR) techniques [37]. Controlling infection and preventing spread through active surveillance of fish health using biosecurity guidelines are recommended to manage EHNV, as there is no commercial vaccine available [38].

3.1.2. Cyprinid Herpesvirus (CyHV)

The cyprinid herpesvirus is classified under the species *Cyprinid herpesvirus 1* (CyHV-1), within the genus *Cyprinivirus* of the *Alloherpesviridae* family, which belongs to the order *Herpesvirales* [39]. Two other major cyprinid species viruses have been identified, *Cyprinid herpesvirus 2* (CyHV-2, also called goldfish haematopoietic necrosis virus) and *Cyprinid herpesvirus 3* (CyHV-3, also referred to as koi herpesvirus (KSV)) [40,41]. In 2003, CyHV-2 was first detected in Western Australia; it originated from goldfish (*Carassius auratus*) that had been cleared from quarantine and subsequently transmitted the virus to domestic fish through the live fish trade [24]. Previous studies have highlighted that CyHV-2 exemplifies how aquatic pathogens from the ornamental fish trade could infiltrate and spread within farmed and wild populations, emphasising its critical significance for risk assessments of megalocytivirus infection in live ornamental fish imported into Australia [24,42]. CyHV-3 is distinguishable from CyHV-1, the causative agent of carp pox, and CyHV-2 based on differences in the clinical signs, host range, antigenic properties, growth characteristics, cytopathic effects (CPEs) in cell culture, and DNA sequencing [43]. CyHV-3 has been detected in goldfish (*C. auratus*) and crucian carp (*Carassius carassius*) [44–46]. Infected fish typically exhibit severe gill necrosis, pale gills and skin, and skin haemorrhages, with the disease progression varying according to the environmental conditions, accelerating in summer and slowing in winter [47]. PCR and ELISA have been employed to detect CyHV viral DNA and viral antigen (Ag) in faeces [48]. To control the endemic spread of CyHV2 in domestic farms, the Australian Government Department of Agriculture, Fisheries and Forestry revoked the requirement for goldfish exported to Australia to be certified free of CyHV2 [24].

3.1.3. Tasmanian Atlantic Salmon Reovirus (TSRV)

Tasmanian Atlantic salmon reovirus (TSRV) is a member of the genus *Aquareovirus*, belonging to the family *Reoviridae*. TSRV possesses an 11-segment dsRNA genome, similar to those of viruses in the genera *Mycoreovirus* and *Rotavirus* [49]. In 1990, TSRV was first isolated in Tasmania from *S. salar* [25]. A previous study reported that TSRV is non-pathogenic; however, it still has an impact on fish production [50]. According to Carlile [25], there was increased mortality in *S. salar* infected with TSRV, especially when the temperature increased above 18 °C; thus, the incidence was higher during summer. The temperature conditions of Tasmanian waters have historically been favourable for the growth of Atlantic salmon; however, as water temperatures have risen due to climate change, the susceptibility of Atlantic salmon to clinical infections with TSRV has also increased, as noted in 2014 [51]. Infected Atlantic salmon exhibited petechial haemorrhages on their scales, peritoneal surfaces, and visceral fats. There is also evidence of hepatic necrosis with subacute non-suppurative myocarditis and pericarditis, as well as acute posterior uveitis [52]. In 2009, a total of 144 fish (12–13 fish per site) were collected from nine sites in two regions of Tasmania, the Tamar River and Southeast Tasmania, during late spring to early summer. The collected samples were tested via qPCR, which showed a sensitivity and specificity of 95.2% for detecting TSRV, with a prevalence ranging from 6% to 22% in both regions [50].

3.1.4. Pilchard Orthomyxovirus (POMV)

Pilchard orthomyxovirus (POMV) is a single-stranded RNA virus within the *Orthomyxoviridae* family, consisting of eight viral segments that encode ten putative proteins [53]. Although phylogenetic analysis identified infectious salmon anaemia virus (ISAV) as its closest relative, a comparison of six major proteins encoded by the two viruses revealed significant divergence [54]. POMV virus was first identified in 1998 as an incidental finding in wild pilchards (*Sardinops sagax*) collected from waters off the South Australian coast [54]. In 2012, it was subsequently discovered in Atlantic salmon from different fish farms in southern Tasmania [27]. Therefore, there are concerns regarding this virus within the Atlantic salmon industry, and it is classified as a notifiable finfish disease under the regulations of the World Organization for Animal Health (WOAH) [53]. The clinical condition caused by POMV, recently designated salmon orthomyxoviral necrosis (SON), is characterised by symptoms such as lethargy, darkened skin pigmentation, and petechial haemorrhages on ventral regions of the body [55]. The viral load is most prominent in the head kidney and heart, especially during the early stages of infection. However, even if fish recovered from the infection, the virus was still detectable in some tissues and remained at relatively high levels in the gills [56]. Additionally, virus-infected melanomacrophages were found in different tissues, indicating a host immune response against the virus [55]. Diagnostic tools such as reverse transcriptase real-time TaqMan polymerase chain reaction (RT-qPCR) and conventional reverse transcriptase nested PCR (RT-nPCR) were developed to detect POMV in infected fish [54].

3.1.5. Nervous Necrosis Virus (NNV)

Nervous necrosis viruses (NNVs) are classified within the genus *Betanodavirus* and the family *Nodaviridae* [57–60]. NNV is a small non-enveloped spherical or icosahedral single-stranded RNA virus that can be transmitted either horizontally or vertically [61]. NNVs encode two single-stranded positive-sense RNAs, RNA1 and RNA2, which encode an RNA-dependent RNA polymerase and a structural capsid protein, respectively [61]. NNV has been officially reported in New South Wales, the Northern Territory, Queensland, South Australia, Tasmania, and Western Australia [28]. It is also known to cause viral encephalopathy and retinopathy (VER), which is one of the most significant pathologies of affected fish species [62]. In 1980, VER was first described in larval barramundi (*L. calcarifer*) or Asian sea bass in Australia, and its impact on hatchery production was attributed to a neurological disease [62,63]. Additional signs observed in fish infected with NNV include anorexia, abnormal swimming patterns, fish resting belly-up due to a loss of equilibrium, sporadic protrusion of the head from the water, colour changes such as becoming lighter, blindness, abrasions, over-inflated swim bladders with vacuolation of central nervous tissues, especially the retina, and the presence of crystalline arrays or aggregates of intracytoplasmic inclusions in brain tissues [28,31]. NNV can be detected using validated real-time reverse transcriptase quantitative polymerase chain reaction (RT-qPCR) assays or virus isolation in striped snakehead (SSN-1) cell culture [36].

3.1.6. Infectious Spleen and Kidney Necrosis Virus (ISKNV)

Infectious spleen and kidney necrosis virus (ISKNV) is a species within the *Megalocytivirus* genus, which belongs to the *Iridoviridae* family [64]. *Megalocytivirus* is the most recent addition to the five genera within the *Iridoviridae* family, which consists of large, enveloped, double-stranded DNA viruses [65]. The international trade of live ornamental fish, both freshwater and marine, has been identified as a potential source for the spread of megalocytiviruses, particularly the ISKNV-like virus [66,67]. While megalocytiviruses have not been found in wild fish populations in Australia, several ISKNV-like viruses have been detected in imported ornamental fish and this has caused disease in native species on one occasion [68]. For example, strains such as dwarf gourami *Iridovirus* (DGIV) have been isolated from ornamental fish [67,69–71]; notably, DGIV was found in dwarf gourami imported to Australia that later died in commercial aquaria [69]. In 2011, various gourami species imported from multiple countries were found to carry DGIV in both quarantine and post-quarantine facilities [72]. Furthermore, ornamental fish species imported to Australia between 2012 and 2013 showed high mortality rates in quarantine, which was associated with infections with an ISKNV-like virus [73]. Most recently, DGIV was detected at an Australian ornamental fish farm breeding exotic species [72]. The clinical signs of ISKNV infection in these fish included high mortality rates (50–100%), lethargy, poor feeding, respiratory distress, colour changes, exophthalmos, abdominal swelling, and microscopic signs like hypertrophied basophilic cells in haematopoietic tissues, with some affected cells appearing amoeboid [38]. TaqMan qPCR has emerged as a widely used diagnostic tool for screening ISKNV [68].

3.1.7. Bohle Iridovirus (BIV)

Bohle iridovirus (BIV), classified under the genus *Ranavirus* in the family *Iridoviridae*, was initially isolated in Australia from the ornate burrowing frog (*Limnodynastes ornatus*) [74]. Ranavirus is characterised by large, double-stranded DNA virions measuring approximately 150–180 nm in diameter and possessing genomes ranging from 150 to 170 kb [75]. Since its discovery, BIV has been found to infect a variety of hosts, including amphibians, reptiles, and fish such as tilapia (*O. mossambicus*) [76]. The first case of BIV infection in tilapia fry was documented in an aquatic disease research facility in North Queensland, Australia, where it caused 100% mortality [77]. Affected fish displayed erratic, corkscrew-like swimming behaviour, a symptom that led to the disease being referred to as "spinning tilapia" (ST) syndrome [77]. Given Australia's dependence on robust aquatic populations, the emergence of BIV presents a significant threat to both local aquaculture industries and international trade, especially if the virus spreads to commercially valuable species. At present, no vaccines or targeted treatments exist for BIV, and its diagnosis largely relies on molecular methods, such as PCR, and histopathological examinations [35,78].

3.1.8. Wamena Virus (WV)

Wamena virus (WV), belonging to the family *Iridoviridae*, is generally classified within the genus *Ranavirus* [79]. In Australia, WV was first isolated from a python that had been illegally imported from Irian Jaya [80]. This virus is known to cause severe diseases in fish, amphibians, and snakes, although its distribution appears to be largely confined to the Australasian region [79]. While direct reports of widespread outbreaks within Australia are limited, early detections combined with ecological similarities across regions highlight a credible risk of viral introduction and spread.

3.2. *Pathogenic Viruses in Prawn*

Nine prawn viruses were identified via the search process. Table 2 lists the nomenclature, source, and year of publication for the nine identified viruses.

Table 2. Pathogenic viruses reported in the literature from prawn species found in Australia.

Pathogenic Virus	Viral Genus	Viral Family	Genome Type	Susceptible Prawn Species	Origin	Year of Study	Impact Level	References
IHHNV	*Penstylhamaparvovirus*	*Parvoviridae*	ssDNA	*P. monodon*	Northern Territory and Queensland	July 2018 and October 2020	High	[14,20]
GAV	*Okavirus*	*Roniviridae*	ssRNA	*P. monodon*	Queensland	July 2018 and October 2020	Moderate	[14,20]
MrNV	Unclassified	*Nodaviridae*	ssRNA	*Giant freshwater prawn Macrobrachium rosenbergii*	Queensland	During January and March 2020	Moderate	[81]
WSSV	*Whispovrius*	*Nimaviridae*	dsDNA	Penaeid shrimps	South-Eastern Queensland	2017	Very high	[82]
When-2	Unclassified	Unclassified	ssRNA	*P. monodon*	Wenzhou in Zhejiang province, China	Between July 2018 and October 2020	Unknown	[20]
MVB	*Nucleopolyhedrovirus*	*Baculoviridae*	dsDNA	*P. monodon*	Australia	*	Unknown	[83]
MoV	*Wenrivirus*	*Phenuiviridae*	ssRNA	*P. monodon* and Kuruma shrimp (*Penaeus japonicus*)	Eastern Australia	*	Moderate	[84]
HPV	*Densoviruses*	*Parvoviridae*	ssDNA	*P. monodon*, *P. japonicus*, Brown tiger prawn (*P. esculentus*), Indian white prawn (*P. indicus*), *P. merguiensis*	Moreton Bay Gulf of Carpentaria	2005	Moderate	[85]
YHV-7	*Unclassified*	*Unclassified*	-	*P. monodon*	Joseph Bonaparte Gulf in northern Australia	November 2012	High	[86]

* Year the study was performed was not mentioned. Table 2 is an overview of the key viral pathogens affecting prawn species in Australian aquaculture, including the genome type, host specificity, and geographical distribution. Pathogens such as WSSV and IHHNV are marked as high impact due to their widespread occurrence and severe economic implications.

3.2.1. Infectious Hypodermal and Haematopoietic Necrosis Virus (IHHNV)

Infectious hypodermal and haematopoietic necrosis virus (IHHMV) is currently classified as a potential *Penaeus stylirostris* densovirus (PstDNV), belonging to the genus *Penstylhamaparvovirus* within the subfamily *Hamaparvovirinae* of the *Parvoviridae* family [87,88]. IHHNV is the smallest known penaeid shrimp virus and is characterised by a single-stranded DNA genome approximately 4.0 kb in size with a non-enveloped icosahedral nucleocapsid [89,90]. Two known IHHNV genotypes have been shown to be pathogenic to either Pacific white prawn (*Penaeus* (*Litopenaeus*) *vannamei*) or *P. monodon* [91]. The clinical manifestations in prawns infected with IHHNV include slow floating to the water surface, cannibalism, poor hatching and survival of larvae, opaque abdominal musculature, white to buff lesions on the carapace, runt deformity syndrome, cuticular roughness, cuticular deformities, and mottling of the shell at the abdominal shell plate. Additionally, histologically, virus-infected cells present characteristic intranuclear eosinophilic and haloed Cowdry type A inclusion bodies [91]. Standard diagnostic methods used to detect IHHNV in prawns include histopathological examination, PCR analysis, and ELISA tests [92–94].

3.2.2. Gill-Associated Virus (GAV)

Gill-associated viruses (GAVs) are classified as novel members of the family *Roniviridae*, within the genus *Okavirus* [95,96]. GAV belongs to the yellow head virus (YHV) complex and is also referred to as YHV-2 [14]. GAV is characterised by four major structural proteins (~170, 135, 67, and 22 kDa), with the 135 kDa protein being glycosylated; its genome consists of a single-stranded RNA (ssRNA) molecule exceeding 22 kb in length [97,98]. GAV is usually found in eastern Australia and is known to cause chronic infection in both farmed and healthy broodstock *P. monodon* [14]. The virus that emerged in 1996 was less virulent, causing only mid-crop mortality syndrome in farmed *P. monodon* in Australia [99]. The prevalence of GAV infection in healthy *P. monodon* broodstock and farmed prawn in Australia has been reported to be close to 100%. While viruses can be transmitted through injection or exposure to moribund prawn via oral routes, outbreaks in ponds are most likely driven by the amplification of viral loads due to environmental stress [100]. The other four known viral genotypes were exclusively detected in healthy *P. monodon* in Asia, where they are prevalent in various regions but not associated with disease [101]. The pathological signs observed due to this virus include reddening of the body and appendages, pink to yellow coloration of the gills, and the necrosis of lymphoid organs [102]. According to Nobel et al. (2020), GAV infection is affected by sex, since female prawns have a higher prevalence of infection (50.2%) compared to their male counterparts (41.8%). Thus, one of the factors affecting the selection of females for breeding is that GAV can be transferred vertically and presented on egg surfaces. In addition, selecting host families with lower viral infection loads will likely lead to lower GAV prevalence [103]. The diagnostic methods for detecting GAV in clinical samples include in situ hybridisation (ISH), RT-nested PCR, and RT-qPCR [104–107].

3.2.3. Macrobrachium Rosenbergii Nodavirus (MrNV)

Macrobrachium rosenbergii nodavirus (MrNV), a member of the *Nodaviridae* family [108], is a non-enveloped, icosahedral virus (27 nm in diameter) with a genome consisting of two positive-sense single-stranded RNA fragments: RNA 1 and RNA 2 [109]. It causes white muscle disease/white tail disease (WTD) in giant freshwater prawn (*Macrobrachium rosenbergii*) [108] and was first reported on Guadeloupe Island in 1997 [110] and later in Australia [111]. It is known to be a limiting factor in crustacean species development, especially in *M. rosenbergii* larvae and juveniles, where it had a 100% mortality rate. *Mr*NV targets the haemocytes and myonuclei of the prawn's lower abdomen before spreading. Infection occurs when a virus binds to a host cell receptor through the caveolin-mediated endocytosis pathway. This virus is spread through the haemolymph circulatory system and is found in almost every organ except the eyestalks and hepatopancreas [112]. Several diagnostic techniques have been developed for detecting MrNV, including a sandwich enzyme-linked immunosorbent assay (S-ELISA), dot blot hybridisation, in situ hybridisation, and RT-PCR [113].

3.2.4. White Spot Syndrome Virus (WSSV)

White spot syndrome virus (WSSV) has been assigned to the newly established genus *Whispovirus* within the family *Nimaviridae* on the basis of its distinctive morphology and genomic composition [114–116]. WSSV is a large, enveloped, double-stranded DNA virus that is known to cause panzootic white spot disease (WSD), which affects prawn aquaculture [117]. The first reported cases of WSD date back to 1992 in mainland China and Taiwan, and it had spread throughout Asia by the following decade [82]. Australia remained free from WSD until 2016, when there was a WSSV-associated disease outbreak in a prawn farm near Brisbane, southeastern Queensland. The clinical signs of the disease included reddish pink and white calcified spots on the prawns [82]. According to the Department of Agriculture (2019), prawns infected with WSSV showed white spots embedded within the exoskeleton, a loosened carapace, and anorexia, since the GIT was empty, along with the delayed clotting of haemolymph and excessive fouling of the gills [118]. WSD has caused large economic losses due to the high mortality of penaeids through the ingestion of infected tissues or direct contact with infected individuals. Furthermore, WSSV is likely transmitted vertically through infected females to their offspring [82]. For the diagnosis of WSSV, PCR has been shown to be a highly accurate method, even in asymptomatic prawns [119].

3.2.5. Whenzhou (Syn. Wenzhou) Shrimp Virus-2 (When-2)

Whenzhou (syn. wenzhou) shrimp virus-2 (When-2) was first officially identified in an unclassified segmented virus 3 RNA sequence of *P. monodon* in Wenzhou, Zheijiang Province, China [120]. Most recently, an uncharacterised When-2 viral genome was also identified in *P. monodon* stocks in Australia [121]. A study reported the total prevalence of When-2 to be 2.9% (220/7472), with the majority of positive detections occurring on the East Coast of Queensland (199 out of 220 samples), indicating its potential presence in Australian stocks [20]. However, the impact of this virus on the health and productivity of *P. monodon* has yet to be thoroughly investigated [20]. Additionally, the detection of *Hepandenovirus* in the replicative phase of GAV suggests that similar techniques could be employed to identify viruses with DNA genomes [121]. Notably, quantitative PCR is the most widely used technique for the detection of When-2 [20].

3.2.6. Monodon Baculovirus (MBV)

Monodon baculovirus (MBV), also referred to as nuclear polyhedrosis virus (NPV), is classified within the genus *Nucleopolyhedrovirus* of the *Baculoviridae* family [122]. NPV has a double-stranded circular DNA genome of 80–160 kb pairs and a rod-shaped, enveloped particle that is often occluded within proteinaceous bodies [123]. In Australia, NPV has been detected in both cultured *P. monodon* and wild *P. merguiensis* [124], but it is not thought to be pathogenic, as it is usually contracted in hatcheries where prawns are cultured under sub-optimal conditions. The primary target organs of NPV are the hepatopancreas and anterior midgut [125]. Prawns affected by NPV are lethargic, anorexic, and foul smelling; dysfunction in the hepatopancreas, anterior midgut, and other organs is also observed. In addition, co-infection with bacteria can also occur, resulting in hypertrophied nuclei containing spherical occlusion bodies [123,126]. Good management practices, such as washing fertilised eggs and nauplii with clean seawater, have been recommended to control MBV or avoid its spread within the cultured stock [127]. Furthermore, rapid molecular detection methods, including PCR and DNA hybridisation-based genomic probes (either in situ or dot blot), have been developed for the early identification of MBV in prawn samples [128–130].

3.2.7. Mourilyan Virus (MoV)

Mourilyan virus (MoV) is classified within the family *Phenuiviridae*, genus *Wenrivirus*, and order *Bunyavirale* [131]. MoV is an enveloped negative-sense ssRNA bunyavirus that is 85–100 nm in diameter and contains four segmented genomes [132]. The virus is known for infecting *P. monodon* and *P. japonicus*, which are both farmed in eastern Australia [133,134].

Prawns infected with MoV show damage to mesodermal and ectodermal tissues, accompanied by the formation of enveloped, ovoid-shaped particles [84]. In *P. japonicus* reared in tanks after grow-out in farm ponds, increased MoV infection loads are correlated with higher mortality rates [135]. Additionally, recent experimental challenge studies have validated the pathogenicity of the virus [84]. The detection of MoV in prawn samples is primarily conducted using the TaqMan qRT-PCR assay [136].

3.2.8. Hepatopancreatic Parvovirus (HPV)

Hepatopancreatic parvovirus (HPV) is classified within the *Parvoviridae* family based on its virion characteristics [137]. HPV is a non-enveloped, icosahedral virus with an average diameter ranging from 22 to 24 nm. The HPV virus genome is a single linear DNA strand approximately 6 kb in size [138]. HPV was first reported in 1984 in wild *P. merguiensis* and *P. indicus* from Singapore [139]. In Australia, the earliest detection occurred in 1985, when HPV was identified in *Penaeus esculentus* from Moreton Bay and the Gulf of Carpentaria [140]. Subsequently, HPV was found in *P. merguiensis* in 1989 and later in *P. monodon* and *P. japonicus* [141–143]. This virus is associated with severe outbreaks, where it causes mortality rates of up to 50–100% as well as stunting [142,144,145]. Furthermore, HPV infection has been linked to reduced growth rates in juvenile prawns [144]. qPCR has been demonstrated to be a valuable tool for accurately assessing the HPV levels in broodstock [146].

3.2.9. Yellow Head Virus Genotype 7 (YHV7)

Yellow head virus genotype 7 (YHV7) is a recently identified strain within the YHV complex. YHV7 was first detected in *P. monodon* from the Joseph Bonaparte Gulf (JBG) in northern Australia [86]. Phylogenetic analysis of the open reading frame 1b (ORF1b) gene revealed that YHV-7 is the seventh distinct genotype in the YHV complex and has the highest genetic similarity to YHV-1 [86]. The pathogenicity of YHV-7 was confirmed through injection, co-habitation, and feeding in *P. monodon*, which resulted in approximately 60% cumulative mortality at 28 days post-infection [147]. Additionally, YHV-7 has also been associated with sporadic disease outbreaks in pond-reared *P. monodon* [148]. To detect YHV-7 in prawn populations, sensitive and specific diagnostic methods, including TaqMan qRT-PCR and conventional nested PCR assays targeting the ORF1b gene, have been developed [148].

3.3. *Comparative Insights on Viruses in Fish and Prawn*

A comparative assessment of the viruses affecting Australian aquaculture reveals distinct patterns in the emergence, endemicity, and trade risk. Among finfish pathogens, EHNV, TSRV, and BIV are long-established and endemic, causing localised but significant losses [23,52,77]. In contrast, NNV and POMV are increasingly widespread and linked to severe disease outbreaks, particularly in hatchery systems. CyHV-2, though introduced via the ornamental trade, exemplifies the risks posed by imported live animals. ISKNV and WV, while not yet established, represent emerging threats with high incursion potential.

In prawn aquaculture, WSSV is the most economically disruptive pathogen, with its 2016 outbreak causing over AUD 23 million in damages [149]. Endemic viruses such as IHHNV and GAV remain prevalent in wild-sourced broodstock, complicating control efforts [150,151]. YHV-7, MoV, and MrNV are emerging concerns due to their virulence and limited diagnostic capacity [136,148,152]. Other detected viruses, including When-2, MBV, and HPV, vary in impact but highlight the growing complexity of disease surveillance.

3.4. Impacts of Viral Diseases

In the next section, we will explore the economic impacts of viral diseases on Australian fish and prawn aquaculture, with an emphasis on the direct and indirect costs, trade implications, and mitigation strategies. A summary comparing the economic impact of major viruses is presented Table 3.

Table 3. Comparative summary of the economic impact of major viral diseases in fish and prawn aquaculture in Australia.

Virus	Year of Major Outbreak	Affected Species	Direct Economic Losses	Indirect Economic Costs	Trade and Market Impact	Socioeconomic Implications	References
WSSV	2016	*P. monodon*	Estimated AUD 23.5 million loss (2016–2017); loss of 25 million prawns; single farm loss > AUD 1 million despite AUD 400 k support	Restocking delays, poor feed conversion, downtime, higher production cost	Export bans, market uncertainty, compliance costs, price volatility	AUD 24 million farm gate loss (~25% of sector); 100% mortality within 3–10 days; income loss and debt in farmers	[147,149, 153–161]
NNV	Recurrent (no fixed date)	Barramundi (larvae and juveniles)	Insolvency of first Barramundi hatchery in Australia	High larval mortality; delays in production cycles	Market instability due to supply chain disruption	Disruption of breeding; financial loss for hatcheries	[162,163]
TSRV	Endemic (concerns resurfacing recently)	*S. salar*	Not quantified; concerns emerging from association with diseased fish	Potential productivity loss under stress conditions; diagnostic and surveillance costs	No documented trade bans, but growing industry concern	Industry-driven monitoring and management; vertical and horizontal transmission risks	[26]

3.4.1. Direct Economic Losses

Viral diseases such as TSRV, KHV, and WSSV have had devastating impacts on Australian aquaculture production [26,162]. A notable outbreak occurred in December 2016, when WSSV spread in commercial *P. monodon* prawn farms along the Logan River in Queensland, northeast Australia. This led to the complete loss of prawn production in the region, either directly due to the virus or indirectly through quarantine measures that resulted in the destruction of affected farms [153]. This outbreak had a significant and immediate financial impact on prawn farming in the region, directly affecting five farming families and resulting in the destruction of stock across grow-out ponds, hatcheries, and breeding programmes, with losses estimated at AUD 23.5 million for the 2016–2017 period [154]. Additionally, two farms that managed breeding programmes for *P. monodon* in their private hatcheries experienced a complete loss of spawning stocks, severely disrupting their genetic development capabilities [154], while Gold Coast Marine Aquaculture, one of the Australia's largest prawn farming enterprises, reported a loss of 25 million prawns due to mandatory destocking under national eradication efforts [155]. The first farm to test positive for WSSV reported stock losses exceeding AUD 1 million, despite receiving an AUD 400,000 government support package to mitigate the financial impact [156].

Beyond prawns, the broader aquaculture industry also faces challenges, as Barramundi farming, which initially thrived as an alternative, was undermined by nodavirus outbreaks that caused high larval and juvenile mortality, significant financial losses, and the insolvency of the country's first Barramundi hatchery [162,163]. The direct financial losses primarily stem from mass mortalities among farmed aquatic species, which not only reduce overall yields but also disrupt farming cycles.

3.4.2. Indirect Economic Costs

The indirect economic costs resulting from viral infections in fish and prawn aquaculture in Australia extend beyond the immediate loss of stock. Infected fish and prawns often exhibit poor feed conversion ratios, requiring farmers to spend more on feed for diminished output. Farmers also face significant financial strain from the delays in production cycles caused by the need to restock and rebuild populations, which can take months or even years, leading to lost revenue and disrupted supply chains [157,158]. This inefficiency translates into higher production costs and lower profitability [159]. Restrictions on the movement and trade of aquatic products imposed during outbreaks further exacerbate the economic losses by limiting market access and increasing compliance costs [164]. Additionally, disease outbreaks can cause significant disruptions to production cycles. Farmers may need to halt operations to manage and eradicate diseases, leading to delays in restocking and rebuilding populations [165]. For example, the emergence of white spot syndrome in Queensland's prawn farms in 2016 led to an immediate cessation of operations to treat the disease. Eradication efforts included culling infected stock and implementing strict biosecurity measures, resulting in delays in restocking and prolonged production downtime [149].

3.4.3. Impacts on Trade and Market Access

Viral diseases also disrupt international trade by triggering trade restrictions and a loss of consumer confidence. Countries importing Australian aquaculture products may impose stringent sanitary requirements or outright bans on exports from regions experiencing outbreaks. For instance, the presence of WSSV in Australian prawns has previously led to temporary trade bans, significantly affecting export revenues and damaging Australia's reputation as a reliable supplier [153,160]. Beyond direct trade restrictions, viral outbreaks also create market uncertainty. Consumers, wary of potential health risks, may shift their preferences away from products originating from affected regions, even after the outbreak is controlled. Exporters may also face increased costs related to compliance with new sanitary regulations, including certifications, testing, and biosecurity audits, which can be resource-intensive and time-consuming processes [166]. Outbreaks can reduce the availability of high-quality products, leading to price volatility and reduced competitiveness against imported alternatives. This dynamic can erode market share and profitability for Australian aquaculture producers in the long term.

3.4.4. Socioeconomic Implications

Viral diseases have profound socioeconomic implications for Australia's aquaculture industry, particularly fish and prawn farming. The white spot disease outbreak in Queensland's Logan River region in 2016 resulted in an estimated farm gate loss of approximately AUD 24 million, representing around 25% of the country's aquaculture prawn sector at that time [161]. Such outbreaks can be associated with high mortality rates, with WSSV causing up to 100% mortality in infected prawns within 3–10 days, severely impacting production [167]. These losses place a significant financial strain on farmers, many of whom struggle with income loss and debt due to production failures [154]. Additionally, disease outbreaks disrupt trade and markets; for example, following the detection of WSSV in 2016, movement restrictions and heightened biosecurity measures were imposed, affecting supply chains and international exports [149]. In response, the Australian government implemented strict movement restrictions and decontamination procedures to control outbreaks, aiming to mitigate the long-term industry damage [168].

3.5. Biosecurity Policies and Solutions for Managing Viral Diseases

Viral diseases are a significant threat to Australian fish and prawn aquaculture, causing severe economic losses and threatening sector sustainability. The implementation of robust biosecurity measures is essential for the future growth of aquaculture, as ensuring the health of aquatic species is essential for meeting the global demand for food [169]. Australia's biosecurity focuses on animal health, public health, and the integration of best management practices at all the production stages—from broodstock facilities to exports [170]. National strategies such as AQUAPLAN and AQUAVETPLAN have been central to this effort [171]. These strategies were first implemented in 1987 to address aquatic animal diseases [172] and later evolved through multiple revisions [173–175]. AQUAPLAN, developed by the Department of Agriculture, Fisheries and Forestry, outlines national priorities for aquatic animal health, while AQUAVETPLAN provides technical response guidelines during disease outbreaks [176,177]. Moreover, Animal Health Australia (AHA) plays a key role in the emergency response coordination and cost-sharing arrangements between the industry and government [178].

3.5.1. Mitigation Strategies

Investments in disease prevention, early detection, and effective management are crucial to minimising the economic impacts of viral diseases. The development of robust biosecurity protocols, vaccination programmes, and genetic improvements for disease-resistant strains has shown promise in reducing the losses in aquaculture globally [179], but these measures are not well documented in Australia. Advancements in molecular diagnostics and next-generation sequencing have enhanced our ability to detect and monitor viral pathogens in aquaculture systems [14,180]. Ongoing investment in research, particularly in the development of RNAi-based therapies for viral diseases, may have an impact on the long-term sustainability of aquaculture [181]. Early detection allows for swift intervention, minimising the spread of disease and reducing the overall economic impact. Additionally, international collaborations and knowledge-sharing initiatives provide Australian aquaculture stakeholders with access to cutting-edge research and best practices in disease management [182]. Moreover, implementing comprehensive insurance schemes and financial support mechanisms for farmers affected by outbreaks can also help to mitigate economic losses. AHA has been working with industry sectors and the government to determine the emergency response and cost-sharing arrangements for future aquatic emergencies [183]. However, although AHA's efforts to establish emergency response frameworks are crucial, implementing equitable cost-sharing and compensation remains a challenge.

3.5.2. Quarantine and Import Controls

Quarantine and import controls are the first line of defence against the introduction of viral pathogens into Australian aquaculture. The Biosecurity Act 2015 provides a robust framework for screening aquatic species and related products [184,185]. Diagnostic tests, such as PCR, are conducted during this period for high-sensitivity pathogen detection [14]. Australia also restricts the importation of certain feed ingredients and other biological materials to minimise the contamination risks. For example, regulations on frozen prawn imports are crucial to mitigating threat of WSSV, which devastated several farms in Queensland during a 2016 outbreak [186]. The Federal Government almost entirely relied on an at-border testing programme for risk mitigation in November/December 2016 [187]. However, this reliance on border testing revealed vulnerabilities, prompting updated risk analyses and surveillance protocols.

3.5.3. Surveillance and Early Detection

DAFF emphasises that surveillance is essential for early disease detection, demonstrating Australia's disease status, and maintaining market access [188]. Advancements in diagnostic technologies have further enhanced early detection capabilities [189]. Australian biotech company Genics developed Shrimp MultiPath2.0, a technology capable of detecting 18 shrimp diseases and identifying genetic variations in a single test. Additionally, the Commonwealth Scientific and Industrial Research Organization (CSIRO) validated PCR tests for detecting WSSV, confirming their high accuracy in samples from apparently healthy prawns [119]. On the other hand, a non-invasive, environmental DNA (eDNA) analysis technique has become a valuable tool for detecting viral pathogens in water samples [190] and allows for continuous monitoring without disturbing aquaculture stocks. While the technical capabilities are effective, ensuring consistent adoption across farms and maintaining long-term funding remain hurdles.

3.5.4. Farm-Level Biosecurity Measures

Farm-level biosecurity practices are critical to minimising the risk of viral transmission within and between aquaculture facilities. Hatcheries often implement water filtration systems and use treated water to reduce the contamination risks [191]. Regular cleaning and disinfection of tanks, nets, and feeding equipment can eliminate viral particles and prevent their spread. To reduce the risk of introducing viruses into aquatic environments, farms should obtain disease-free stock and feed that is free from pathogens [168]. Training and raising awareness among farm staff about biosecurity practices ensure consistent implementation and adherence. These practices, when consistently implemented, significantly lower the risk of viral outbreaks and contribute to farm sustainability [179,192].

3.5.5. Emergency Response and Contingency Planning

Australia's biosecurity framework includes various emergency response plans to address viral disease outbreaks [192]. In the event of an outbreak, affected farms are immediately isolated to prevent contact with surrounding facilities. The culling of infected stock is another standard measure during severe outbreaks, followed by the safe disposal of carcasses. The farms are disinfected, and all the equipment is sanitised to eliminate residual pathogens. Temporary bans on the movement of live aquatic animals and related materials have also been enforced in affected areas [193]. During the 2016 WSSV outbreak in Queensland, rapid response teams worked with local farmers to limit the virus's spread, demonstrating the importance of coordinated action [153,185]. However, contingency planning must also address post-outbreak recovery support for affected farmers.

3.5.6. Research and Innovation

The Cooperative Research Centre for Developing Northern Australia (CRCNA) has conducted studies to understand the impact of pathogens on prawn health and survival, with a particular focus on environmental factors, including water quality and temperature [194]. Additionally, James Cook University is advancing the use of eDNA sampling methods, which serve as early detection tools for pathogens on prawn farms, potentially reducing biosecurity costs and preventing outbreaks [190]. Furthermore, treatment with chemicals, like trichlorfon, has been validated for controlling white spot disease, enhancing biosecurity protocols on Australian prawn farms and contributing to effective management and prevention strategies that are used to safeguard the Australian aquaculture industry [195]. Additionally, the CSIRO in Australia developed Shrimp MultiPath technology, which helps to detect 13 commercially significant prawn diseases, including IHHNV and WSSV [196]. However, broader industry uptake of these solutions is needed to ensure system-wide resilience.

3.5.7. Building Resilience Through Education and Training

To enhance the national biosecurity capacity, the Australian government, in collaboration with Charles Sturt University, established the Biosecurity Training Centre in 2022, offering professional training and specialised biosecurity programmes aimed at building sector-wide expertise [197]. Complementing this initiative, the National Biosecurity Training Hub provides an accessible online platform with a wide range of aquatic biosecurity training resources, although maintaining high levels of participation across remote farming communities remains a challenge [198]. To further strengthen biosecurity awareness, the CRCNA developed a biosecurity training video specifically for aquaculture, aimed at improving on-farm biosecurity practices in northern Australia [199]. Additionally, the Australian government provides guidelines for creating farm-specific biosecurity plans, emphasising staff training to manage disease prevention and outbreaks [192]. To further support industry needs, the FRDC offers an Aquatic Animal Health and Biosecurity Training Scheme to support the industry's health management training needs [200]. Together, these initiatives represent a comprehensive approach, though ongoing challenges related to engagement, consistency, and regional implementation highlight the need for continuous improvement.

3.6. Knowledge Gaps and Recommendations

Viral diseases pose a major threat to the economic viability and environmental sustainability of the Australian fish and prawn industries, as they can lead to stock losses, increased operational costs, export restrictions, and reduced industry confidence. Efforts to address these challenges are hindered by key knowledge gaps, including the lack of early diagnostic tools, insufficient understanding of viral transmission dynamics, and inadequate assessment of economic impacts.

To safeguard the industry's sustainability, priority should be given to developing rapid, accessible molecular diagnostics and establishing a coordinated national surveillance framework. Strengthening biosecurity through advanced water treatment systems and selective breeding for genetic resistance is equally critical. Targeted research into the effects of environmental changes, particularly temperature fluctuations, on viral emergence and transmission is urgently needed. Understanding the temperature-dependent viral dynamics is vital for refining predictive models and adapting disease mitigation strategies under climate change pressures.

Finally, integrating stakeholder engagement, economic resilience planning, and support for small-scale producers into biosecurity responses will be key to strengthening industry-wide collaboration. Through these actions, Australia can effectively manage the economic and ecological consequences of viral threats, ensuring the long-term sustainability of its aquaculture sectors.

Supplementary Materials: The following supporting information can be downloaded at https://www.mdpi.com/article/10.3390/v17050692/s1, Table S1: PRISMA checklist.

Author Contributions: Conceptualisation: M.M.R. and S.S.; investigation: M.M.R. and S.S.; funding acquisition: S.S. and S.T.; formal analysis: M.M.R.; data analysis: M.M.R., S.S. and S.T.; data curation: M.M.R. and S.S.; supervision: S.S. and S.T.; writing (original draft preparation): M.M.R. and S.S.; writing (review and editing): M.M.R., S.S., B.S., M.J.U., M.A.B.S. and S.S. All authors have read and agreed to the published version of the manuscript.

Funding: This research received no external funding.

Data Availability Statement: No external link is applicable for the data.

Acknowledgments: Sarker is the recipient of an Australian Research Council Discovery Early Career Researcher Award (grant number DE200100367) funded by the Australian Government. The Australian Government had no role in the study design, data collection and analysis, decision to publish, or preparation of the manuscript.

Conflicts of Interest: The authors declare no conflict of interest.

References

1. Trebilco, R.; Fischer, M.; Hunter, C.; Hobday, A.; Thomas, L.; Evans, K. Australia state of the environment 2021: Marine, independent report to the Australian Government Minister for the Environment. *Commonw. Aust. Canberra* **2021**, *10*. [CrossRef]
2. Fisheries, F. *Aquaculture Division*; Food and Agriculture Organization of the United Nations: Rome, Italy, 2022.
3. Ritchie, H. The World Now Produces More Seafood from Fish Farms Than Wild Catch. Our-WorldinData.org. 2019. Available online: https://ourworldindata.org/rise-of-aquaculture (accessed on 9 September 2024).
4. Gui, J.-F. Chinese wisdom and modern innovation of aquaculture. *Water Biol. Secur.* **2024**, *3*, 100271. [CrossRef]
5. Garlock, T.; Asche, F.; Anderson, J.; Bjørndal, T.; Kumar, G.; Lorenzen, K.; Ropicki, A.; Smith, M.D.; Tveterås, R. A global blue revolution: Aquaculture growth across regions, species, and countries. *Rev. Fish. Sci. Aquac.* **2020**, *28*, 107–116. [CrossRef]
6. CSIRO. Protecting Marine Biodiversity. Available online: https://www.csiro.au/en/about/facilities-collections/Collections/ANFC/Marine-biodiversity (accessed on 9 September 2024).
7. Henriksson, P.J.G.; Troell, M.; Banks, L.K.; Belton, B.; Beveridge, M.C.M.; Klinger, D.H.; Pelletier, N.; Phillips, M.J.; Tran, N. Interventions for improving the productivity and environmental performance of global aquaculture for future food security. *One Earth* **2021**, *4*, 1220–1232. [CrossRef]
8. Patterson, A.D.; Gonzalez, F.J.; Idle, J.R. Xenobiotic metabolism: A view through the metabolometer. *Chem. Res. Toxicol.* **2010**, *23*, 851–860. [CrossRef]
9. Crane, M.; Hyatt, A. Viruses of fish: An overview of significant pathogens. *Viruses* **2011**, *3*, 2025–2046. [CrossRef]
10. Saravanan, K.; Praveenraj, J.; Kiruba-Sankar, R.; Devi, V.; Biswas, U.; Kumar, T.S.; Sudhagar, A.; El-Matbouli, M.; Kumar, G. Co-Infection of infectious hypodermal and hematopoietic necrosis virus (IHHNV) and white spot syndrome virus (WSSV) in the wild crustaceans of Andaman and Nicobar Archipelago, India. *Viruses* **2021**, *13*, 1378. [CrossRef]
11. Vallejos-Vidal, E.; Reyes-López, F.E.; Sandino, A.M.; Imarai, M. Sleeping with the enemy? The current knowledge of piscine Orthoreovirus (PRV) immune response elicited to counteract infection. *Front. Immunol.* **2022**, *13*, 768621. [CrossRef]
12. Gui, J.-F.; Tang, Q.; Li, Z.; Liu, J.; De Silva, S.S. *Aquaculture in China: Success Stories and Modern Trends*; John Wiley & Sons: Hoboken, NJ, USA, 2018.
13. Zhang, Q.-Y.; Ke, F.; Gui, L.; Zhao, Z. Recent insights into aquatic viruses: Emerging and reemerging pathogens, molecular features, biological effects, and novel investigative approaches. *Water Biol. Secur.* **2022**, *1*, 100062. [CrossRef]
14. Arbon, P.; Martinez, M.A.; Garrett, M.; Jerry, D.; Condon, K. Determining patterns of tissue tropism for IHHNV, GAV and YHV-7 infection in giant black tiger shrimp (*Penaeus monodon*) using real-time RT-qPCR. *Aquaculture* **2024**, *584*, 740680. [CrossRef]
15. Kralik, P.; Ricchi, M. A basic guide to real time PCR in microbial diagnostics: Definitions, parameters, and everything. *Front. Microbiol.* **2017**, *8*, 108. [CrossRef] [PubMed]
16. Kibenge, F.; Kibenge, M.; Montes de Oca, M.; Godoy, M. Parvoviruses of Aquatic Animals. *Pathogens* **2024**, *13*, 625. [CrossRef] [PubMed]
17. Mugimba, K.K.; Byarugaba, D.K.; Mutoloki, S.; Evensen, Ø.; Munang'andu, H.M. Challenges and solutions to viral diseases of finfish in marine aquaculture. *Pathogens* **2021**, *10*, 673. [CrossRef] [PubMed]
18. Mondal, H.; Chandrasekaran, N.; Mukherjee, A.; Thomas, J. Viral infections in cultured fish and shrimps: Current status and treatment methods. *Aquac. Int.* **2022**, *30*, 227–262. [CrossRef]
19. Subasinghe, R.; Alday-Sanz, V.; Bondad-Reantaso, M.G.; Jie, H.; Shinn, A.P.; Sorgeloos, P. Biosecurity: Reducing the burden of disease. *J. World Aquac. Soc.* **2023**, *54*, 397–426. [CrossRef]

20. Arbon, P.; Condon, K.; Martinez, M.A.; Jerry, D. Molecular detection of six viral pathogens from Australian wild sourced giant black tiger shrimp (Penaeus monodon) broodstock. *Aquaculture* **2022**, *548*, 737651. [CrossRef]
21. Page, M.J.; McKenzie, J.E.; Bossuyt, P.M.; Boutron, I.; Hoffmann, T.C.; Mulrow, C.D.; Shamseer, L.; Tetzlaff, J.M.; Akl, E.A.; Brennan, S.E. The PRISMA 2020 statement: An updated guideline for reporting systematic reviews. *BMJ* **2021**, *372*, n71. [CrossRef]
22. Viswanathan, M.; Ansari, M.; Berkman, N.; Chang, S.; Hartling, L.; McPheeters, M.; Treadwell, J. *Assessing the Risk of Bias of Individual Studies in Systematic Reviews of Health Care Interventions. Agency for Healthcare Research and Quality Methods Guide for Comparative Effectiveness Reviews*; AHRQ Methods for Effective Health Care; AHRQ: Rockville, MD, USA, 2012.
23. Becker, J.A.; Gilligan, D.; Asmus, M.; Tweedie, A.; Whittington, R.J. Geographic distribution of epizootic haematopoietic necrosis virus (EHNV) in freshwater fish in south eastern Australia: Lost opportunity for a notifiable pathogen to expand its geographic range. *Viruses* **2019**, *11*, 315. [CrossRef]
24. Becker, J.A.; Tweedie, A.; Rimmer, A.; Landos, M.; Lintermans, M.; Whittington, R.J. Incursions of Cyprinid herpesvirus 2 in goldfish populations in Australia despite quarantine practices. *Aquaculture* **2014**, *432*, 53–59. [CrossRef]
25. Carlile, G.A. *Epidemiological, Molecular and Pathogenesis Studies of Atlantic Salmon Aquareovirus (TSRV)*; School of Veterinary Science, University of Melbourne: Melbourne, Australia, 2013.
26. Zainathan, S.C.; Knowles, G. Tasmanian Atlantic salmon reovirus. In *Aquaculture Pathophysiology*; Elsevier: Amsterdam, The Netherlands, 2022; pp. 277–283.
27. Costa, V.A.; Holmes, E.C. Diversity, evolution, and emergence of fish viruses. *J. Virol.* **2024**, *98*, e00118–e00124. [CrossRef]
28. Toffan, A.N.N.A. Viral Encephalopathy and Retinopathy (VER). In *Fish Viruses and Bacteria: Pathobiology and Protection*; CABI: Wallingford, UK, 2017; pp. 128–146.
29. Rimmer, A.E.; Becker, J.A.; Tweedie, A.; Lintermans, M.; Landos, M.; Stephens, F.; Whittington, R.J. Detection of dwarf gourami iridovirus (Infectious spleen and kidney necrosis virus) in populations of ornamental fish prior to and after importation into Australia, with the first evidence of infection in domestically farmed Platy (*Xiphophorus maculatus*). *Prev. Vet. Med.* **2015**, *122*, 181–194. [CrossRef] [PubMed]
30. Ariel, E.; Owens, L.; Moody, N. A barramundi bioassay for iridovirus refractory to cell culture. In *Diseases in Asian Aquaculture II*; Fish Health Section, Asian Fisheries Society: Manila, PH, USA, 1995; pp. 355–367.
31. Faheem, M.; Azmat, H.; Swar, S.O.; Khaliq, S.; Hoseinifar, S.H. Some important viral diseases of farmed fish. *Vet. Pathobiol. Public Health* **2021**, 472–480. [CrossRef]
32. Langdon, J. Experimental transmission and pathogenicity of epizootic haematopoietic necrosis virus (EHNV) in redfin perch, *Perca fluviatilis* L., and 11 other teleosts. *J. Fish Dis.* **1989**, *12*, 295–310. [CrossRef]
33. Sahoo, P.; Goodwin, A. Viruses of freshwater finfish in the Asian–Pacific region. *Indian J. Virol.* **2012**, *23*, 99–105. [CrossRef]
34. Becker, J.A.; Tweedie, A.; Gilligan, D.; Asmus, M.; Whittington, R.J. Experimental infection of Australian freshwater fish with epizootic haematopoietic necrosis virus (EHNV). *J. Aquat. Anim. Health* **2013**, *25*, 66–76. [CrossRef]
35. Whittington, R.; Becker, J.; Dennis, M. Iridovirus infections in finfish–critical review with emphasis on ranaviruses. *J. Fish Dis.* **2010**, *33*, 95–122. [CrossRef]
36. Tandanus, T.; Schmida, G. Redfin Perch and the EHN Virus. Available online: https://www.finterest.au/stories/redfin-perch-and-the-ehn-virus (accessed on 10 September 2024).
37. Hyatt, A. Diagnosis and Classification of Epizootic Haematopoietic Necrosis Virus (EHNV). Available online: https://www.frdc.com.au/sites/default/files/products/1989-086-DLD.pdf (accessed on 10 September 2024).
38. Schipp, M. Aquatic Animal Diseases Significant to Australia: Identification Field Guide, 5th ed. Available online: https://www.agriculture.gov.au/sites/default/files/documents/field-guide-5th-edition.pdf (accessed on 10 September 2024).
39. Nagata, J.; Takita, K.; Kasai, H. Long-term detection of cyprinid herpesvirus 1 genome and tumorous transcriptome in common carp (*Cyprinus carpio*) infected with the virus. *Aquaculture* **2024**, *592*, 741158. [CrossRef]
40. Aoki, T.; Hirono, I.; Kurokawa, K.; Fukuda, H.; Nahary, R.; Eldar, A.; Davison, A.J.; Waltzek, T.B.; Bercovier, H.; Hedrick, R.P. Genome Sequences of Three Koi Herpesvirus Isolates Representing the Expanding Distribution of an Emerging Disease Threatening Koi and Common Carp Worldwide. *J. Virol.* **2007**, *81*, 5058–5065. [CrossRef]
41. Davison, A.J.; Kurobe, T.; Gatherer, D.; Cunningham, C.; Korf, I.; Fukuda, H.; Hedrick, R.P.; Waltzek, T.B. Comparative Genomics of Carp Herpesviruses. *J. Virol.* **2013**, *87*, 2908–2922. [CrossRef]
42. Kurita, J.; Nakajima, K. Megalocytiviruses. *Viruses* **2012**, *4*, 521–538. [CrossRef]
43. Hedrick, R.P.; Gilad, O.; Yun, S.C.; McDowell, T.S.; Waltzek, T.B.; Kelley, G.O.; Adkison, M.A. Initial isolation and characterization of a herpes-like virus (KHV) from koi and common carp. *Bull. Fish. Res. Agency* **2005**, *2*, V7.
44. El-Matbouli, M.; Saleh, M.; Soliman, H. Detection of cyprinid herpesvirus type 3 in goldfish cohabiting with CyHV-3-infected koi carp (Cyprinus carpio koi). *Vet. Rec.* **2007**, *161*, 792.
45. Hedrick, R.P.; Waltzek, T.B.; McDowell, T.S. Susceptibility of koi carp, common carp, goldfish, and goldfish × common carp hybrids to cyprinid herpesvirus-2 and herpesvirus-3. *J. Aquat. Anim. Health* **2006**, *18*, 26–34. [CrossRef]

46. Sadler, J.; Marecaux, E.; Goodwin, A. Detection of koi herpes virus (CyHV-3) in goldfish, *Carassius auratus* (L.), exposed to infected koi. *J. Fish Dis.* **2008**, *31*, 71. [CrossRef] [PubMed]
47. Omori, R.; Adams, B. Disrupting seasonality to control disease outbreaks: The case of koi herpes virus. *J. Theor. Biol.* **2011**, *271*, 159–165. [CrossRef]
48. McColl, K.A. Review of the Literature on Cyprinid Herpesvirus 3 (CyHV-3) and Its Disease. 2013. Available online: https://pestsmart.org.au/wp-content/uploads/sites/3/2020/06/McColl_KHVLitReview.pdf (accessed on 20 September 2024).
49. Zainathan, S.C.; Carson, J.; Crane, M.S.J.; Williams, L.M.; Hoad, J.; Moody, N.J.; Gudkovs, N.; Crameri, S.; Hyatt, A.D.; Young, J. Preliminary characterization of Tasmanian aquareovirus (TSRV) isolates. *Arch. Virol.* **2017**, *162*, 625–634. [CrossRef]
50. Zainathan, S.C. *Detection of Aquareovirus in Farmed Tasmanian Atlantic Salmon (Salmo salar)*; University of Tasmania: Hobart, Australia, 2012.
51. Carlile, G.; East, I.; McColl, K.; Ellard, K.; Browning, G.; Crane, M.S.J. The spatial and temporal variation of the distribution and prevalence of Atlantic salmon reovirus (TSRV) infection in Tasmania, Australia. *Prev. Vet. Med.* **2014**, *116*, 214–219. [CrossRef]
52. Zainathan, S.; Carlile, G.; Carson, J.; McColl, K.; Crane, M.S.J.; Williams, L.; Hoad, J.; Moody, N.; Aiken, H.; Browning, G. Development and application of molecular methods (PCR) for detection of Tasmanian Atlantic salmon reovirus. *J. Fish Dis.* **2015**, *38*, 739–754. [CrossRef]
53. Samsing, F.; Alexandre, P.; Rigby, M.; Taylor, R.S.; Chong, R.; Wynne, J.W. Transcriptome response of atlantic salmon (*Salmo salar*) to a new piscine orthomyxovirus. *Pathogens* **2020**, *9*, 807. [CrossRef]
54. Mohr, P.G.; Crane, M.S.J.; Hoad, J.; Williams, L.M.; Cummins, D.; Neave, M.J.; Shiell, B.; Beddome, G.; Michalski, W.P.; Peck, G.R. Pilchard orthomyxovirus (POMV). I. Characterisation of an emerging virus isolated from pilchards Sardinops sagax and Atlantic salmon Salmo salar. *Dis. Aquat. Org.* **2020**, *139*, 35–50. [CrossRef]
55. Godwin, S.E.; Morrison, R.N.; Knowles, G.; Cornish, M.C.; Hayes, D.; Carson, J. Pilchard orthomyxovirus (POMV). II. Causative agent of salmon orthomyxoviral necrosis, a new disease of farmed Atlantic salmon Salmo salar. *Dis. Aquat. Org.* **2020**, *139*, 51–68. [CrossRef] [PubMed]
56. Samsing, F.; Rigby, M.; Tengesdal, H.K.; Taylor, R.S.; Farias, D.; Morrison, R.N.; Godwin, S.; Giles, C.; Carson, J.; English, C.J. Seawater transmission and infection dynamics of pilchard orthomyxovirus (POMV) in Atlantic salmon (*Salmo salar*). *J. Fish Dis.* **2021**, *44*, 73–88. [CrossRef] [PubMed]
57. Peducasse, S.; Castric, J.; Thiery, R.; Jeffroy, J.; Le Ven, A.; Laurencin, F.B. Comparative study of viral encephalopathy and retinopathy in juvenile sea bass Dicentrarchus labrax infected in different ways. *Dis. Aquat. Org.* **1999**, *36*, 11–20. [CrossRef] [PubMed]
58. Mu, Y.; Lin, K.; Chen, X.; Ao, J. Diagnosis of nervous necrosis virus in orange-spotted grouper, Epinephelus coioides, by a rapid and convenient RT-PCR method. *Acta Oceanol. Sin.* **2013**, *32*, 88–92. [CrossRef]
59. Jia, P.; Jia, K.-T.; Yi, M.-S. Complete genome sequence of a fish nervous necrosis virus isolated from sea perch (*Lateolabrax japonicus*) in China. *Genome Announc.* **2015**, *3*, e00048-15. [CrossRef]
60. Pascoli, F.; Serra, M.; Toson, M.; Pretto, T.; Toffan, A. Betanodavirus ability to infect juvenile European sea bass, Dicentrarchus labrax, at different water salinity. *J. Fish Dis.* **2016**, *39*, 1061–1068. [CrossRef]
61. Zhang, Y.; Dong, F.; Xing, J.; Tang, X.; Sheng, X.; Chi, H.; Zhan, W. Characterization of Nervous Necrosis Virus (NNV) Nonstructural Protein B2 and Its Enhancement on Virus Proliferation. *Viruses* **2022**, *14*, 2818. [CrossRef]
62. Bandín, I.; Souto, S. Betanodavirus and VER disease: A 30-year research review. *Pathogens* **2020**, *9*, 106. [CrossRef]
63. Nakai, T.; Sugaya, T.; Nishioka, T.; Mushiake, K.; Yamashita, H. Current knowledge on viral nervous necrosis (VNN) and its causative betanodaviruses. *Isr. J. Aquac.-Bamidgeh* **2009**, *61*, 198–207. [CrossRef]
64. Guo, C.-J.; Wu, Y.-Y.; Yang, L.-S.; Yang, X.-B.; He, J.; Mi, S.; Jia, K.-T.; Weng, S.-P.; Yu, X.-Q.; He, J.-G. Infectious spleen and kidney necrosis virus (a fish iridovirus) enters Mandarin fish fry cells via caveola-dependent endocytosis. *J. Virol.* **2012**, *86*, 2621–2631. [CrossRef]
65. Jancovich, J. Family iridoviridae. Virus taxonomy: Ninth report of the international committee on taxonomy of viruses. *J. Gen. Virol.* **2012**, *98*, 193–210.
66. Whittington, R.; Chong, R. Global trade in ornamental fish from an Australian perspective: The case for revised import risk analysis and management strategies. *Prev. Vet. Med.* **2007**, *81*, 92–116. [CrossRef] [PubMed]
67. Jeong, J.B.; Kim, H.Y.; Jun, L.J.; Lyu, J.H.; Park, N.G.; Kim, J.K.; Do Jeong, H. Outbreaks and risks of infectious spleen and kidney necrosis virus disease in freshwater ornamental fishes. *Dis. Aquat. Org.* **2008**, *78*, 209–215. [CrossRef]
68. Mohr, P.G.; Moody, N.J.; Williams, L.M.; Hoad, J.; Cummins, D.M.; Davies, K.R.; Crane, M.S. Molecular confirmation of infectious spleen and kidney necrosis virus (ISKNV) in farmed and imported ornamental fish in Australia. *Dis. Aquat. Org.* **2015**, *116*, 103–110. [CrossRef]
69. Go, J.; Lancaster, M.; Deece, K.; Dhungyel, O.; Whittington, R. The molecular epidemiology of iridovirus in Murray cod (Maccullochella peelii peelii) and dwarf gourami (Colisa lalia) from distant biogeographical regions suggests a link between trade in ornamental fish and emerging iridoviral diseases. *Mol. Cell. Probes* **2006**, *20*, 212–222. [CrossRef]

70. Sudthongkong, C.; Miyata, M.; Miyazaki, T. Iridovirus disease in two ornamental tropical fresh-water fishes: African lampeye and dwarf gourami. *Dis. Aquat. Org.* **2002**, *48*, 163–173. [CrossRef]
71. Anderson, I.; Prior, H.; Rodwell, B.; Harris, G. Iridovirus-like virions in imported dwarf gourami (*Colisa lalia*) with systemic amoebiasis. *Aust. Vet. J.* **1993**, *70*, 66–67. [CrossRef]
72. Becker, J.R.A.; Tweedie, A.; Landos, M.; Lintermans, M.; Whittington, R. *Aquatic Animal Health Subprogram: Surveys of Ornamental Fish for Pathogens of Quarantine Significance*; Fisheries Research and Development Corporation: Canberra, Australia, 2013; Project Number 2009/044.
73. Nolan, D.; Stephens, F.; Crockford, M.; Jones, J.; Snow, M. Detection and characterization of viruses of the genus Megalocytivirus in ornamental fish imported into an Australian border quarantine premises: An emerging risk to national biosecurity. *J. Fish Dis.* **2015**, *38*, 187–195. [CrossRef]
74. Hick, P.M.; Subramaniam, K.; Thompson, P.; Whittington, R.J.; Waltzek, T.B. Complete genome sequence of a *Bohle iridovirus* isolate from ornate burrowing frogs (*Limnodynastes ornatus*) in Australia. *Genome Announc.* **2016**, *4*, e00632-16. [CrossRef]
75. Chinchar, V.G.; Hick, P.; Ince, I.A.; Jancovich, J.K.; Marschang, R.; Qin, Q.; Subramaniam, K.; Waltzek, T.B.; Whittington, R.; Williams, T. ICTV virus taxonomy profile: Iridoviridae. *J. Gen. Virol.* **2017**, *98*, 890–891. [CrossRef]
76. Sunarto, A.; Grimm, J.; McColl, K.A.; Ariel, E.; Nair, K.K.; Corbeil, S.; Hardaker, T.; Tizard, M.; Strive, T.; Holmes, B. Bioprospecting for biological control agents for invasive tilapia in Australia. *Biol. Control* **2022**, *174*, 105020. [CrossRef]
77. Machimbirike, V.I.; Jansen, M.D.; Senapin, S.; Khunrae, P.; Rattanarojpong, T.; Dong, H.T. Viral infections in tilapines: More than just tilapia lake virus. *Aquaculture* **2019**, *503*, 508–518. [CrossRef]
78. Shin, Y.; Kwon, T.; Seo, J.; Kim, T. Oral immunization of fish against iridovirus infection using recombinant antigen produced from rice callus. *Vaccine* **2013**, *31*, 5210–5215. [CrossRef]
79. Marsh, I.; Whittington, R.; Hyatt, A.; Chisholm, O. Rapid differentiation of Australian, European and American ranaviruses based on variation in major capsid protein gene sequence. *Mol. Cell. Probes* **2002**, *16*, 137–151. [CrossRef]
80. Hyatt, A.; Williamson, M.; Coupar, B.; Middleton, D.; Hengstberger, S.; Gould, A.; Selleck, P.; Wise, T.; Kattenbelt, J.; Cunningham, A. First identification of a ranavirus from green pythons (*Chondropython viridis*). *J. Wildl. Dis.* **2002**, *38*, 239–252. [CrossRef]
81. Costa, V.A.; Geoghegan, J.L.; Holmes, E.C.; Harvey, E. Genetic Reassortment between Endemic and Introduced Macrobrachium rosenbergii Nodaviruses in the Murray-Darling Basin, Australia. *Viruses* **2022**, *14*, 2186. [CrossRef]
82. Oakey, H.J.; Smith, C.S. Complete genome sequence of a white spot syndrome virus associated with a disease incursion in Australia. *Aquaculture* **2018**, *484*, 152–159. [CrossRef]
83. Rajendran, K.; Makesh, M.; Karunasagar, I. Monodon baculovirus of shrimp. *Indian J. Virol.* **2012**, *23*, 149–160. [CrossRef]
84. Cowley, J.A. Mourilyan virus pathogenicity in kuruma shrimp (*Penaeus japonicus*). *J. Fish Dis.* **2020**, *43*, 1401–1407. [CrossRef]
85. La Fauce, K.A.; Layton, R.; Owens, L. TaqMan real-time PCR for detection of hepatopancreatic parvovirus from Australia. *J. Virol. Methods* **2007**, *140*, 10–16. [CrossRef]
86. Mohr, P.G.; Moody, N.J.; Hoad, J.; Williams, L.M.; Bowater, R.O.; Cummins, D.M.; Cowley, J.A.; Crane, M.S. New yellow head virus genotype (YHV7) in giant tiger shrimp Penaeus monodon indigenous to northern Australia. *Dis. Aquat. Org.* **2015**, *115*, 263–268. [CrossRef] [PubMed]
87. Stanley, J.; Bisaro, D.; Briddon, R.; Brown, J.; Fauquet, C.; Harrison, B.; Rybicki, E.; Stenger, D. Geminivridae: Eighth Report of the ICTV on Virus Taxonomy. In *Virus Taxonomy-Eighth Report of the International Committee on Taxonomy of Viruses*; Elsevier: San Diego, CA, USA, 2005; pp. 301–306.
88. ICTV. Taxon Name: *Penstylhamaparvovirus decapod1*. Available online: https://ictv.global/taxonomy/taxondetails?taxnode_id=202304232&taxon_name=Penstylhamaparvovirus%20decapod1 (accessed on 16 October 2024).
89. Kim, J.; Kim, H.; Nguyen, V.; Park, B.; Choresca, C.; Shin, S.; Han, J.; Jun, J.; Park, S. Genomic sequence of infectious hypodermal and hematopoietic necrosis virus (IHHNV) KLV-2010-01 originating from the first Korean outbreak in cultured Litopenaeus vannamei. *Arch. Virol.* **2012**, *157*, 369–373. [CrossRef] [PubMed]
90. Mari, J.; Bonami, J.-R.; Lightner, D. Partial cloning of the genome of infectious hypodermal and haematopoietic necrosis virus, an unusual parvovirus pathogenic for penaeid shrimps; diagnosis of the disease using a specific probe. *J. Gen. Virol.* **1993**, *74*, 2637–2643. [CrossRef] [PubMed]
91. Hsieh, C.Y.; Chuang, P.C.; Chen, L.C.; Tu, C.; Chien, M.S.; Huang, K.C.; Kao, H.F.; Tung, M.C.; Tsai, S.S. Infectious hypodermal and haematopoietic necrosis virus (IHHNV) infections in giant freshwater prawn, *Macrobrachium rosenbergii*. *Aquaculture* **2006**, *258*, 73–79. [CrossRef]
92. Infectious Hypodermal and Haematopoietic Necrosis (IHHN) (Also Known as Infection with Penaeus Stylirostris Densovirus [PstDNV]). Aquatic Animal Diseases Significant to Australia: Identification Field Guide 4th Edition. Available online: https://www.agriculture.gov.au/sites/default/files/sitecollectiondocuments/animal-plant/aquatic/field-guide/4th-edition/crustaceans/ihhn.pdf? (accessed on 20 October 2024).
93. Rai, P.; Safeena, M.P.; Krabsetsve, K.; La Fauce, K.; Owens, L.; Karunasagar, I. Genomics, molecular epidemiology and diagnostics of infectious hypodermal and hematopoietic necrosis virus. *Indian J. Virol.* **2012**, *23*, 203–214. [CrossRef]

94. Cowley, J.A.; Rao, M.; Coman, G.J. Real-time PCR tests to specifically detect IHHNV lineages and an IHHNV EVE integrated in the genome of Penaeus monodon. *Dis. Aquat. Org.* **2018**, *129*, 145–158. [CrossRef]
95. Lightner, D.V.; Redman, R.M.; Pantoja, C.R.; Tang, K.; Noble, B.; Schofield, P.; Mohney, L.; Nunan, L.; Navarro, S. Historic emergence, impact and current status of shrimp pathogens in the Americas. *J. Invertebr. Pathol.* **2012**, *110*, 174–183. [CrossRef]
96. Pechenik, J. *Invertebrates*; McGraw Hill: Singapore, 2000; Volume 193.
97. Wongteerasupaya, C.; Sriurairatana, S.; Vickers, J.E.; Akrajamorn, A.; Boonsaeng, V.; Panyim, S.; Tassanakajon, A.; Withyachumnarnkul, B.; Flegel, T. Yellow-head virus of Penaeus monodon is an RNA virus. *Dis. Aquat. Org.* **1995**, *22*, 45–50. [CrossRef]
98. Nadala, E.C.B., Jr.; Tapay, L.M.; Loh, P.C. Yellow-head virus: A rhabdovirus-like pathogen of penaeid shrimp. *Dis. Aquat. Org.* **1997**, *31*, 141–146. [CrossRef]
99. Spann, K.; Cowley, J.; Walker, P.; Lester, R. A yellow-head-like virus from Penaeus monodon cultured in Australia. *Dis. Aquat. Org.* **1997**, *31*, 169–179. [CrossRef]
100. Walker, P.J.; Winton, J.R. Emerging viral diseases of fish and shrimp. *Vet. Res.* **2010**, *41*, 51. [CrossRef] [PubMed]
101. Wijegoonawardane, P.K.; Cowley, J.A.; Phan, T.; Hodgson, R.A.; Nielsen, L.; Kiatpathomchai, W.; Walker, P.J. Genetic diversity in the yellow head nidovirus complex. *Virology* **2008**, *380*, 213–225. [CrossRef] [PubMed]
102. Australian Government Department of Agriculture. Infection with Gill-Associated Virus (GAV). 2019. Available online: https://www.agriculture.gov.au/sites/default/files/documents/gill_associated_virus_disease.pdf (accessed on 12 November 2024).
103. Noble, T.H.; Coman, G.J.; Wade, N.M.; Thomson, P.C.; Raadsma, H.W.; Khatkar, M.S.; Guppy, J.L.; Jerry, D.R. Genetic parameters of Gill-associated virus infection and body weight under commercial conditions in black tiger shrimp, Penaeus monodon. *Aquaculture* **2020**, *528*, 735580. [CrossRef]
104. Cowley, J.A.; Cadogan, L.C.; Wongteerasupaya, C.; Hodgson, R.A.; Boonsaeng, V.; Walker, P.J. Multiplex RT-nested PCR differentiation of gill-associated virus (Australia) from yellow head virus (Thailand) of *Penaeus monodon*. *J. Virol. Methods* **2004**, *117*, 49–59. [CrossRef]
105. Cowley, J.A.; Dimmock, C.M.; Spann, K.M.; Walker, P.J. Detection of Australian gill-associated virus (GAV) and lymphoid organ virus (LOV) of Penaeus monodon by RT-nested PCR. *Dis. Aquat. Org.* **2000**, *39*, 159–167. [CrossRef]
106. Tang, K.F.-J.; Spann, K.M.; Owens, L.; Lightner, D.V. In situ detection of Australian gill-associated virus with a yellow head virus gene probe. *Aquaculture* **2002**, *205*, 1–5. [CrossRef]
107. Noble, T.; Stratford, C.; Wade, N.; Cowley, J.; Sellars, M.; Coman, G.; Jerry, D. PCR testing of single tissue samples can result in misleading data on gill-associated virus infection loads in shrimp. *Aquaculture* **2018**, *492*, 91–96. [CrossRef]
108. Low, C.F.; Md Yusoff, M.R.; Kuppusamy, G.; Ahmad Nadzri, N.F. Molecular biology of Macrobrachium rosenbergii nodavirus infection in giant freshwater prawn. *J. Fish Dis.* **2018**, *41*, 1771–1781. [CrossRef]
109. Senapin, S.; Jaengsanong, C.; Phiwsaiya, K.; Prasertsri, S.; Laisutisan, K.; Chuchird, N.; Limsuwan, C.; Flegel, T.W. Infections of MrNV (Macrobrachium rosenbergii nodavirus) in cultivated whiteleg shrimp Penaeus vannamei in Asia. *Aquaculture* **2012**, *338*, 41–46. [CrossRef]
110. Arcier, J.-M.; Herman, F.; Lightner, D.V.; Redman, R.M.; Mari, J.; Bonami, J.-R. A viral disease associated with mortalities in hatchery-reared postlarvae of the giant freshwater prawn Macrobrachium rosenbergii. *Dis. Aquat. Org.* **1999**, *38*, 177–181. [CrossRef]
111. Owens, L.; La Fauce, K.; Juntunen, K.; Hayakijkosol, O.; Zeng, C. Macrobrachium rosenbergii nodavirus disease (white tail disease) in Australia. *Dis. Aquat. Org.* **2009**, *85*, 175–180. [CrossRef] [PubMed]
112. Chen, K.F.; Tan, W.S.; Ong, L.K.; Zainal Abidin, S.A.; Othman, I.; Tey, B.T.; Lee, R.F.S. The Macrobrachium rosenbergii nodavirus: A detailed review of structure, infectivity, host immunity, diagnosis and prevention. *Rev. Aquac.* **2021**, *13*, 2117–2141. [CrossRef]
113. Hayakijkosol, O.; Burgess, G.; La Fauce, K.; Owens, L. The complete sequence of the Australia recognizate of Macrobrachium rosenbergii nodavirus which causes white tail disease. *Aquaculture* **2012**, *366*, 98–104. [CrossRef]
114. Mayo, M. Virus taxonomy-houston 2002. *Arch. Virol.* **2002**, *147*, 1071–1076.
115. Mayo, M. A summary of taxonomic changes recently approved by ICTV. *Arch. Virol.* **2002**, *147*, 1655–1656. [CrossRef]
116. Vlak, J.M.; Bonami, J.-R.; Flegel, T.W.; Kou, G.-H.; Lightner, D.V.; Lo, C.-F.; Loh, P.C.; Walker, P.J. Nimaviridae, a new virus family infecting aquatic invertebrates. In Proceedings of the XII International Congress of Virology, Paris, France, 27 July–1 August 2002.
117. Pradeep, B.; Rai, P.; Mohan, S.A.; Shekhar, M.S.; Karunasagar, I. Biology, host range, pathogenesis and diagnosis of white spot syndrome virus. *Indian J. Virol.* **2012**, *23*, 161–174. [CrossRef]
118. Department of Agriculture, Water and the Environment. Infection with White Spot Syndrome Virus (WSSV). 2019. Available online: https://www.agriculture.gov.au/sites/default/files/documents/infection-white-spot-syndrome-virus.pdf (accessed on 15 November 2024).
119. CSIRO. White Spot Syndrome Virus Diagnostic Test Validation. Available online: https://www.csiro.au/en/about/facilities-collections/acdp/white-spot-test-validation (accessed on 25 November 2024).

120. Li, C.-X.; Shi, M.; Tian, J.-H.; Lin, X.-D.; Kang, Y.-J.; Chen, L.-J.; Qin, X.-C.; Xu, J.; Holmes, E.C.; Zhang, Y.-Z. Unprecedented genomic diversity of RNA viruses in arthropods reveals the ancestry of negative-sense RNA viruses. *eLife* **2015**, *4*, e05378. [CrossRef]
121. Huerlimann, R.; Wade, N.M.; Gordon, L.; Montenegro, J.D.; Goodall, J.; McWilliam, S.; Tinning, M.; Siemering, K.; Giardina, E.; Donovan, D. De novo assembly, characterization, functional annotation and expression patterns of the black tiger shrimp (Penaeus monodon) transcriptome. *Sci. Rep.* **2018**, *8*, 13553. [CrossRef]
122. Lightner, D.V.; Redman, R.M. A baculovirus-caused disease of the penaeid shrimp, Penaeus monodon. *J. Invertebr. Pathol.* **1981**, *38*, 299–302. [CrossRef]
123. Ahamed, A.S.; Haq, M.; Banu, M.N.; Tiwary, C.; Sedhuraman, R. An update of major DNA shrimp viruses-White Spot Syndrome virus (WSSV), Monodon Baculovius (MBV), Infectious Hypodermal and Hematopoietic Necrosis virus (IHHNV) and Hepatopancreatic Parvovirus (HPV): A Review. *Imp. J. Interdiscip. Res.* **2016**, *3*, 1–15.
124. Doubrovsky, A.; Paynter, J.; Sambhi, S.; Atherton, J.; Lester, R. Observations on the Ultrastracture of Baculovirus in Australian Penaeus monodon and Penaeus merguiensis. *Mar. Freshw. Res.* **1988**, *39*, 743–749. [CrossRef]
125. Lightner, D.; Redman, R.; Bell, T.; Brock, J. Detection of IHHN virus in *Penaeus stylirostris* and *P. vannamei* imported into Hawaii. *J. World Maric. Soc.* **1983**, *14*, 212–225. [CrossRef]
126. Government of Canada. Monodon Baculovirus (MBV) Disease of Penaeid Shrimp. 2018. Available online: https://www.dfo-mpo.gc.ca/science/aah-saa/diseases-maladies/mbvsp-eng.html (accessed on 5 December 2024).
127. NSW Government. NSW Government Invests $21 Million in the Future for Clarence Prawn Fishers and Farmers. 2023. Available online: https://www.dpi.nsw.gov.au/about-us/media-centre/releases/2023/general2/nsw-government-invests-$21-million-in-the-future-for-clarence-prawn-fishers-and-farmers (accessed on 22 December 2024).
128. Belcher, C.R.; Young, P.R. Colourimetric PCR-based detection of monodon baculovirus in whole Penaeus monodon postlarvae. *J. Virol. Methods* **1998**, *74*, 21–29. [CrossRef] [PubMed]
129. Vickers, J.; Paynter, J.; Spradbrow, P.; Lester, R. An impression smear method for rapid detection of Penaeus monodon-type baculovirus (MBV) in Australian prawns. *J. Fish Dis.* **1993**, *16*, 507. [CrossRef]
130. Poulos, B.T.; Mari, J.; Bonami, J.-R.; Redman, R.; Lightner, D.V. Use of non-radioactively labeled DNA probes for the detection of a baculovirus from Penaeus monodon by in situ hybridization on fixed tissue. *J. Virol. Methods* **1994**, *49*, 187–193. [CrossRef]
131. Abudurexiti, A.; Adkins, S.; Alioto, D.; Alkhovsky, S.V.; Avšič-Županc, T.; Ballinger, M.J.; Bente, D.A.; Beer, M.; Bergeron, É.; Blair, C.D. Taxonomy of the order Bunyavirales: Update 2019. *Arch. Virol.* **2019**, *164*, 1949–1965. [CrossRef]
132. Cowley, J. Bunyaviruses of crustaceans. In *Aquaculture Virology*; Elsevier: Amsterdam, The Netherlands, 2016; pp. 489–503.
133. Cowley, J.A.; McCulloch, R.J.; Rajendran, K.; Cadogan, L.C.; Spann, K.M.; Walker, P.J. RT-nested PCR detection of Mourilyan virus in Australian Penaeus monodon and its tissue distribution in healthy and moribund prawns. *Dis. Aquat. Org.* **2005**, *66*, 91–104. [CrossRef]
134. Cowley, J.A.; McCulloch, R.J.; Spann, K.M.; Cadogan, L.; Walker, P.J. Preliminary molecular and biological characterisation of Mourilyan virus (MoV): A new bunya-related virus of penaeid prawns. In *Diseases in Asian Aquaculture V, Proceedings of the 5th Symposium on Diseases in Asian Aquaculture, Gold Coast, Australia, 24–28 November 2002*; Asian Fisheries Society: Manila, PH, USA, 2005; pp. 113–124.
135. Sellars, M.J.; Keys, S.J.; Cowley, J.A.; McCulloch, R.J.; Preston, N.P. Association of Mourilyan virus with mortalities in farm pond-reared Penaeus (Marsupenaeus) japonicus transferred to maturation tank systems. *Aquaculture* **2006**, *252*, 242–247. [CrossRef]
136. Rajendran, K.; Cowley, J.A.; McCulloch, R.J.; Walker, P.J. A TaqMan real-time RT-PCR for quantifying Mourilyan virus infection levels in penaeid shrimp tissues. *J. Virol. Methods* **2006**, *137*, 265–271. [CrossRef]
137. Bonami, J.-R.; Mari, J.; Poulos, B.T.; Lightner, D.V. Characterization of hepatopancreatic parvo-like virus, a second unusual parvovirus pathogenic for penaeid shrimps. *J. Gen. Virol.* **1995**, *76*, 813–817. [CrossRef]
138. Safeena, M.P.; Rai, P.; Karunasagar, I. Molecular biology and epidemiology of hepatopancreatic parvovirus of penaeid shrimp. *Indian J. Virol.* **2012**, *23*, 191–202. [CrossRef] [PubMed]
139. Chong, Y.; Loh, H. Hepatopancreas chlamydial and parvoviral infections of farmed marine prawns in Singapore. *Singap. Vet. J.* **1984**, *8/9*, 51–56.
140. Paynter, J. Prawn virus from juvenile Penaeus esculentus. *Second. Aust. Prawn Semin.* **1985**, *1985*. Available online: https://cir.nii.ac.jp/crid/1570291225526464256 (accessed on 27 March 2025).
141. Roubal, F.; Paynter, J.; Lester, R. Electron microscopic observation of hepatopancreatic parvo-like virus (HPV) in the penaeid prawn, Penaeus merguiensis de Man, from Australia. *J. Fish Dis.* **1989**, *12*, 199. [CrossRef]
142. Lightner, D. *A Handbook of Pathology and Diagnostic Procedures for Diseases of Penaeid Shrimp*; World Aquaculture Society: Baton Rouge, LA, USA, 1996.
143. Spann, K.; Adlard, R.; Hudson, D.; Pyecroft, S.; Jones, T.; Voigt, M. Hepatopancreatic parvo-like virus (HPV) of *Penaeus japonicus* cultured in Australia. *Dis. Aquat. Org.* **1997**, *31*, 239–241. [CrossRef]

144. Flegel, T.; Thamavit, V.; Pasharawipas, T.; Alday-Sanz, V. Statistical correlation between severity of hepatopancreatic parvovirus infection and stunting of farmed black tiger shrimp (*Penaeus monodon*). *Aquaculture* **1999**, *174*, 197–206. [CrossRef]
145. Lightner, D.; Redman, R. A parvo-like virus disease of penaeid shrimp. *J. Invertebr. Pathol.* **1985**, *45*, 47–53. [CrossRef]
146. Pollak, N.M.; Fais, O.; Kristoffersen, J.; Phuthaworn, C.; Knibb, W.; Macdonald, J. Rapid sample preparation and low-resource molecular detection of hepatopancreatic parvoviruses (HPV) by recombinase polymerase amplification lateral flow detection assay in shrimps (*Fenneropenaeus merguiensis*). *PLoS ONE* **2022**, *17*, e0276164. [CrossRef]
147. Moody, N.J.G.; Crane, M. Aquatic Animal Health Subprogram: Determining the Susceptibility of Australian Penaeus Monodon and P. Merguiensis to Newly Identified Enzootic (YHV7) and Exotic (YHV8 and YHV10) Yellow Head Virus (YHV) Genotypes. 2016. Available online: https://www.frdc.com.au/sites/default/files/products/2015-005-DLD.pdf (accessed on 24 December 2024).
148. Cowley, J.A.; Rao, M.; Mohr, P.; Moody, N.J.; Sellars, M.J.; Crane, M.S. TaqMan real-time and conventional nested PCR tests specific to yellow head virus genotype 7 (YHV7) identified in giant tiger shrimp in Australia. *J. Virol. Methods* **2019**, *273*, 113689. [CrossRef]
149. White Spot Disease of Prawns Queensland Response 2016-17 Scenario Planning Advisory Panel Report. Available online: https://www.daf.qld.gov.au/__data/assets/pdf_file/0010/1237285/wssv-advisory-panel-report.pdf (accessed on 27 March 2025).
150. Aquatic Biosecurity Risk Review: Prohibited Matter Listing of Infectious Hypodermal and Haematopoietic Necrosis Virus (IHHNV). Available online: https://www.dpi.nsw.gov.au/biosecurity/aquatic-biosecurity-risk-review-prohibited-matter-listing-of-infectious-hypodermal-and-haematopoietic-necrosis-virus-ihhnv (accessed on 27 March 2025).
151. Sellars, M.J.; Rao, M.; O'Leary, Z.; Wood, A.; Degnan, B.M.; Cowley, J.A. Reduced loads of pre-existing Gill-associated virus (GAV) infection in juvenile Penaeus monodon injected with single or multiple GAV-specific dsRNAs. *Aquaculture* **2014**, *434*, 272–276. [CrossRef]
152. Glanville, R.; Neville, P.; Walker, P. White Spot Disease of Prawns Queensland Response 2016–17 Scenario Planning Advisory Panel Report. Available online: https://www.agriculture.gov.au/sites/default/files/documents/infection-macrobrachium-rosenbergii-nodavirus.pdf (accessed on 1 January 2025).
153. Knibb, W.; Le, C.; Katouli, M.; Bar, I.; Lloyd, C. Assessment of the origin of white spot syndrome virus DNA sequences in farmed Penaeus monodon in Australia. *Aquaculture* **2018**, *494*, 26–29. [CrossRef]
154. FRDC. Economic Impact of 2016 White Spot Disease Outbreak. Available online: https://www.frdc.com.au/sites/default/files/2021-07/2016-267-Project-Summary-Economic-Impact.pdf (accessed on 1 January 2025).
155. ABC. Prawn White Spot Disease: What Is It? Available online: https://www.abc.net.au/news/rural/2017-03-16/what-is-white-spot-disease-in-prawns/8359476 (accessed on 2 January 2025).
156. ABC. White Spot Disease: Emergency Funding for Farmers and Wider Industry. Available online: https://www.abc.net.au/news/2017-01-26/white-spot-emergency-funding-compensation-barnaby-joyce/8214548 (accessed on 4 January 2025).
157. Bondad-Reantaso, M.G.; Subasinghe, R.P.; Arthur, J.R.; Ogawa, K.; Chinabut, S.; Adlard, R.; Tan, Z.; Shariff, M. Disease and health management in Asian aquaculture. *Vet. Parasitol.* **2005**, *132*, 249–272. [CrossRef]
158. ACIL-Allen Consulting. Case Study: Aquaculture Feed and Prawn Breeding. Available online: https://www.csiro.au/-/media/About/Files/Impact-case-studies/Full-Reports/ACIL-Allen_CSIROs-Impact-and-Value_Aquaculture-and-Prawn-Breeding_2014.pdf (accessed on 12 December 2024).
159. Department of Agriculture, Fisheries and Forestry. Implementation of Changes to Import Conditions for Prawns and Prawn Products (Excluding Dried Prawns and Shelf Stable Prawns) for Human Consumption. 2023. Available online: https://www.agriculture.gov.au/biosecurity-trade/import/industry-advice/2023/177-2023 (accessed on 15 December 2024).
160. Diggles, B. Identifying and Addressing the Biosecurity Risks to Australia Associated with Imported Prawns and Seafood Products. Available online: https://www.aph.gov.au/DocumentStore.ashx?id=bbdb978d-5ddd-4499-b540-eb1ea2499e45&subId=510046 (accessed on 17 October 2024).
161. FRDC. An Impact Assessment of FRDC Investment in 2016-266: A plan for the Australian Prawn Farming Industry's Initial Response to the White Spot Disease Incident in Summer 2016–17. Available online: https://www.frdc.com.au/sites/default/files/inline-files/BCA%2016-17%20FRDC%20Project%202016-266%20White%20Spot%20Final_0.pdf (accessed on 13 September 2024).
162. Department of Primary Industries. White Spot Disease. Available online: https://www.daf.qld.gov.au/business-priorities/biosecurity/animal-biosecurity-welfare/animal-health-pests-diseases/list-animal-pest-disease/white-spot-disease (accessed on 13 October 2024).
163. BL, M.; Owens, L. Viral diseases of fish and shellfish in Australian mariculture. *Fish Pathol.* **1998**, *33*, 193–200.
164. Lightner, D. Virus diseases of farmed shrimp in the Western Hemisphere (the Americas): A review. *J. Invertebr. Pathol.* **2011**, *106*, 110–130. [CrossRef]

165. Fish Health Management. Fisheries Victoria Research Report Series-No. 32 December 2005. Available online: https://vfa.vic.gov.au/aquaculture/murray-cod-aquaculture/fish-health-management?utm_source.com (accessed on 23 August 2024).
166. Department of Agriculture, Fisheries and Forestry. Biosecurity Conditions for Imported Prawns. Available online: https://www.agriculture.gov.au/about/news/biosecurity-imported-prawns (accessed on 12 June 2024).
167. Moody, N.; Mohr, P. Australian and New Zealand Standard Diagnostic Procedure (ANZSDP) for White Spot Syndrome Virus. 2022. Available online: https://www.agriculture.gov.au/sites/default/files/sitecollectiondocuments/animal/ahl/ANZSDP-White-spot-syndrome-virus.pdf (accessed on 30 December 2024).
168. Animal and Plant Pests and Disease. White Spot Disease. Available online: https://www.outbreak.gov.au/current-outbreaks/white-spot-disease (accessed on 10 January 2025).
169. Hine, M.; Adams, S.; Arthur, J.R.; Bartley, D.; Bondad-Reantaso, M.G.; Chávez, C.; Clausen, J.H.; Dalsgaard, A.; Flegel, T.; Gudding, R.; et al. Improving Biosecurity: A Necessity for Aquaculture Sustainability. 2010. Available online: https://researchportal.murdoch.edu.au/esploro/outputs/conferencePaper/Improving-biosecurity-A-necessity-for-aquaculture/991005543480907891#file-0 (accessed on 21 August 2024).
170. Kumar, S.; Paul, T. Biosafety and Biosecurity for Sustainable Aquaculture Development. In *Advances in Fisheries Biotechnology*; Springer: Berlin/Heidelberg, Germany, 2022; pp. 453–463.
171. Scarfe, A.D.; Palić, D. Aquaculture biosecurity: Practical approach to prevent, control, and eradicate diseases. In *Aquaculture Health Management*; Elsevier: Amsterdam, The Netherlands, 2020; pp. 75–116.
172. Bernoth, E.; Ernst, I.; Wright, B. National aquatic animal health plans: The Australian experience. *Rev. Sci. Tech. (Int. Off. Epizoot.)* **2008**, *27*, 71–88. [CrossRef]
173. AUS-DAFF. *AQUAPLAN—Australia's National Strategic Plan for Aquatic Animal Health 1998–2003*; Australian Department of Agriculture, Fisheries and Forestry (AU-DAFF): Canberra, Australia, 1988; 34p.
174. AUS-DAFF. *AQUAPLAN 2005–2010—Australia's National Strategic Plan for Aquatic Animal Health*; Australian Department of Agriculture, Fisheries and Forestry (AU-DAFF): Canberra, Australia, 2005; 52p.
175. AUS-DA. *AQUAPLAN 2014–2019—Australia's National Strategic Plan for Aquatic Animal Health*; Australian Department of Agriculture (AU-DA): Canberra, Australia, 2014; 29p.
176. FRDC. Aquaplan. Available online: https://www.agriculture.gov.au/sites/default/files/documents/aquaplan-2022-2027.pdf (accessed on 15 January 2025).
177. Department of Agriculture, Fisheries and Forestry. AQUAVETPLAN. Available online: https://www.agriculture.gov.au/agriculture-land/animal/aquatic/aquavetplan (accessed on 27 December 2024).
178. Animal Health Australia. Emergency Animal Disease Response Agreement. Available online: https://animalhealthaustralia.com.au/eadra/#:~:text=Animal%20Health%20Australia%20(AHA)%20manages,response%20if%20an%20outbreak%20occurs (accessed on 27 August 2024).
179. Hishamunda, N.; Ridler, N.; Martone, E. Policy and Governance in Aquaculture: Lessons Learned and Way Forward. In *FAO Fisheries and Aquaculture Technical Paper*; FAO: Rome, Italy, 2014; Available online: https://www.proquest.com/docview/1541489093?pq-origsite=gscholar&fromopenview=true&sourcetype=Scholarly%20Journals (accessed on 23 July 2024).
180. Mishra, S.; Seshagiri, B.; Rathod, R.; Sahoo, S.N.; Choudhary, P.; Patel, S.; Behera, D.K.; Ojha, D.K.; Jena, A.; Namburu, P.K. Recent advances in fish disease diagnosis, therapeutics, and vaccine development. *Front. Aquac. Biotechnol.* **2023**, 115–145. [CrossRef]
181. Alam, M.S.; Islam, M.N.; Das, M.; Islam, S.F.; Rabbane, M.G.; Karim, E.; Roy, A.; Alam, M.S.; Ahmed, R.; Kibria, A.S.M. RNAi-based therapy: Combating shrimp viral diseases. *Viruses* **2023**, *15*, 2050. [CrossRef]
182. Department of Agriculture, Fisheries and Forestry. The Network of Aquaculture Centres in Asia–Pacific. Available online: https://www.agriculture.gov.au/agriculture-land/fisheries/aquaculture/collaboration (accessed on 19 August 2024).
183. Animal Health Australia. Aquatic Health and Biosecurity. Available online: https://animalhealthaustralia.com.au/aquatic-health/ (accessed on 13 September 2024).
184. Biosecurity Australia. *Generic Import Risk Analysis Report for Prawns and Prawn Products*; Biosecurity Australia: Canberra, Australia, 2009; p. 305.
185. Department of Agriculture, Fisheries and Forestry. Legislation. Available online: https://www.agriculture.gov.au/biosecurity-trade/policy/legislation#biosecurity-legislation (accessed on 12 November 2024).
186. Walker, P.J.; Mohan, C. Viral disease emergence in shrimp aquaculture: Origins, impact and the effectiveness of health management strategies. *Rev. Aquac.* **2009**, *1*, 125–154. [CrossRef]
187. Inspector-General of Biosecurity. Submission to Inspector-General of Biosecurity Review. Available online: https://www.researchgate.net/profile/Dr-B-K-Diggles/publication/320602800_DigsFish_Submission_to_the_Inspector-General_of_Biosecurity%E2%80%99s_review_of_the_circumstances_leading_to_the_2017_suspension_of_uncooked_prawn_imports_into_Autralia_and_the_biosecurity_considerations_rel/links/59efe4d90f7e9baeb26ace23/DigsFish-Submission-to-the-Inspector-General-of-Biosecuritys-review-of-the-circumstances-leading-to-the-2017-suspension-of-uncooked-prawn-imports-into-Australia-and-the-biosecurity-considerations-rele.pdf (accessed on 23 August 2024).

188. Department of Agriculture, Fisheries and Forestry. Disease Surveillance and Reporting. Available online: https://www.agriculture.gov.au/agriculture-land/animal/aquatic/reporting (accessed on 19 August 2024).
189. Global Seafood Alliance. Australian Biotech Company Launches Early Detection Technology for Shrimp Diseases. Available online: https://www.globalseafood.org/advocate/australian-biotech-company-launches-early-detection-technology-for-shrimp-diseases/#:~:text=Genics,%20an%20Australian%20biotech%20firm,%20launched%20Shrimp%20MultiPath2.,just%20one%20pathogen%20per%20sample (accessed on 21 July 2024).
190. James Cook University. JCU Project to Help Safeguard Prawn Industry. Available online: https://www.jcu.edu.au/news/releases/2024/june/jcu-project-to-help-safeguard-prawn-industry#https://www.sciencedirect.com/science/article/pii/S2666765723000303 (accessed on 23 August 2024).
191. FRDC. Water Treatment to Control Influent Water Biosecurity Risk on Australian Prawn Farms. Effectiveness and Impacts on Production Ponds. Available online: https://www.frdc.com.au/project/2017-238 (accessed on 21 July 2024).
192. Sub-Committee on Aquatic Animal Health. Available online Aquaculture Farm Biosecurity Plan. Available online: https://www.agriculture.gov.au/sites/default/files/documents/aquaculture-farm-biosecurity-plan-generic.pdf (accessed on 23 July 2024).
193. Department of Agriculture, Fisheries and Forestry. AQUAVETPLAN. Disease Strategy for White Spot Disease. Available online: https://www.agriculture.gov.au/sites/default/files/sitecollectiondocuments/animal-plant/aquatic/aquavetplan/white-spot.pdf (accessed on 7 May 2024).
194. CRCNA. Demonstrating the Impact of Prawn Viruses on Prawn Aquaculture Production. Available online: https://crcna.com.au/projects/demonstrating-impact-prawn-viruses-prawn-aquaculture-production/ (accessed on 27 August 2024).
195. FRDC. Trichlorfon Strategy Improves Prawn Farm Biosecurity. Available online: https://www.frdc.com.au/trichlorfon-strategy-improves-prawn-farm-biosecurity#:~:text=As%20a%20preventive,%20water%20is,a%20white%20spot%20disease%20outbreak (accessed on 27 December 2024).
196. Hatchery International. New Disease Management Tech for Australia's Prawn Farmers. Available online: https://www.hatcheryinternational.com/new-disease-management-tech-for-australias-prawn-farmers/ (accessed on 13 September 2024).
197. Department of Agriculture, Fisheries and Forestry. Biosecurity Training Centre. Available online: https://www.agriculture.gov.au/biosecurity-trade/policy/australia/biosecurity-reform/biosecurity-training-centre (accessed on 23 November 2024).
198. Farmbiosecurity. New National Biosecurity Training Hub Launched. Available online: https://www.farmbiosecurity.com.au/new-national-biosecurity-training-hub-launched/?utm_source (accessed on 3 January 2025).
199. CRCNA. Northern Australia Aquaculture Biosecurity Training Video. Available online: https://crcna.com.au/resources/publications/northern-australia-aquaculture-biosecurity-training-video-2/ (accessed on 11 October 2024).
200. FRDC. Aquatic Animal Health & Biosecurity. Available online: https://www.frdc.com.au/aquatic-animal-health-biosecurity (accessed on 14 December 2024).

MDPI AG
Grosspeteranlage 5
4052 Basel
Switzerland
Tel.: +41 61 683 77 34

Viruses Editorial Office
E-mail: viruses@mdpi.com
www.mdpi.com/journal/viruses

www.ingramcontent.com/pod-product-compliance
Lightning Source LLC
Chambersburg PA
CBHW040749250926
51850CB00007BB/1445
* 9 7 8 3 7 2 5 8 7 6 8 9 1 *